Tomoko M. Nakanishi • Martin O`Brien
Keitaro Tanoi
Editors

Agricultural Implications of the Fukushima Nuclear Accident (III)

After 7 Years

Editors
Tomoko M. Nakanishi
Graduate School of Agricultural
and Life Sciences
The University of Tokyo
Bunkyo-ku, Tokyo, Japan

Martin O`Brien
Graduate School of Agricultural
and Life Sciences
The University of Tokyo
Bunkyo-ku, Tokyo, Japan

Keitaro Tanoi
Graduate School of Agricultural
and Life Sciences
The University of Tokyo
Bunkyo-ku, Tokyo, Japan

https://doi.org/10.1007/978-981-13-3218-0

Foreword

Seven years have passed since the Fukushima Daiichi Nuclear Power Plant accident caused by the Great East Japan Earthquake in March 2011. Fukushima Prefecture was severely damaged by radioactive contamination. The contamination disrupted agriculture in Fukushima Prefecture, which is a major industry in that Prefecture. In the Graduate School of Agricultural and Life Sciences, the University of Tokyo, many faculty members initiated research activities in their specific fields, such as in soil, crops, livestock, fish, and wildlife, in cooperation with other research organizations such as the Fukushima Agricultural Technology Centre. Our research results not only aided the recovery in agriculture, forestry, and fisheries but also helped the general public understand the extent and the implications of the contamination. We held the first public meeting to report our initial research results on radioactive contamination in the environment on November 11, 2011. The research has continued, and the 14th public meeting was held on November 25, 2017. The research results have been published in scientific journals and books. Internationally, two books reporting the results of research undertaken in Fukushima were published by Springer in 2013 and 2016, and the contents of both books can be downloaded freely.

The decontamination of agricultural land and residential dwellings has continued for 7 years, and the evacuation order zone has gradually decreased in area. Potassium fertilization of crops to exclude the absorption of radioactive cesium is also continuing, as has the monitoring of radioactivity of agricultural, forestry, and fishery products and the inspection of entire crops of rice. Most of the contaminated area in Fukushima Prefecture is covered by forests, and decontamination work has only been carried out in limited areas close to dwellings; forests throughout the Prefecture remain contaminated. In areas of Fukushima Prefecture where the evacuation order has been lifted, local residents are gradually returning to their homes, but they have had difficulties on returning. These difficulties may be from uncertainty on the future effects of the environmental contamination. Because it will take a considerable long time for decontamination by the natural decay of radioactive cesium, long-term monitoring of food and continued study of the spatiotemporal dynamics of radioactive cesium in the forest ecosystem will be required.

It will take a considerable amount of time for agriculture, forestry, and fishery in Fukushima Prefecture to recover and return to the pre-accident conditions. The Graduate School of Agricultural and Life Sciences, the University of Tokyo will continue to research and aid the recovery of these industries and local communities into the future.

Graduate School of Agricultural and Life Sciences
The University of Tokyo
Bunkyo-ku, Tokyo, Japan

Takeshi Tange

Preface

Over seven years have passed since the Fukushima Nuclear Power Plant accident occurred. With a focus on Fukushima Prefecture, the recovery of regions contaminated with radioactive fallout is continuing, and the citizens are returning to their homes. For the agricultural industries in Fukushima Prefecture, the government is recommending farming methods that exclude radioactive materials, e.g., potassium fertilization during crop cultivation. At the same time, the authorities in Fukushima Prefecture have been conducting radioactivity inspections of all agricultural products prior to sale. Especially in the case of rice, the entire Fukushima rice crop of about 10 million bags (30 kg of rice per bag) has been monitored every year. From this inspection, the number of contaminated rice bags has been decreasing from 71 bags in 2012 to zero in 2015, and thereafter no contaminated rice bags have been detected. Because of the abovementioned countermeasures, all agricultural products on the market are now below the threshold levels for radiation exposure.

Since more than 80% of the contaminated region was related to agriculture, the Graduate School of Agricultural and Life Sciences, the University of Tokyo created an independent team immediately after the accident consisting of about 40–50 academic faculty who are specialists in soil, crops, wild and domestic animals, fisheries, forestry, etc. Since the study of the fallout behavior is so closely related to the agricultural environment itself, a multidisciplinary approach was needed. Therefore, faculty members entered the contaminated areas together and performed collaborative research to understand what was happening to the radioactive materials in the agricultural environment. Some of the basic questions we had included: how does radioactive material move within the contaminated soil, and how is it taken up by crops? Can contaminated forests affect agricultural land, etc.?

Although we have been able to accumulate a large amount of data over the past 7 years, data on any one topic is still relatively low. For example, in the case of cereals or rice, we can only harvest the crop once per year, resulting in one data set per year. Therefore, it will take many years to obtain sufficient data to be able to understand the persistence of the fallout in the agricultural environment and the long-term impact on agriculture. We have decided to continue our research for the foreseeable future so that the general public, farmers and other stakeholders can have a much better understanding of the effect of radioactive contamination on agriculture.

The nuclear accident in Fukushima was the first nuclear plant accident in the Asian monsoon region, and because Japan shares a similar agricultural environment to other Asian countries (e.g., rice cultivation in paddy fields), we have an important role to disseminate the findings of our research to the relevant stakeholders around the world, especially those concerned with nuclear fallout and agriculture in the Asian region. The motivation behind publishing this third book was the large interest shown by readers in our previous two books published by Springer Japan in 2013 and 2016 (downloaded 124,000 and 58,500 times, respectively, as of July 2018). For readers interested in the effect the nuclear accident had on the agricultural environment in Fukushima Prefecture, the current book summarizes the latest research undertaken by our faculty.

Tokyo, Japan Tomoko M. Nakanishi

Contents

1 **An Overview of Our Research** 1
Tomoko M. Nakanishi

2 **Transfer of Radiocesium to Rice in Contaminated Paddy Fields** 9
Keisuke Nemoto and Naoto Nihei

3 **Cesium Translocation in Rice** 15
Keitaro Tanoi, Tatsuya Nobori, Shuto Shiomi, Takumi Saito, Natsuko I. Kobayashi, Nathalie Leonhardt, and Tomoko M. Nakanishi

4 **Absorption of Radioceasium in Soybean** 27
Naoto Nihei and Shoichiro Hamamoto

5 **An Observational Study of Pigs Exposed to Radiation** 35
Junyou Li, Chunxiang Piao, Hirohiko Iitsuka, Masanori Ikeda, Tomotsugu Takahashi, Natsuko Kobayashi, Atsushi Hirose, Keitaro Tanoi, Tomoko Nakanishi, and Masayoshi Kuwahara

6 **A Composting System to Decompose Radiocesium Contaminated Baled Grass Silage** 51
Takahiro Yoshii, Tairo Oshima, Saburo Matsui, and Noboru Manabe

7 **Weathered Biotite: A Key Material of Radioactive Contamination in Fukushima** 59
Toshihiro Kogure, Hiroki Mukai, and Ryosuke Kikuchi

8 **Radiocesium Accumulation in Koshiabura (*Eleutherococcus sciadophylloides*) and Other Wild Vegetables in Fukushima Prefecture** 77
Naoto Nihei and Keisuke Nemoto

9 **The Transition of Radiocesium in Peach Trees After the Fukushima Nuclear Accident** 85
Daisuke Takata

10 **Application of the Artificial Annual Environmental Cycle and Dormancy-Induced Suppression of Cesium Uptake in Poplar** 95
Yusaku Noda, Tsutomu Aohara, Shinobu Satoh, and Jun Furukawa

11 **Radiocesium Contamination in Forests and the Current Situation of Growing Oak Trees for Mushroom Logs** 107
Natsuko I. Kobayashi, Ryosuke Ito, and Masaya Masumori

12 **Radiocesium Dynamics in Wild Mushrooms During the First Five Years After the Fukushima Accident** 123
Toshihiro Yamada

13 **The Spatial Distribution of Radiocesium Over a Four-Year Period in a Forest Ecosystem in North Fukushima After the Nuclear Power Station Accident** 141
Masashi Murakami, Takahiro Miyata, Natsuko Kobayashi, Keitaro Tanoi, Nobuyoshi Ishii, and Nobuhito Ohte

14 **Parallel Measurement of Ambient and Individual External Radiation in Iitate Village, Fukushima** 153
Yoichi Tao, Muneo Kanno, Soji Obara, Shunichiro Kuriyama, Takaaki Sano, and Katsuhiko Ninomiya

15 **Mobility of Fallout Radiocesium Depending on the Land Use in Kasumigaura Basin** 165
Shuichiro Yoshida, Sho Shiozawa, Naoto Nihei, and Kazuhiro Nishida

16 **Challenges of Agricultural Land Remediation and Renewal of Agriculture in Iitate Village by a Collaboration Between Researchers and a Non-profit Organization** 177
Masaru Mizoguchi

17 **Radiocesium Contamination on a University Campus and in Forests in Kashiwa City, Chiba Prefecture, a Suburb of Metropolitan Tokyo** 191
Kenji Fukuda

18 **The State of Fisheries and Marine Species in Fukushima: Six Years After the 2011 Disaster** 211
Nobuyuki Yagi

19 Visualization of Ion Transport in Plants 221
Ryohei Sugita, Natsuko I. Kobayashi, Atsushi Hirose, Keitaro Tanoi, and Tomoko M. Nakanishi

20 ^{90}Sr Analysis Using Inductively Coupled Plasma Mass Spectrometry with Split-Flow Injection and Online Solid-Phase Extraction for Multiple Concentration and Separation Steps 233
Makoto Furukawa and Yoshitaka Takagai

Chapter 1
An Overview of Our Research

Tomoko M. Nakanishi

Abstract Immediately after the Fukushima nuclear plant accident (FNPA), 40–50 researchers at the Graduate School of Agricultural and Life Sciences, the University of Tokyo, analyzed the behavior of the radioactive materials in the environment, including agricultural farmland, forests, rivers, etc., because more than 80% of the contaminated land was related to agriculture. Since then, a large number of samples collected from the field were measured for radiation levels at our faculty. A feature of the fallout was that it has hardly moved from the original point contaminated. The fallout was found as scattered spots on all surfaces exposed to the air at the time of the accident. The adsorption onto clay particles, for example, has become firm with time so that it is now difficult to be removed or absorbed by plants. ^{137}Cs was found to bind strongly to fine clay particles, weathered biotite, and to organic matter in the soil, therefore, ^{137}Cs has not mobilized from mountainous regions, even after heavy rainfall. In the case of farmland, the quantity of ^{137}Cs in the soil absorbed by crop plants was small, and this has been confirmed by the real-time imaging experiments in the laboratory. The downward migration of ^{137}Cs in soil is now estimated at 1–2 mm/year. The intake of ^{137}Cs by trees occurred via the bark, not from the roots since the active part of the roots is generally deep within the soil where no radioactive materials exist. The distribution profile of ^{137}Cs within trees was different among species. The overall findings of our research is briefly summarized here.

Keywords ^{137}Cs · Fukushima nuclear plant accident · Agriculture · Soil · Plant · Forest

T. M. Nakanishi (✉)
Graduate School of Agricultural and Life Sciences, The University of Tokyo, Bunkyo-ku, Tokyo, Japan
e-mail: atomoko@mail.ecc.u-tokyo.ac.jp

T. M. Nakanishi et al. (eds.), *Agricultural Implications of the Fukushima Nuclear Accident (III)*, https://doi.org/10.1007/978-981-13-3218-0_1

1.1 General Features of the Fallout

When the fallout from the nuclear power plant between Fukushima and Chernobyl was compared, the total radioactivity released into the environment by FNPA was estimated at 770,000 TBq, which is approximately 15% of that released by the Chernobyl accident. The radioactive nuclides released by FNPA contained 21% ^{131}I (half-life: 8 days), 2.3% ^{134}Cs (half-life: 2 years) and 1.9% ^{137}Cs (half-life: 30 years). The remaining nuclides in the environment are ^{134}Cs and ^{137}Cs, and the ratio of which has changed from roughly 1:1 in 2011 to 0.12:1 in 2017.

Since the accident occurred in late winter, the only crop in the fields was wheat. The relevant feature, with regards to the fallout, is that the radioactive Cs remained at the initial contact site and had not moved since, therefore, this would imply that Cs will be difficult to remove from fields. When the radiograph of any materials exposed to the air at the time of the accident was taken, the contamination was found as scattered spots on all the surfaces investigated, including soil particles and plant material.

Today, there are no contaminated agricultural products on the market, and researchers are starting to turn their attention to the situation in the forests. At the time of the accident, most of the radioactive material was trapped in leaves located high in the evergreen trees and in the bark of these trees; therefore, the radioactivity was relatively low on the forest floor. In the past few years the contaminated leaves of evergreen trees have fallen to the ground and along with the decomposition process of the litter, ^{137}Cs has gradually moved to the soil and become firmly adsorbed by soil particles.

Since most of the radioactive Cs is adsorbed to fine clay or organic matter in the soil, radioactivity was not detected in the water itself flowing out from the mountain. A simple filtration of the water was effective to remove the radioactive fine particles suspended in water. In the forests, no biological concentration of ^{137}Cs was found in any specific animal along the food chain.

1.2 Radioactivity Measurement

Most of the radioactivity measurement and imaging was performed by the Isotope Facility for Agricultural Education and Research in our faculty. Two academic faculty members and two technicians continue to measure all samples collected from the field, as well as samplers generated from laboratory experiments. Approximately 300 samples are measured per month using two Ge counters and several hundred samples are measured using a Na(Tl)I counter with an automatic sample changer (Figs. 1.1 and 1.2). The number of the samples measured in each month does not mean the number of the samples actually collected. The number of the sample is dependent on the activity level of the sample collected, since when the radioactivity level of the sample is low, it required a longer time of the measurement, therefore, only small number of the sample was able to measure in the month.

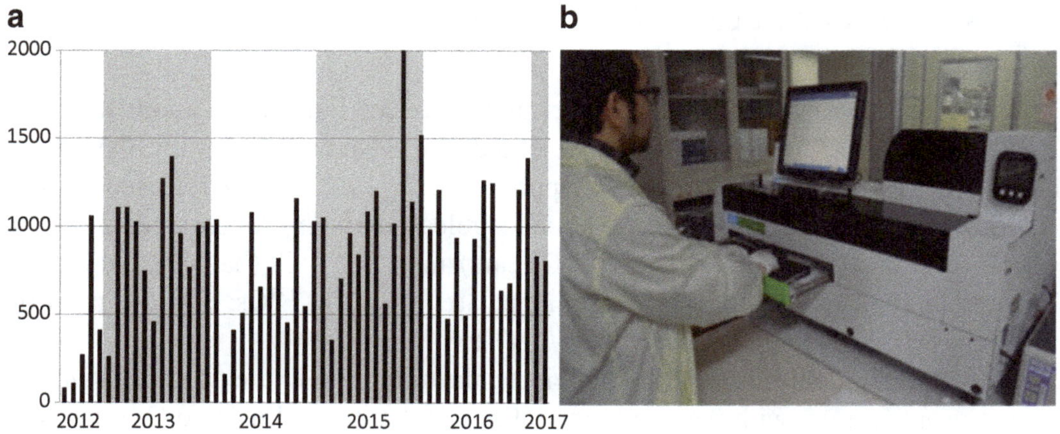

Fig. 1.1 Radioactivity measurement of samples. (**a**) The number of samples measured each month using a Na(Tl)I counter. (**b**) Picture of the Na(Tl)I counter

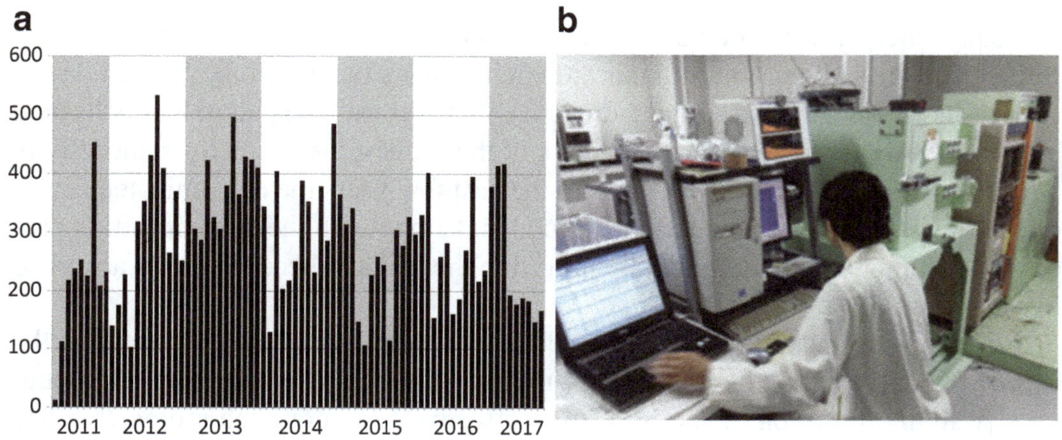

Fig. 1.2 Radioactivity measurement of samples. (**a**) The number of samples measured each month using a Ge counter. (**b**) Picture of the Ge counter

1.3 A Brief Summary of Our Findings

1.3.1 Soil

1.3.1.1 Vertical Migration of Radiocesium

To measure the vertical migration of radiocesium in the soil, a vinyl chloride cylinder was placed in a borehole in the soil. A scintillation counter covered with a lead collimator with a slit window was inserted into the pipe to measure the radioactivity vertically along the borehole. About 2 months after the accident, the vertical radiocesium (^{134}Cs and ^{137}Cs) concentration in the top 0–15 cm layer of soil was measured in an undisturbed paddy field. Approximately 96% of the radiocesium was found within the top 0–5 cm layer. The measurement was repeated every few months to record the downward movement of the radiocesium. The radiocesium movement in soil was very fast during the first 2–3 months and then the speed was drastically reduced, indicating that the adsorption of the radiocesium to soil particles had become stronger with time, indifferent to the amount of rainfall. The speed of the downward movement of the radiocesium is now much slower than immediately after the accident, at about 1–2 mm/year.

The radioactivity of the surface soil at the bottom of a pond was measured periodically. Radioactivity had gradually decreased with time, except for one pond, where run-off water from the city flowed into the pond. Water had been used to decontaminate concrete and other surfaces in the city after the accident, and radiocesium in this contaminated water had moved to the bottom of the pond.

1.3.1.2 ^{137}Cs Adsorption Site

One study determined where radiocesium is adsorbed on soil. The soil was separated according to particle size and an autoradiograph of each fraction was taken. It was found that radiocesium was adsorbed by the fine clay and organic matter but not by the larger components of the soil, such as gravel and sand. It was an important finding which led to the development of an efficient and practical decontamination protocol for farmland.

To determine the kind of clay mineral adsorbing the radiocesium, eight mineral species were prepared and an adsorption/desorption experiment was carried out using a small quantity of ^{137}Cs tracer. It was found that the weathered biotite (WB) sorbed ^{137}Cs far more readily and firmly than the other clay minerals. The WB sample was then cut into pieces by a focused ion beam and the radioactivity of each piece was measured by an imaging plate to know the distribution of ^{137}Cs. To our surprise, each separated fraction of WB showed similar radioactivity per area, suggesting the uniform distribution of radiocesium within the clay piece. This finding has completely changed our understanding of the adsorbed site of clay minerals. The layered shape of the clay was reported to have a loosed edge due to weathering, known as a frayed edge site, where ^{137}Cs was selectively fixed, therefore, ^{137}Cs was expected to be fixed at the margin of the clay. It was also suggested that the adsorption behavior of ^{137}Cs was different when the quantity was very small, i.e. radiotracer level.

1.3.1.3 ^{133}Cs and ^{137}Cs

To compare ^{137}Cs distribution with that of ^{133}Cs, which is a stable nuclide, an agricultural field (3.6 × 30 m) in Iitate-village was selected. In this field, the total radiocesium activity was 5000 Bq/kg, which corresponded to about 10^{-3} μg/kg of ^{137}Cs, whereas the concentration of stable ^{133}Cs was about 7 mg/kg. The stable ^{133}Cs was derived from the minerals in the field and the ^{137}Cs was derived from the fallout. Though there was a high correlation between total ^{137}Cs distribution and that of exchangeable ^{137}Cs, it was found that the extraction ratio, exchangeable ^{137}Cs/total ^{137}Cs, was higher than that of stable Cs. Since this extraction ratio is expected to be the same between stable and radioactive Cs in the future, when equilibrium is attained, the higher extraction ratio of radioactive Cs suggested that the exchangeable radioactive Cs is still moving toward the stable state, which could be interpreted that the fixing process of fallout nuclide is still proceeding.

1.3.2 Plants

1.3.2.1 Rice and Soybean

To study the transfer factor of ^{137}Cs from soil to rice, the relationship between the radioactivity in the soil and that in plants was measured. But there was a reciprocal correlation between the K concentration in soil and ^{137}Cs concentration in plants, which suggested that applying K to soil prevents rice from becoming contaminated. Actually, when an optimum amount of K was not supplied in K deficient fields, radiocesium content in rice plants was high. Although the natural abundance of Cs compared to K is only 1/1000, it was interesting that the competition between the two ions in plant absorption was observed.

When a single grain of rice was sliced and placed on an imaging plate to obtain the radiograph, it was shown that ^{137}Cs accumulated in both the hull and in the cereal germ of the grain. To study the distribution of the radioactive Cs in more detail, the micro-autography method was employed which was developed by our faculty. After slicing the grain, the film emulsion was painted on the surface of the glass to produce a thin film. The film was exposed to radiation from the sample and then developed to obtain a micro-radiograph. Examining the micro-radiograph under a microscope showed that ^{137}Cs accumulated around the plumule and radicle, suggesting that Cs was not incorporated into the newly developing tissue itself but accumulated in the surrounding tissue of the meristem, similar to the phenomenon that the meristem is generally protected and free from heavy metals or viruses.

Radiocesium accumulation in soybean seed tends to be higher than that of rice grain. One of the reasons is that a soybean seed does not have an albumen. In the case of a rice grain, radiocesium accumulates in the embryo and not in the albumen. The soybean seed itself develops into a cotyledon, a kind of embryo, therefore it contains a high amount of minerals. Another difference between rice and soybean plants is that, in the case of a rice, radiocesium absorption occurs before ear emergence and out of the total radiocesium amount absorbed, 10–20% was accumulated in the seeds. However, in the case of a soybean plant, half of the radiocesium accumulated in the seed is taken up during pod formation, and out of the total radiocesium absorbed about 42% accumulated in the seeds.

1.3.2.2 Fruit Trees

Generally, in the case of the trees, radioactive Cs moved directly from the surface of the bark to the inside of the trunk. To understand how radioactive Cs is transferred to the inner part of a fruit tree in the following season, a contaminated peach tree was transplanted to a non-contaminated site after removing twigs, leaves and fine roots. Then, 1 year later, all of the newly developed tissue, including the fruits, was harvested and the radioactivity was measured. Only 3% of the radioactive Cs had moved the following year to the newly developing tissue, including roots. That means that 97% of the radioactive Cs that had accumulated inside the tree did not move. In the case of the fruit tissue, about 0.6% of the radioactive Cs that accumulated inside the tree had moved and accumulated in the fruit.

1.3.3 Forests and Animals

1.3.3.1 Forests

In the mountain forests, leaves were only present on evergreen trees and these needle-like leaves were highly contaminated due to the fallout because the accident occurred in late winter. However, even these needle-like leaves received high amounts of radioactive material and prevented the fallout from moving to the forest floor. Therefore, the radioactivity of the soil under the deciduous trees without leaves was higher than the soil under the evergreen trees. In the case of the evergreen trees, leaves located higher on the trunk of the trees were more contaminated than those located lower on the trunk and the trunk itself was highly contaminated. Though the amount of radioactivity moved into the heartwood was different along the height of the tree, the contamination inside the tree was not due to the radioactive Cs transport from the roots. Since the radioactive Cs was only at the surface of the soil, it was not possible for the active roots to absorb Cs. The active part of the roots for most trees is at least 20–30 cm below the surface of the soil and at this depth, there was no radioactive cesium. In the past few years the contaminated leaves of evergreen trees have fallen to the ground and along with the decomposition process of the litter, ^{137}Cs has gradually moved into the soil and then firmly adsorbed by soil particles.

Mushrooms can be found growing in forests in mountainous regions all over Japan, however, the radioactivity of the mushrooms growing in the forest has not drastically decreased with time. Some of the mushrooms harvested more than 300 km from the site of the accident were found to accumulate ^{137}Cs only, indicating that they are still accumulating the global fallout from nuclear weapons testing that occurred during the 1960s. Since the half-life of ^{137}Cs is 30 years, it is much longer than that of ^{134}Cs (half-life of 2 years), all of the ^{134}Cs in the global fallout in the 1960s has decayed after 50 years. This means when only ^{137}Cs was detected in mushrooms, the ^{137}Cs found was not from the Fukushima nuclear accident. In the case of the fallout from the Fukushima nuclear accident, the initial radioactivity ratio of ^{137}Cs to ^{134}Cs was the same in 2011.

The river water flowing from the mountains show very low radioactivity (less than 10 Bq/l). It was also found that the water itself flowing out from the mountain had low radioactivity and the radioactivity was removed after filtering out the suspended radioactive clay in the water. The amount of the radioactive Cs flowing out from the mountain was in the order of 0.1% of the total fallout amount per year.

1.3.3.2 Animals

Contaminated haylage was supplied to dairy cattle and the radioactivity of the milk was measured. It was found that radioactive Cs was detected in the milk soon after the contaminated feed was supplied. After radioactivity levels in the milk reached a plateau after 2 weeks, the non-contaminated feed was fed to the cattle and the radioactivity in the milk decreased and became close to the background level after 2 weeks. Similar results were found for animal meat, indicating that when contaminated animals are identified, it is possible to decontaminate them by feeding non-contaminated feeds. The biological half-life of ^{137}Cs was estimated to be less than 100 days because of the animal's metabolism, whereas the physical half-life of ^{137}Cs is 30 years.

At the time of the accident, radioactive Cs contaminated every surface exposed to the air, and this also included the feathers of birds. Male bush warblers were captured in a highly contaminated area of the Abukuma highlands in 2011, and it was found that the feathers were contaminated with ^{137}Cs. The accident occurred just as these birds had started molting, therefore, they had a limited home range in the highlands, which was close to the site of the accident. This contamination of feathers was not removed by washing. However, in the following year, no radioactivity was found on the feathers of the bush warbler caught in the same area.

1.4 Decontamination Trial

The most effective and efficient way to prevent radioactive Cs uptake in crops is to apply K fertilizer on farmland. Since the soil in agricultural land is a very important natural resource, the removal of the soil surface cannot be compensated by simply replacing it with non-contaminated soil. The best way to decontaminate farmland is to eliminate only the contaminated particles in the soil. Radioactive Cs was only found to be adsorbed firmly on the fine clay component of soil. Therefore, introducing water into a contaminated field and mixing it well with the surface soil (about 5 cm in depth), the soil components precipitate and the suspended fine clay particles in the water can be drained off into an adjacent ditch in the field. Thus, more than 80% of the radioactivity in the field was removed.

1.5 Conclusion

The behavior of the radioactive Cs emitted from the nuclear accident was different from that of so-called macroscopic Cs chemistry we know. Because the amount of Cs deposited on leaves was so small and carrier-free, the nuclides seem to behave like radio-colloids, or as if they were electronically adsorbed onto the tissue.

Through our activities, many scientific findings have been accumulated. The results of our research introduced above are only a small portion of our total findings since the Fukushima nuclear accident occurred.

References

Nakanishi TM (2018) Agricultural aspects of radiocontamination induced by the Fukushima nuclear accident – a survey of studies of the University of Tokyo Agricultural Department (2011–2016). Proc Jpn Acad Ser B 94:20–34

Nakanishi TM, Tanoi K (eds) (2013) Agricultural implications of the Fukushima nuclear accident. Springer, Tokyo

Nakanishi TM, Tanoi K (eds) (2016) Agricultural implications of the Fukushima nuclear accident. The first three years. Springer, Tokyo

Chapter 2
Transfer of Radiocesium to Rice in Contaminated Paddy Fields

Keisuke Nemoto and Naoto Nihei

Abstract Rice contaminated with high concentrations of radiocesium was found in some local areas after the nuclear accident in Fukushima Prefecture in 2011. Here we discuss the issues of cultivating rice in contaminated areas through our field experiments. The transfer of radiocesium to commercial rice has been artificially down-regulated by potassium fertilizer in radiocesium-contaminated areas in Fukushima. Since 2012, we have continued to cultivate rice experimentally in paddy fields under conventional fertilizer to trace the annual change of radiocesium uptake. The radiocesium concentration in rice cultivated under conventional fertilizer has seen almost no change since 2013. One of the reasons for this is that radiocesium fixation in soil has hardly progressed in these paddy fields.

Keywords Paddy field · Radiocesium · Rice

2.1 Radiocesium in the Paddy Field Ecosystem

The Fukushima Daiichi Nuclear Power Plant Accident in March 2011 caused extensive radiation exposure to fields in Fukushima Prefecture. A large proportion of the released radiation consisted of two radionuclides, namely ^{137}Cs and ^{134}Cs. ^{137}Cs is of most concern because of its long half-life (30.2 years), and thus a long-term problem for agriculture.

One of the most important agricultural products produced in Fukushima Prefecture is rice, which accounts for 40% of total food production from this prefecture. Rice contaminated with high concentrations of radiocesium was found in some local areas after the nuclear accident, and thus it was necessary to take immediate measures to reduce radiocesium uptake in rice. As an aquatic plant, rice has developed specific physiological and ecological characteristics to take up nutrients, and the ecosystem of the paddy field has also its own unique characteristics regarding

K. Nemoto (✉) · N. Nihei
Graduate School of Agricultural and Life Sciences, The University of Tokyo, Bunkyo-ku, Tokyo, Japan
e-mail: unemoto@mail.ecc.u-tokyo.ac.jp

T. M. Nakanishi et al. (eds.), *Agricultural Implications of the Fukushima Nuclear Accident (III)*, https://doi.org/10.1007/978-981-13-3218-0_2

material cycle, such as nitrogen, phosphorus, potassium, etc. Therefore, research conducted in the aftermath of the Chernobyl nuclear accident related to agriculture was not applicable to the situation in Fukushima. Here we discuss the issues of cultivating rice in contaminated fields through a series of experiments carried out in Fukushima Prefecture.

2.2 Transfer of Radiocesium to Rice in 2011 (After the Accident)

In 2011, cultivation of rice was restricted in the areas where the soil contained 5000 Bq/kg of radiocesium. The concentration of radiocesium in rice produced outside the restricted areas was low, and the governor of Fukushima Prefecture announced that the rice cultivated in Fukushima Prefecture was safe in the fall of 2011. However, in the fall of 2011 after this announcement, rice with radiocesium concentration exceeding the provisional regulation level of 500 Bq/kg was found in the northern part of the Abukuma highland in Fukushima Prefecture. Strangely, there were cases where rice cultivated in one paddy field contained several hundred Bq/kg of radiocesium and rice cultivated in an adjacent field had radiocesium concentrations below the detection limit. It was difficult to infer the reason for such a wide variation in radiocesium uptake, even after consulting the literature related to the Chernobyl accident.

2.3 Experimental Cultivation in 2012

Because rice had been detected with over 500 Bq/kg of radiocesium in 2011, rice cultivation was restricted in many regions in 2012. However, local municipalities began 'experimental cultivation'. The basic purpose of these experiments were to investigate the effect of applying potassium fertilizer which was thought to reduce radiocesium uptake. Potassium is one of the three major nutrients for plants along with nitrogen and phosphorus, and it is usually applied as a compound fertilizer containing nitrogen and phosphorus, not as a straight fertilizer. Although plants do not require cesium as a nutrient, it is inadvertently taken up instead of potassium because both elements share similar chemical characteristics. Generally, this 'accidental uptake' of cesium can occur more frequently as there is less the exchangeable potassium (extracted with 1 mol L^{-1} ammonium acetate) that plants can absorb in the soil.

Thus, in the experimental cultivation in 2012, a sufficient amount of potassium fertilizer was applied to the paddy fields where the radiocesium-contaminated rice was found in 2011. Rice was experimentally cultivated in these paddy fields after applying fertilizer to confirm whether radiocesium concentration in the grain was lower than the maximum limit for shipment (<100 Bq/kg, adopted in 2012). When the concentration was confirmed to be lower than the maximum limit in this experiment, the prefecture allowed farmers to cultivate rice commercially from the following year (2013) with a prerequisite that potassium fertilizer would be applied.

The result of the experiment in 2012 demonstrated that applying potassium fertilizer thoroughly can reduce the transfer of radiocesium to rice effectively even in radiocesium-contaminated areas (Ministry of Agriculture, Forestry and Fisheries, and Fukushima Prefecture 2014). Indeed, brown rice with radiocesium levels over 100 Bq/kg were detected in only 71 bags (0.0007%) out of whole commercial rice (a total of about 10 million bags) produced in Fukushima in 2012, due to the thorough application of potassium fertilizer all over the prefecture (Nihei et al. 2015).

2.4 The Experimental Cultivation in Oguni, Date City

As stated above, the reason rice had a high concentration of radiocesium is now better understood, and this knowledge helped to pave the way for the resumption of rice cultivation in 2013. However, not all problems were resolved by the experimental cultivation. When uptake of radiocesium is down-regulated by the application of potassium as was seen in the experimental cultivation, it becomes difficult to identify the specific reason why the radiocesium concentration of rice differs for each paddy field within an area. Furthermore, to decide the timing of ceasing potassium application in the future, it will be important to ensure that radiocesium remaining in fields will not be absorbed by crops under conventional fertilizer use. Hence, it is necessary to trace the annual change of radiocesium uptake in rice in the natural paddy field ecosystem under conventional fertilizer use.

Oguni in Date City is a district where rice with radiocesium over 500 Bq/kg was harvested in 2011. Date City and the local community of Oguni appreciated the importance of the research into radiocesium uptake in the natural paddy field ecosystem and gave their support for our project. In this way, our research group in the University of Tokyo, Koyama group (agricultural economics) in the University of Fukushima, and Gotoh group (pedology) in Tokyo University of Agriculture collaborated and carried out the experimental cultivation in Oguni from 2012.

Oguni is located in hilly terrain in the north of the Abukuma highlands, and the Oguni River runs through the center of the district. During periods of water shortage in the basin of the tributaries of the Oguni River, numerous reservoirs are frequently used to supply water to the paddy fields. Sixty paddy fields were selected in different geographical locations encompassing a variety of different local environments. Potassium silicate and zeolite (each 200 g/m^2) were applied to 5 of the 60 paddy fields as a radiocesium reduction measure, whereas the remaining 55 paddy fields were cultivated under conventional fertilizer usage; however, in each of these 55 fields, 6.6 m^2 of land was separated from the rest of the field using corrugated sheets to investigate the effect of applying potassium silicate (200 g/m^2) (Fig. 2.1). In the fall of 2012, we measured the radiocesium concentration in the rice cultivated under

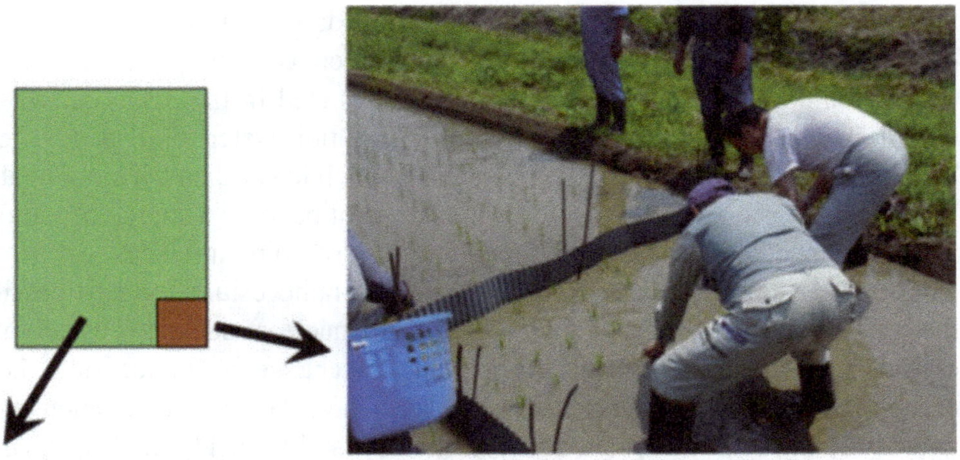

Fig. 2.1 The experiment planning of paddy fields in Oguni

normal fertilizer usage. Forty-one of the 55 paddy fields produced brown rice with less than 100 Bq/kg of radiocesium. Some of these paddy fields had produced rice with several hundred Bq/kg of radiocesium one year earlier. Judging from this result, it seemed that the uptake of radiocesium in rice cultivated in Oguni had decreased over time.

Soils in the paddy fields that produced rice with high radiocesium concentrations typically contained exchangeable potassium less than 10 mg K_2O/100 g, and thus it was confirmed that potassium concentration in paddy fields was an important factor to reduce radiocesium contamination in rice. An interesting finding was that the paddy fields that produced rice with high concentrations of radiocesium were all located in the basin of the tributaries. Actually, rice with a radiocesium concentration of >50 Bq/kg was not produced from paddy fields on the bank of the mainstream, even though potassium concentrations in the soil were low. These paddy fields were located only a few hundred meters from fields in the basin that produced rice with high concentrations of radiocesium. Hence, there is a possibility that some geographical factors increase radiocesium uptake.

2.5 No Decrease of Radiocesium in Rice

After 2012, we continued to cultivate rice continuously in the paddy fields in Oguni with high radiocesium concentrations, in collaboration with the local community, the City, and other Universities (Nemoto 2014). We originally estimated that the uptake of radiocesium in rice would decrease year by year. However, to our surprise, there has been almost no change since 2013. This fact suggests that the amount of radiocesium which is responsible for producing rice with a high radiocesium concentration still exists in the paddy fields without being fixed by the soil. Certainly, there is a possibility that the irrigation water acted as a source of radiocesium because the reservoirs contained 3–4 Bq/L of radiocesium one year after the accident. However, radiocesium in the reservoirs decreased sharply after 2013, and at present, radiocesium >1 Bq/L has not been detected in the irrigation water from Oguni. The inflow of radiocesium via the irrigation water to all paddy fields has been about 100 Bq/m^2 since 2013, and this is much lower than the amount absorbed by the rice. It seems that irrigation water will not become a source of radiocesium for rice at present unless mud at the bottom of reservoirs mobilizes.

Because we believe water is not the source of radiocesium, the next possibility is the soil. Radiocesium deposited on soil after the nuclear accident is usually fixed by clay mineral over time, and thus the amount of exchangeable radiocesium, i.e. radiocesium absorbed by roots decreases with time. To verify this phenomenon, we investigated the fixation of deposited radiocesium on soil in paddy fields in Oguni. One year after the accident (i.e., 2012), about 80% of the deposited radiocesium was fixed by the soil, and the other 20% was exchangeable radiocesium. Surprisingly, the fixation of cesium to soil has not progressed much in 5 years (2012–2016), and about 15% of the radiocesium is still in an exchangeable form which can be absorbed by plants.

Of course, not all paddy fields in Fukushima Prefecture are in the same situation. As stated above, the radiocesium uptake dropped below 100 Bq/kg in three-quarters of the 60 experimental fields in Oguni without any measures to control radiocesium uptake in 2012. Exchangeable radiocesium in some paddy fields has decreased to 5%.

2.6 Summary of the Experiments Performed in Oguni, Date City

These results outlined above raises two questions. First, as mentioned previously, it is necessary to continue applying potassium fertilizer. Indeed, thanks to the municipalities' exhaustive instructions to farmers about applying additional potassium fertilizer, no rice in Fukushima Prefecture with radiocesium over the standard value (100 Bq/kg) has been detected since 2015 (Fukushima Association for Securing Safety of Agricultural Products). However, judging from the result of the experimental cultivation in Oguni, the concentration of radiocesium in rice might increase over the standard value again if we moderate applications of potassium in the paddy fields where cesium does not easily fix to soil, or where radiocesium is mobile in the soil. It is necessary for the Government to take responsibility not only for providing farmers with potassium fertilizer as grant aid but also for its application, in order to complete the control measure of radiocesium.

Secondly, we need to apply our research findings to continually investigate regions where cultivation will be resumed in the future. For example, some farmers returning to regions where the evacuation order has just been lifted will want to cultivate rice again. When considering risks of uptake of radiocesium by rice cultivated in these regions, the data obtained in Oguni will be very important and applicable.

Acknowledgment The authors would like to thank Riona Kobayashi (The University of Tokyo) for her technical assistance.

References

Fukushima Association for Securing Safety of Agricultural Products. https://fukumegu.org/ok/kome/

Ministry of Agriculture, Forestry and Fisheries, and Fukushima Prefecture (2014) Factors causing rice with high radioactive cesium concentration and its countermeasures. http://www.maff.go.jp/j/kanbo/joho/saigai/pdf/kome.pdf

Nemoto K (2014) Transfer of radioactive cesium to rice (Fourth report). http://www.a.u-tokyo.ac.jp/rpjt/event/20141109slide5.pdf

Nihei N, Tanoi K, Nakanishi TM (2015) Inspections of radiocesium concentration levels in rice from Fukushima Prefecture after the Fukushima Dai-ichi Nuclear Power Plant accident. Scientific Reports 5, Article number: 8653-8658 2015/3

Chapter 3
Cesium Translocation in Rice

Keitaro Tanoi, Tatsuya Nobori, Shuto Shiomi, Takumi Saito, Natsuko I. Kobayashi, Nathalie Leonhardt, and Tomoko M. Nakanishi

Abstract To breed a low Cs rice variety, it is important to clarify the mechanism of Cs transport in a plant. In the present report, we found a difference in Cs distribution in rice cultivars using a ^{137}Cs tracer experiment. In addition, the difference was also found in Cs distribution of each leaf position among the same rice cultivars. There has been no report clarifying the molecular mechanism of Cs translocation, nor those of other cations, in plants. Using the rice cultivars, Akihikari and Milyang23, to find the Cs translocation mechanism can contribute to developing crops that contain lower levels of Cs when cultivated in radiocesium contaminated land.

Keywords Breeding · Brown rice · Cesium · Grain · Fukushima Daiichi Nuclear Power Plant Accident · Rice · Translocation

K. Tanoi (✉) · S. Shiomi · N. I. Kobayashi · T. M. Nakanishi
Graduate School of Agricultural and Life Sciences, The University of Tokyo, Bunkyo-ku, Tokyo, Japan
e-mail: uktanoi@g.ecc.u-tokyo.ac.jp

T. Nobori
Graduate School of Agricultural and Life Sciences, The University of Tokyo, Bunkyo-ku, Tokyo, Japan

Department of Plant Microbe Interactions, Max Planck Institute for Plant Breeding Research, Cologne, Germany

T. Saito
Nuclear Professional School, School of Engineering, The University of Tokyo, Tokai-mura, Ibaraki, Japan

N. Leonhardt
Laboratoire de Biologie du Développement des Plantes (LBDP), Institut de Biosciences et Biotechnologies d'Aix-Marseille (BIAM), St Paul lez Durance, France

T. M. Nakanishi et al. (eds.), *Agricultural Implications of the Fukushima Nuclear Accident (III)*, https://doi.org/10.1007/978-981-13-3218-0_3

3.1 Introduction

In March 2011, a 9.1 magnitude earthquake occurred in Eastern Japan, triggering an extremely large tsunami. Consequently, the Tokyo Electric Power Company's Fukushima Daiichi Nuclear Power Plant (TEPCO-FDNPP) was unable to withstand the pressure exerted upon it by both forces resulting in a nuclear meltdown and radioactive contamination of the area surrounding the power plant. The radiocesium isotopes (^{137}Cs and ^{134}Cs) are of most concern for local agriculture because of their relatively long half-lives (^{137}Cs = 30.2 years; ^{134}Cs = 2.06 years).

Because rice is the primary staple food in Japan, we have been particularly concerned over the rice crop in the fallout area. All rice bags produced in Fukushima have been inspected by screening equipment that was specifically designed for 30 kg rice bags (Nihei et al. 2015). Inspections have indicated that, after 2016, no single rice bag had radiocesium concentrations higher than the standard in Japan (100 Bq/kg; Table 3.1). We can confirm, finally, that this rice is safe to consume.

There are many reports supporting the mediation of Cs^+ transport via potassium ion (K^+) channels in root systems (Kim et al. 1998; Qi et al. 2008). In Arabidopsis, AtHAK5 is the most well-known K^+ channel among numerous genes that transport Cs^+ (Qi et al. 2008; Nieves-Cordones et al. 2017; Ishikawa et al. 2017; Rai et al. 2017). Qi et al. (2008) reported that AtHAK5 transports Cs^+ in plants under conditions of low K^+ availability. In rice plants, there have been reports that OsHAK1, expressed in roots under low potassium conditions, is involved in Cs^+ uptake from paddy soils (Nieves-Cordones et al. 2017; Ishikawa et al. 2017; Rai et al. 2017). We grew the *athak5* null mutant on Fukushima soil and determined that the ^{137}Cs in shoots was drastically decreased compared with that observed in wild-type shoots (Fig. 3.1).

When trying to clarify the mechanism of Cs accumulation in grain, Cs absorption by roots is not the only issue to be considered. The incident at the TEPCO-FDNPP occurred in March, which means that paddy soils were contaminated with radiocesium before any rice was planted in May. After planting the rice cultivar, the radiocesium in the paddy soil was absorbed by rice roots, and consequently, translocated to the grains. However, in March 2011, wheat was growing in the field as the nuclear crisis unfolded and leaves of the wheat were contaminated directly by radiocesium. The radiocesium concentrations in wheat grains grown in the same field correlated with wheat leaf mass at the time the fallout occurred, suggesting that ^{137}Cs translocation from leaf to grain was the main pathway for contamination of the wheat product at that time (Fig. 3.2). If a similar incident occurs during the rice growing season, radiocesium contamination directly to rice leaves would have a greater impact on rice grains via translocation; therefore, it is necessary to clarify the Cs translocation mechanism to breed low-Cs rice. However, in contrast to K^+ and Cs^+ absorption in roots, there is no molecular information regarding a transporter that mediates transport of K^+ or Cs^+ in above-ground biomass.

In the present study, we focused on Cs distribution in rice plants and tried to obtain a low-cesium phenotype by analyzing Cs translocation in different rice

Table 3.1 The inspection of all rice in all rice bags performed in Fukushima prefecture

Cultivation year	2012	2013	2014	2015	2016
Inspection period	08/25/2012~07/10/2015	08/22/2013~03/26/2015	08/21/2014~07/20/2016	08/20/2015~02/08/2017	08/24/2016~06/23/2017
Number of total rice bags	10,346,169	11,006,551	11,014,971	10,498,715	10,259,868
Number of rice bags containing 100 Bq/kg	71	28	2	0	0

Nihei et al., Sci. Rep. 2015, web site of Fukushima Association for securing safety of agricultural products

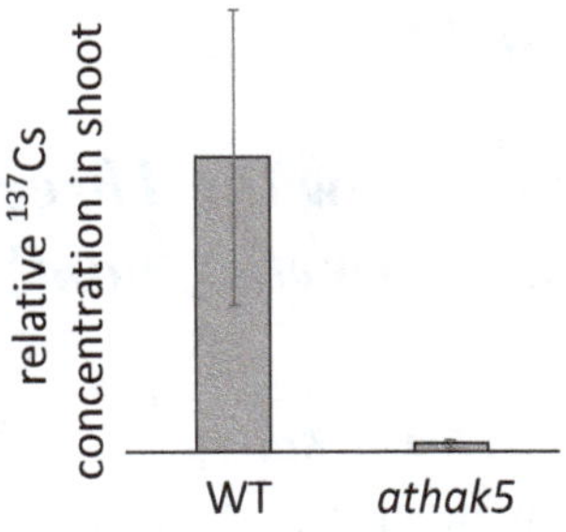

Fig. 3.1 Relative ^{137}Cs concentration in shoot. ^{137}Cs concentrations in shoots were drastically decreased in the *oshak5* null mutant when the plants were grown on the same Fukushima soil with low K^+. The exchangeable K in the soil was 6.7 mg K/100 g soil. Data represents the mean ± standard deviation (Welch's t test: P = 0.0050)

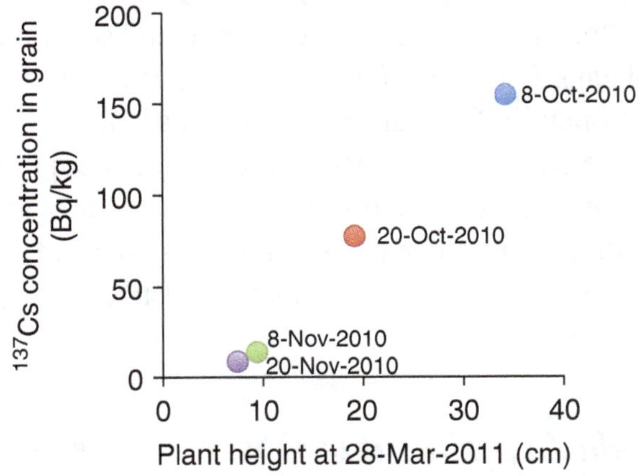

Fig. 3.2 ^{137}Cs concentrations in wheat grains produced in a field in Fukushima. The seeding dates were separated into four different 2010 plantings: October 8th, October 20th, November 8th, and November 20th. After the TEPCO-FDNPP incident occurred, the wheat plant heights were recorded on March 28th, 2011. The radiocesium concentrations of rice grains were measured after maturity at the end of June 2011. These data (in Japanese text) were provided by Arai Y., Nihei N., Takeuchi M. and Endo A., Fukushima Agricultural Technology Centre. The figure was modified from Tanoi (2013)

varieties. We selected three cultivars (Nipponbare, Akihikari, and Milyang23) based on the variation in radiocesium concentrations found in brown rice cultivated in Fukushima paddy fields in 2011 (Ono et al. 2014). Ono et al. studied 30 different cultivars, including 18 Japonica varieties, 2 Javanica varieties, and 10 Indica varieties. Results determined that the Nipponbare brown rice had low radiocesium concentrations, while that of Akihikari had high radiocesium concentrations among the Japonica varieties, and that the Milyang23, an Indica variety, had higher radiocesium concentrations than all other Japonica varieties (Ono et al. 2014). We grew the rice cultivars and separated the plantlets according to organ type (leaf, stem, peduncle, and ear) to measure the Cs concentrations within. Different distributions of Cs were found among the cultivars. These differences could provide a better understanding of both Cs^+ and K^+ translocation in rice plants.

3.2 Materials and Methods

3.2.1 ^{137}Cs Experiment to Grow the Three Rice Cultivars Hydroponically in a Growth Chamber

Seeds of three rice cultivars (*Oryza sativa* L. "Nipponbare," "Akihikari," and "Milyang23") were soaked in water for 2–4 days and then transferred to a floating net in a 0.5 mM $CaCl_2$ solution. After 2 days, the seedlings were transferred to a 2-litre container with modified half-strength Kimura B nutrient solution (pH 5.6; Tanoi et al. 2011). Two-week-old rice seedlings were transferred to 300-ml pots with culture solutions containing ^{137}Cs (non-carrier-added ^{137}Cs; Eckert & Ziegler Isotope Products, Valencia, CA, USA). The plants were grown at 30 °C with a 12 h:12 h light: dark photoperiod. Culture solutions were changed twice per week.

Rice plants in the "heading" stage were collected and separated into organs (leaf, stem, peduncle, and ear). Each leaf number was set as an arbitrary ordinal number of leaves counted acropetally from an incomplete leaf on the main stem. When the grains had matured, we collected the ears and separated them into husk, brown rice, and rachis branch. The weight of each sample was measured after drying at 60 °C for 1 week. The radioactivity of each sample was measured using a well-type NaI(Tl) scintillation counter (ARC-300; Aloka Co., Ltd., Tokyo, Japan).

3.2.2 Paddy Field Experiment to Observe ^{133}Cs Distribution in Grains

Seeds of two rice cultivars (Oryza sativa L. "Akihikari" and "Milyang23") were soaked in water for 2 days and then transferred to a seedbed in a greenhouse mid-April. Approximately 1 month later, the seedlings were transplanted to a paddy field in Tokyo (Institute for Sustainable Agro-ecosystem Services, Graduate School of Agricultural and Life Sciences, The University of Tokyo).

We analyzed ^{133}Cs instead of ^{137}Cs in the present field experiment. The matured rice grains were harvested in October. The grains were separated into husk and brown rice after being dried at 60 °C for more than 24 h. The samples were digested with 60% nitric acid for 3 h using the "Eco-Pre-Vessel system" (ACTAC; Tokyo, Japan). The digested solution was filtered using a 0.20 μm PTFE filter and diluted with deionized water to 5% nitric acid concentration. The concentrations of ^{133}Cs and ^{85}Rb were determined by inductively coupled plasma mass spectrometry using the ICP-MS 7500cx (Agilent Technologies) with ^{115}In as an internal standard. The concentrations of K, Ca, Mg, and Na were determined from the digested solution using an inductively coupled plasma optical emission spectrometry (ICP-OES; Optima 7300 DV, PerkinElmer). We analyzed nine plants for each cultivar.

3.2.3 ^{137}Cs Tracer Experiment Using Juvenile-Phase Rice

To observe the ^{137}Cs distribution in each leaf, we grew two cultivars Akihikari and Milyang23, in 250 ml of modified half-strength Kimura B solution (Tanoi et al. 2011) containing 1.8 kBq of ^{137}Cs at 30 °C for 16 days, until the seedlings had grown the 6th leaf after emergence. The solution was changed every other day. After the 16-day growth period, the shoots of the plants were separated into leaf sheaths and leaf blades for each leaf stage. After measuring the fresh weights, ^{137}Cs activity was measured in the samples using the well-type NaI(Tl) scintillation counter (ARC-300; Aloka Co., Ltd.). There were four replicates for each cultivar. In addition to the ^{137}Cs experiments, we prepared the same culture set without ^{137}Cs, digesting the leaf samples from the culture with 30% nitric acid using the DigiPREP system (GL Science; Tokyo, Japan). Concentrations of potassium (K), calcium (Ca), sodium (Na) and magnesium (Mg) were measured by ICP-OES (Optima 7300 DV; PerkinElmer).

To analyze ^{137}Cs uptake rate by roots, seedlings of Akihikari and Milyang23 cultivars that had grown the 6th leaf (about 16-day-old seedlings) were cultured in 200 ml of modified half-strength Kimura B solution containing 3.7 kBq of ^{137}Cs at 30 °C under lighted conditions for 30 min. After rinsing the root with tap water, the seedlings were washed with ice-cold half-strength Kimura B solution for 10 min. After cutting roots and shoots and measuring the fresh weight, the ^{137}Cs activities of the samples were measured using the well-type NaI(Tl) scintillation counter (ARC-300, Aloka Co., Ltd.). There were three replicates for each cultivar.

3.3 Results and Discussion

We grew our three chosen cultivars (Nipponbare, Akihikari, and Milyang23) in a culture solution containing ^{137}Cs inside a growth chamber. We then analyzed the ^{137}Cs distribution twice, at the heading stage and at the mature stage. When we measured the distribution at the heading stage, we found that the total amount of ^{137}Cs was lowest in Nipponbare and highest in Milyang23 (Fig. 3.3). We then separated the rice shoots into organs. When we analyzed the proportion of ^{137}Cs in shoots, ^{137}Cs concentrations in the ears of Milyang23 were twice as high compared to Nipponbare and Akihikari ears (Fig. 3.3). At that point we decided to focus on the ears, separating them into husks, brown rice, and rachis branches in the mature stage. We found that ^{137}Cs concentrations in husk and rachis branches were nearly the same between Akihikari and Milyang23, but the ^{137}Cs concentration in the brown rice from Akihikari was half that in Milyang23 (Fig. 3.4). The ^{137}Cs distribution suggests that ^{137}Cs translocation activity from leaves to brown rice occurs differently between Akihikari and Milyang23 varieties.

To confirm the different ^{137}Cs accumulation patterns between Akihikari and Milyang23 in field conditions, we grew these two cultivars in a paddy field in Tokyo

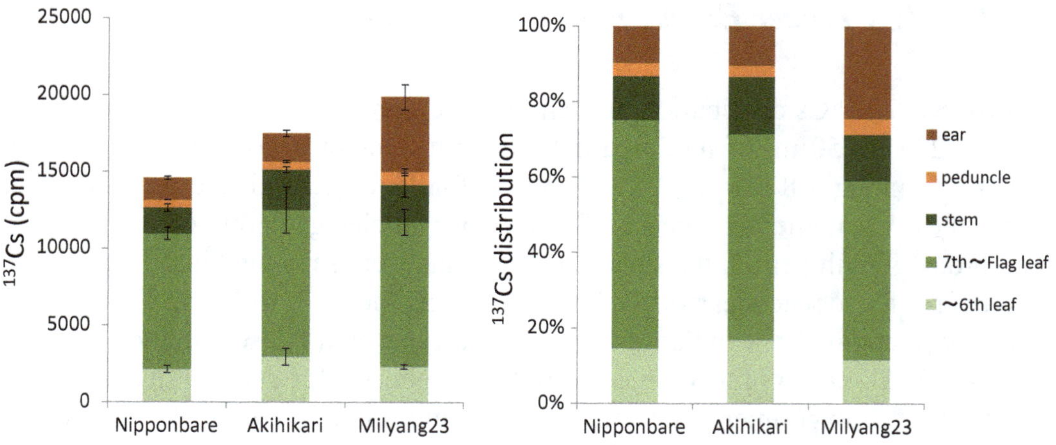

Fig. 3.3 ^{137}Cs amount in rice plants. Left: ^{137}Cs amount in the upper part of the plants. Right: ^{137}Cs distribution pattern in the upper part of the plants. Error bars: standard deviation

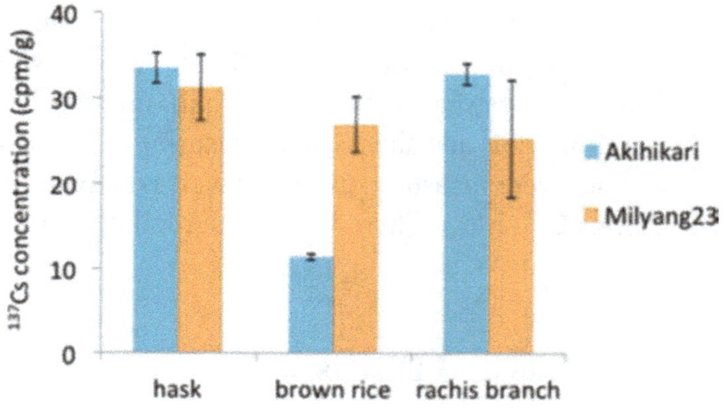

Fig. 3.4 ^{137}Cs concentration in ears of Akihikari and Milyang23. Error bars: standard deviation

(Fig. 3.5). In fact, the ^{137}Cs contamination in the paddy field was so low that the ^{137}Cs in grain was at an undetectable level, and we resorted to measuring ^{133}Cs instead, which confirmed our laboratory results, as described below.

The ^{133}Cs concentrations in brown rice from Akihikari were half those from Milyang23, while ^{133}Cs concentrations in the husks were comparable between the two cultivars (Fig. 3.6). Thus, Milyang23 showed preferential ^{133}Cs accumulation in brown rice over husk compared with Akihikari, which was consistent with our previous laboratory experiments using ^{137}Cs in a hydroponic culture.

There were no similar trends observed between Akihikari and Milyang23 for Rb (Fig. 3.6), K, Na, Ca or Mg concentrations (Table 3.2). The K concentration measured in husks from Milyang23 was double that of husks from Akihikari (Table 3.2). Mineral concentrations in the grain, showing no correlation between K and Cs, suggest that Cs concentrations in grain can be decreased without greatly deteriorating K concentrations simultaneously.

Translocation from old organs to new organs occurs in the juvenile phase. We analyzed 2-week-old plants of Akihikari and Milyang23 using ^{137}Cs. Before carrying

Fig. 3.5 Photos of the paddy field in Tokyo

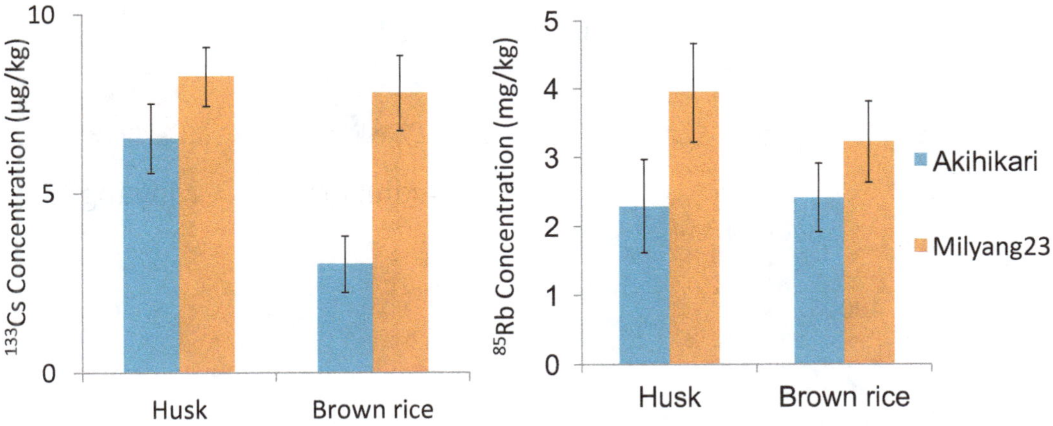

Fig. 3.6 ^{133}Cs and ^{85}Rb concentrations in husk and brown rice of Akihikari and Milyang23. Error bars: standard deviation

out the translocation experiments, we analyzed ^{137}Cs uptake rates in roots and determined that they did not differ between Akihikari and Milyang23 (Fig. 3.7).

Next, we analyzed the ^{137}Cs distribution in young rice plants at the leaf-6 stage. Results indicated that the ^{137}Cs concentration of L6, the newest leaf, was high in Milyang23 and low in Akihikari. On the other hand, the ^{137}Cs concentration of L4B, the oldest leaf blade among the leaves, was high in Akihikari and low in Milyang23 (Fig. 3.8). In general, minerals in leaves are transported via the xylem and phloem, and the phloem contribution is larger in newer leaves. In addition, minerals in old and mature leaves are translocated to new organs via the phloem. These results

Table 3.2 Mineral concentrations in brown rice and husk of Akihikari and Milyang23

	Brown rice		Husk	
	Milyang23	Akihikari	Milyang23	Akihikari
Ca (mg/kg)	0.064	0.095	0.58	0.46
Mg (mg/kg)	0.94	0.97	0.19	0.17
Na (mg/kg)	0.15	0.16	0.39	0.43
K (mg/kg)	1.0	1.1	2.2	1.1

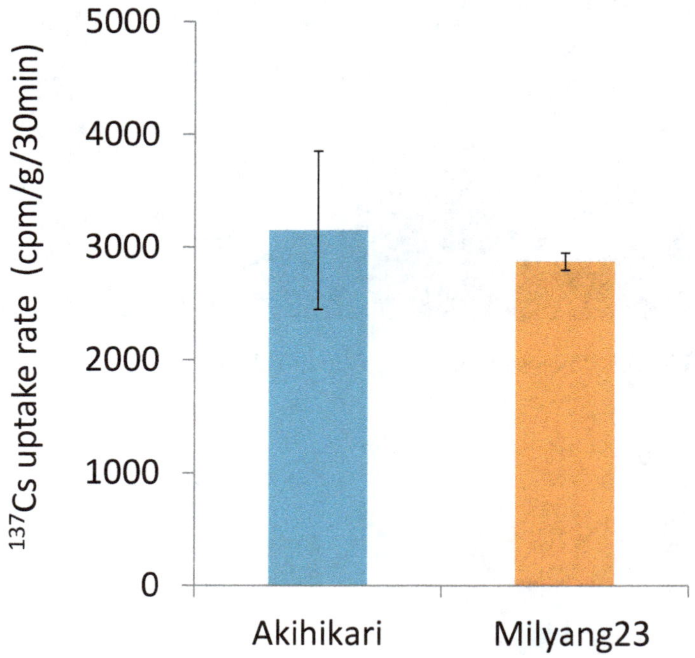

Fig. 3.7 ^{137}Cs uptake rate of Akihikari and Milyang23. Error bars: standard deviation

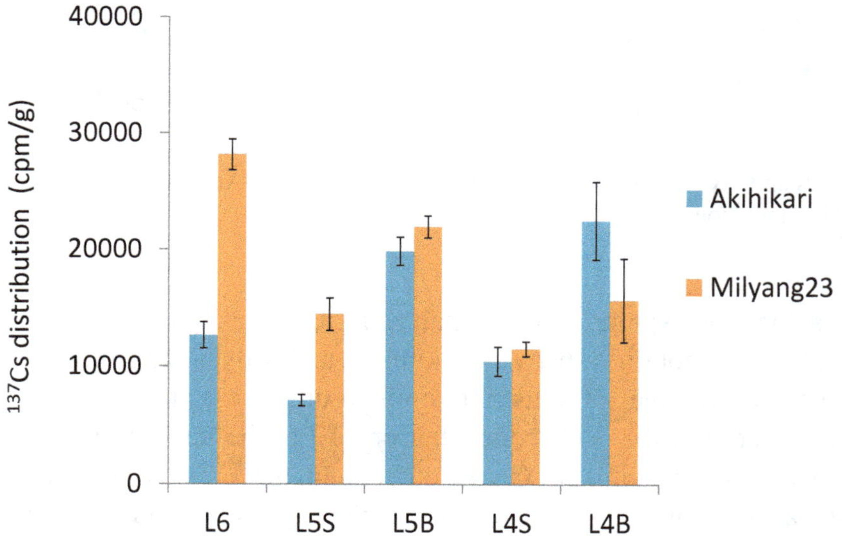

Fig. 3.8 ^{137}Cs amount in leaves of Akihikari and Milyang23. L4, L5 and L6 means 4th leaf, 5th leaf and 6th leaf, respectively. B: blade, S: sheath. L6 is a blade on the 6th leaf whose sheath was too small to collect. Error bars: standard deviation

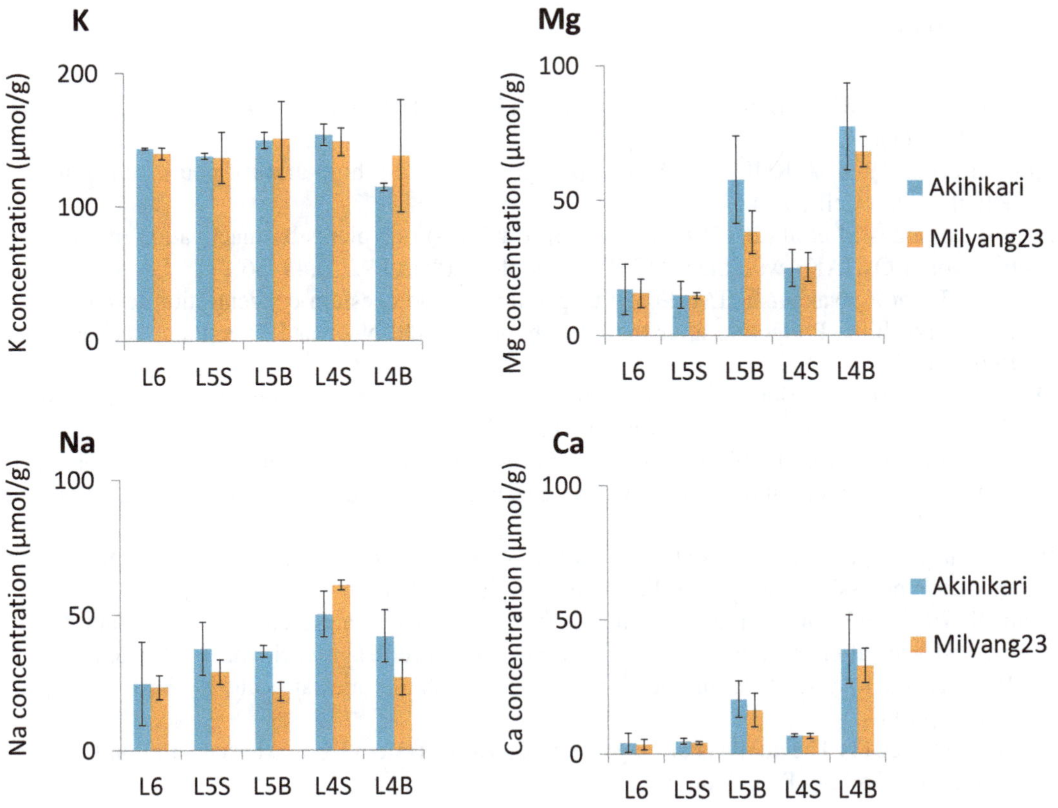

Fig. 3.9 Mineral concentrations in leaves of Akihikari and Milyang23. Error bars: standard deviation

indicate that Cs translocation from old mature leaves to new leaves is more vigorous in Milyang23 than Akihikari in the juvenile phase.

We also analyzed other minerals in the leaves, but there were no differences in mineral distribution between Akihikari and Milyang23 (Fig. 3.9).

To our knowledge, all the transporters reported to mediate Cs^+ transport *in planta* were K^+ channels (Qi et al. 2008; Ishikawa et al. 2017; Nieves-Cordones et al. 2017; Rai et al. 2017); therefore, the candidate transporters involved in Cs translocation should be K^+ channels. Currently, however, there are no reports elaborating on the molecular mechanisms of K^+ translocation in the upper part of a plant. Using the two cultivars, Akihikari and Milyang23, it may be possible to find the translocation system, not only of Cs^+ but also of K^+.

The low-Cs phenotype in the present study is not related to K concentrations. Clarifying the mechanism establishing this phenotype would contribute to breeding low-Cs crops without decreasing K concentrations and, consequently, without lessening the quantity and quality of the grains.

References

Ishikawa S et al (2017) Low-cesium rice: mutation in OsSOS2 reduces radiocesium in rice grains. Sci Rep 7(1):2432

Kim EJ et al (1998) AtKUP1: an Arabidopsis gene encoding high-affinity potassium transport activity. Plant Cell 10(1):51–62

Nieves-Cordones M et al (2017) Production of low-Cs(+) rice plants by inactivation of the K(+) transporter OsHAK1 with the CRISPR-Cas system. Plant J 92(1):43–56

Nihei N, Tanoi K, Nakanishi TM (2015) Inspections of radiocesium concentration levels in rice from Fukushima Prefecture after the Fukushima Dai-ichi Nuclear Power Plant accident. Sci Rep 5:8653

Ono Y et al (2014) Variation in rice radiocesium absorption among different cultivars. Fukushima-Ken Nogyo Sogo Senta Kenkyu Hokoku 48:29–32

Qi Z et al (2008) The high affinity K+ transporter AtHAK5 plays a physiological role in planta at very low K+ concentrations and provides a caesium uptake pathway in Arabidopsis. J Exp Bot 59(3):595–607

Rai H et al (2017) Caesium uptake by rice roots largely depends upon a single gene, HAK1, which encodes a potassium transporter. Plant Cell Physiol 58(9):1486–1493

Tanoi K (2013) Behavior of radiocesium adsorbed by the leaves and stems of wheat plant during the first year after the Fukushima Daiichi Nuclear Power Plant accident. In: Nakanishi TM, Tanoi K (eds) Agricultural implications of the Fukushima nuclear accident. Springer Japan, Tokyo, pp 11–18

Tanoi K et al (2011) The analysis of magnesium transport system from external solution to xylem in rice root. Soil Sci Plant Nutr 57(2):265–271

Chapter 4
Absorption of Radioceasium in Soybean

Naoto Nihei and Shoichiro Hamamoto

Abstract Radioactive materials, primarily radiocesium (^{134}Cs + ^{137}Cs), were released into the environment by the Fukushima Daiichi Nuclear Power Plant accident in March 2011. The percentage of soybean plants that had a concentration of radiocesium over 100 Bq/kg was higher than that of other crops. To examine the reason why the concentration of radiocesium in soybeans was high, its concentration and distribution in seeds were analyzed and compared to rice.

Potassium fertilization is one of the most effective countermeasures to reduce the radiocesium uptake by soybean and nitrogen fertilizer promotes soybean growth. To use potassium and nitrogen fertilizers safely and efficiently, applied potassium behavior in soil and the effect of nitrogen fertilizer on radiocesium absorption in soybean were studied.

Keywords Nitrogen · Seed · Soybean · Potassium · Radiocesium

4.1 Introduction

The Great East Japan Earthquake occurred on March 11, 2011 and it was immediately followed by the nuclear accident at the Fukushima Daiichi Nuclear Power Plant, Tokyo Electric Power Company. Radiocaesium, the dominant nuclide released, was deposited on agricultural lands in Fukushima and its neighboring prefectures, which contaminated the soil and agricultural products.

To revitalize agriculture in the affected regions, the authorities in Fukushima Prefecture have been promoting countermeasures for reducing radiocaesium (RCs) uptake by plants and the remediation of polluted agricultural land. Some of these remediation techniques include the application of potassium (K) fertilizer, plowing

N. Nihei (✉) · S. Hamamoto
Graduate School of Agricultural and Life Sciences, The University of Tokyo,
Bunkyo-ku, Tokyo, Japan
e-mail: anaoto@mail.ecc.u-tokyo.ac.jp

T. M. Nakanishi et al. (eds.), *Agricultural Implications of the Fukushima Nuclear Accident (III)*, https://doi.org/10.1007/978-981-13-3218-0_4

to bury the topsoil, and stripping the topsoil. Emergency environmental radiation monitoring of agricultural products (hereafter, referred to as "monitoring inspections") have been conducted by the Nuclear Emergency Response Headquarters to assess the effectiveness of the countermeasures to keep food safe (Nihei 2016). Approximately 500 food items were monitored, which produced 100,000 data points by the end of March 2016. The monitoring inspections indicated that the percentage of soybean plants with a radiocaesium content of greater than 100 Bq kg^{-1} (fresh weight), was higher compared to other cereal crops (Fig. 4.1); 100 Bq kg^{-1} is the maximum allowable limit of radiocaesium in general foods. Because the cultivation area of soybean plants in Fukushima Prefecture is the second largest after rice, the analysis of RCs uptake by soybean plants is particularly important.

To cultivate soybean after the accident, farmers were recommended to apply potassium fertilizer until the exchangeable potassium (Ex-K; extracted with 1 mol L^{-1} ammonium acetate) is greater than 25 mg K_2O 100 g^{-1} or higher. This recommendation was made because it is known that potassium fertilization is effective for reducing radiocaesium concentration in agricultural crops. However, Ex-K did not increase in the soil for some soybean fields in Fukushima Prefecture after the application of K, resulting in relatively higher RCs concentration in those seeds. Moreover nitrogen fertilizer has a large effect on crop growth, but few studies have examined how nitrogen contributes to the absorption of RCs in soybean.

In this chapter, the reasons why RCs concentration in soybean is higher compared to other crops, potassium behavior in the soil with the low effectiveness of potassium application, and the effect of nitrogen fertilization on RCs absorption in soybean will be discussed.

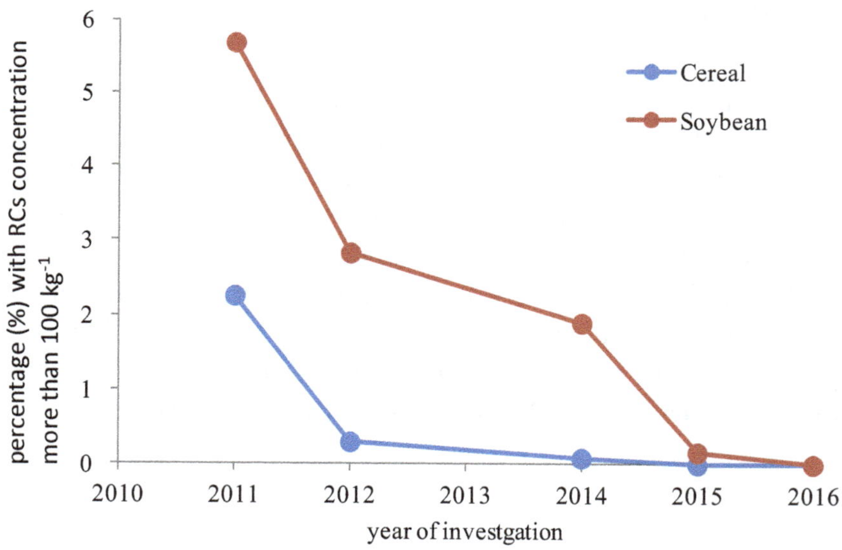

Fig. 4.1 The percentage of soybean plants and cereals with a radiocaesium content exceeding 100 Bq kg^{-1} in the monitoring inspections carried out in Fukushima prefecture

4.2 The Concentration Distribution of Cs in Soybean Seeds

Even though there are some reports of Cs uptake by soybean plants, it is not clear why the concentration of Cs in soybean seed is higher than those in other crops such as rice.

When the concentration distribution of Cs in soybean seeds were analyzed using radioluminography with RCs (Nihei et al. 2017), it was found that Cs was uniformly distributed in the soybean seed, as was potassium, both of which likely accumulated in the cotyledon. The chemical behavior of Cs is expected to be similar to that of K because they are both alkali metal elements and have similar physicochemical properties. Therefore, it is assumed that Cs is also accumulated in the cotyledon like K. In the case of rice grain, the concentration distribution of RCs is localized and rice grains accumulate Cs in the embryo (Sugita et al. 2016), which is only a small part of the rice grain. The different distributions of RCs for rice grain and soybean seed appear to be derived from their seed storage tissues. Soybean seed does not develop its albumen and is therefore called an exalbuminous seed. The cotyledon capacity occupies the largest part of soybean seed. The monitoring inspections measured the edible parts of the crop, i.e., seeds and grains for soybean and rice, respectively. The results suggested that the large capacity of Cs accumulation in soybean seeds is one of the reasons why the concentration of radiocaesium in soybeans was higher than that of rice in the monitoring inspections. In addition, the Cs concentration of each organ and the ratio of absorbed Cs to seeds in mature soybeans were examined. Approximately 40% of absorbed Cs was accumulated in the soybean seeds (Nihei et al. 2018) (Fig. 4.2), while rice grains accumulate only 20% of the entire amount absorbed (Nobori et al. 2016). It is not clear whether the amount of Cs that the soybean plants absorb is larger than that of rice. However, the results from this examination indicates that soybean plants translocate absorbed Cs to its seeds more easily than rice.

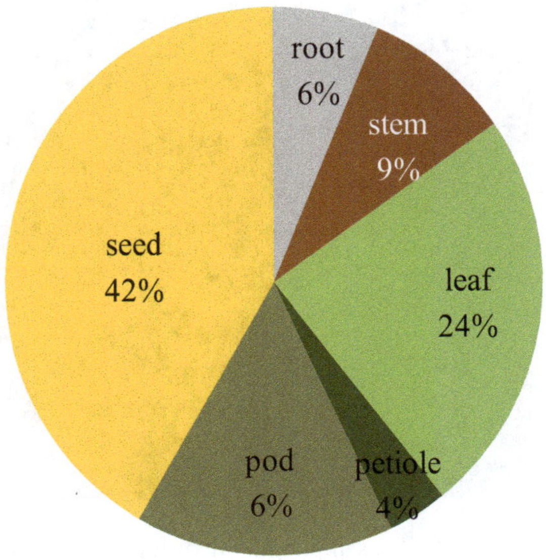

Fig. 4.2 Percentage of ^{137}Cs activity in different organs of mature soybean plants

4.3 Potassium Behavior in the Soil with Low Effectiveness of Potassium Application

It is important to understand the behavior of applied K and the soil characteristics with low Ex-K content to establish efficient techniques to decrease RCs in crops. Therefore, we examined the behavior of K in soil following K fertilizer application (Hamamoto et al. 2018). We tested two types of soil in Fukushima Prefecture (Fig. 4.3). Soil A increases Ex-K with K fertilization (i.e., control treatment), and soil B does not increase Ex-K.

First, a batch experiment was conducted with the two types of soil. After adding KCl (27 mg, 54 mg, 81 mg K 100 g^{-1}) to these soils and culturing for 5 days, Ex-K in soil A increased with the addition of KCl, however, Ex-K in soil B did not increase (Fig. 4.4). There are two reasons for this result: (1) leaching of applied K from the soil and (2) fixation of applied K in the soil. Since this experiment was carried out in a closed system, it was considered that the reason why Ex-K did not increase in soil B was due to the strong adsorption of applied K onto the soil which could not be extracted with ammonium acetate.

Next, a column transport experiment was undertaken to investigate fertilized K behavior in detail (Fig. 4.5). In the repacked soils, the radioisotope tracer, ^{42}K (half-life =12.36 h), was applied to the top 4-cm soil layer, and the soil beneath the top layer was ^{42}K free. Water was applied to the top of the column using a rainfall

Fig. 4.3 Soil A sampling in Fukushima prefecture

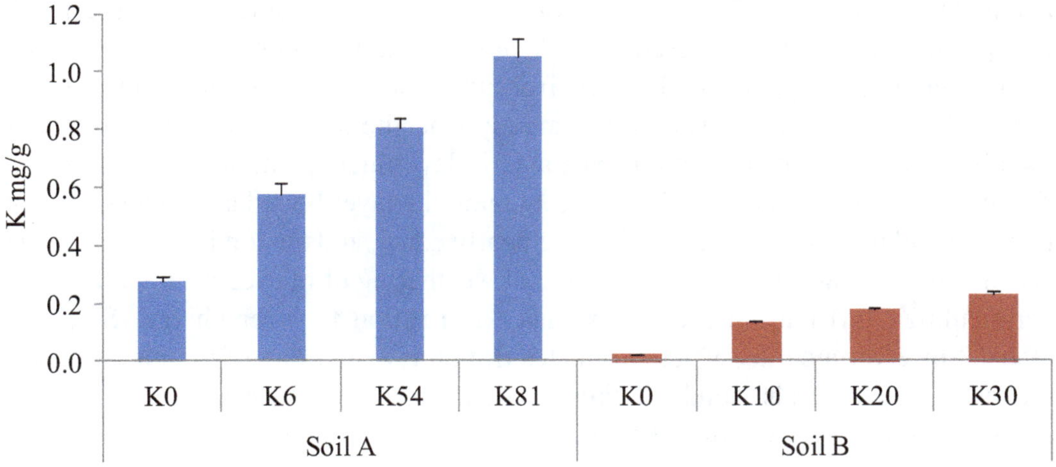

Fig. 4.4 The ratio of exchangeable K to applied K in soils A and B. K0, no K fertilizer. K27, K54, and K81, adding 0.27, 0.54, 0.81 mg K (as KCl) to 1 g soil

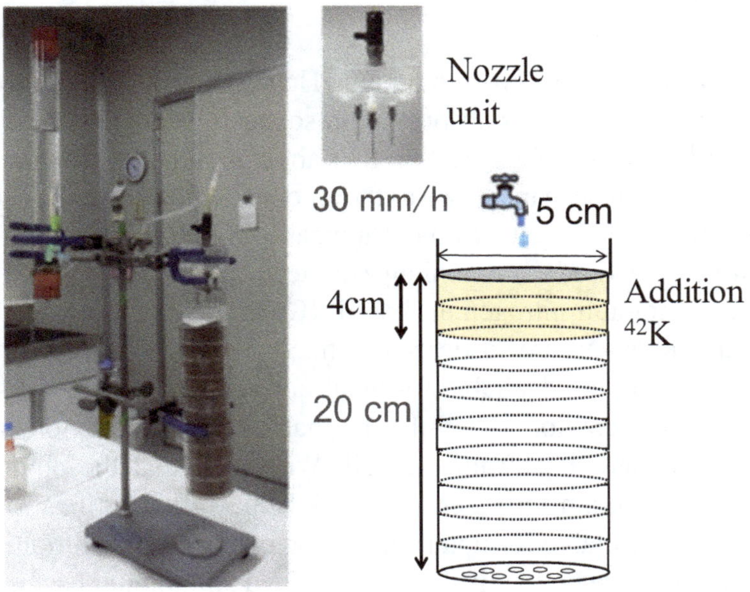

Fig. 4.5 Column transport experiments with soils A and B

simulator connected to a Mariotte's tank. During the experiment, rainfall intensity was maintained at 30 mm/h. After 1 h of 'rainfall', the soil column was horizontally sectioned in 2-cm discs. Samples were taken from each disc, extracted with water (i.e., water-soluble K fraction) and ammonium acetate (i.e., exchangeable K fraction). The ^{42}K activity of each fraction was measured with a semiconductor detector. The ^{42}K obtained by subtracting ^{42}K for ammonium acetate extracts from total ^{42}K was defined as the fixed form. Although 30 mm of water was applied, water content increased almost up to the bottom of the column, however the mobility of the applied ^{42}K was very low in both soils. Only a small quantity of ^{42}K was detected at a soil depth of 4–6 cm after applying ^{42}K to the top 4-cm soil layer in both soils. In soil A, about 75% of ^{42}K retained in each 2-cm disc was the exchangeable form, while in soil B, about 60% of ^{42}K was the fixed form. Again, the findings suggest the soil B can fix a large amount of applied K. Further, from the result of X-ray diffraction (XRD) charts, the clay mineral of soil A was mainly smectite and zeolite, and soil B was mainly vermiculite. In vermiculite such as 2:1 clay mineral, a hollow six-membered ring exists on the tetrahedral silicon sheet facing the layer boundaries. The radius of this void and the potassium ion radius are nearly equal, and the ion is attracted to the six-membered ring. Moreover, the degree of weathering of micaceous minerals (i.e., vermiculitization) may affect the extent of K fixation to layer charge (Sawhney 1970). Since vermiculite, which originates from the weathering of granite, is one of the major clay minerals found in Fukushima, Japan, special attention is needed when K application is used to reduce RCs transfer to crops in such soils.

4.4 The Effect of Nitrogen Fertilization on RCs Absorption in Soybean

Nitrogen (N) has a large effect on crop growth. However, few studies have examined how nitrogen contributes to RCs absorption in soybean. Focusing on this point, we studied the effect of nitrogen fertilizers on RCs absorption in soybean seedlings. The RCs concentration in soybean increased as the amount of nitrogen fertilizer increased. The different forms of nitrogen treatment increased the RCs concentration of soybean in the following order: ammonium sulfate > ammonium nitrate > calcium nitrate. Hence, ammonium-nitrogen increased RCs absorption more than nitrate.

Geometrically, RCs ions are adsorbed and fixed strongly to the clay mineral, and these ions are probably not available for plant uptake. However, because the ionic radius of the ammonium ion is similar to that of the cesium ion, ammonium exchanged and released RCs from the soil. We found that the amount of RCs extracted by the ammonium-fertilizer increased the day after fertilization, and thus, RCs would become available for uptake by soybean plants. In addition, the ammonium and cesium ions are both univalent cations, and ammonium has been found to restrict cesium absorption in hydroponics (Tensyo et al. 1961). However, with the soil was used in the current study, ammonium ions were oxidized to nitrate ions during cultivation. Therefore, we suggest that ammonium fertilizer promotes the activity of RCs in soybean without restricting it.

Soybean cultivation area in Fukushima Prefecture is the second largest after rice cultivation. Therefore, to assist the recovery and revitalization of agriculture in the contaminated regions, it is important to develop agricultural techniques that inhibit Cs accumulation in soybean plants. One technique is the use of K fertilizer, which has been used in contaminated regions. Further studies will be needed to develop more efficient techniques, such as the examination of the mechanism of Cs accumulation in soybean seeds and the improvement of soybean varieties able to alleviate the absorption of radiocaesium.

References

Hamamoto S, Eguchi T, Kubo K, Nihei N, Hirayama T, Nishimura T (2018) Adsorption and transport behaviors of potassium in vermiculitic soils. Radioisotope. RADIOISOTOPES 67:93–100

http://www4.pref.fukushima.jp/nougyou-centre/news/kenkyuuseikasen_h25/h25_housyanou_09.pdf

Nihei N (2016) Monitoring inspection for radioactive substances in agricultural, livestock, forest and fishery products in Fukushima prefecture. In: Nakanishi TM, Tanoi K (eds) Agricultural implications of the Fukushima nuclear accident: the first three year. Springer, Tokyo, pp 11–22

Nihei N, Sugiyama A, Ito Y, Onji T, Kita K, Hirose A, Tanoi K, Nakanishi TM (2017) The concentration distributions of Cs in soybean seeds. Radioisotopes 66:1–8

Nihei N, Tanoi K, Nakanishi TM (2018) Effect of different Cs concentrations on overall plant growth and Cs distribution in soybean. Plant Prod Sci:1–6

Nobori T, Kobayashi NI, Tanoi K, Nakanishi TM (2016) Alteration in caesium behavior in rice caused by the K, phosphorous, and nitrogen deficiency. J Radioanal Nucl Chem 307:1941–1943

Sawhney B (1970) Potassium and cesium ion selectivity in relation to clay mineral structure. Clay Clay Min 18:47–52

Sugita R, Hirose A, Kobayashi NI, Tanoi K, Tomoko M (2016) Nakanishi imaging techniques for radiocaesium in soil and plants. In: Nakanishi TM, Tanoi K (eds) Agricultural implications of the Fukushima nuclear accident: the first three year. Springer, Tokyo, pp 247–263

Tensyo K, Yeh KL, Mitsui S (1961) The uptake of strontium and cesium by plants from soil with special reference to the unusual cesium uptake by lowland rice and its mechanism. Soil Plant Food 6:176

Chapter 5
An Observational Study of Pigs Exposed to Radiation

Junyou Li, Chunxiang Piao, Hirohiko Iitsuka, Masanori Ikeda, Tomotsugu Takahashi, Natsuko Kobayashi, Atsushi Hirose, Keitaro Tanoi, Tomoko Nakanishi, and Masayoshi Kuwahara

Abstract On June 28, 2011, 26 pigs were rescued from the alert area, 17 km northwest of the Fukushima Daiichi Nuclear Power Plant, where radiation levels were approximately 1.9–3.8 μSv/h. The pigs were transferred outside of the radiation alert area to the Animal Resource Science Center (ARSC), The University of Tokyo. It was confirmed by the farm owner that the pigs were never fed radiation-contaminated concentrate and they had access to uncontaminated groundwater (http://www.env.go.jp/press/press.php?serial=16323) while living inside the radiation alert area; however, radiocesium was detected in the rescued pigs' organs, testis/ovary, spleen, liver, kidney, psoas major, urine, and blood, within nine months after the nuclear disaster. Radiocesium levels in samples collected in early January 2012 were significantly lower than those collected in either early or late September 2011, indicating a continuing decrease in radiation levels over that duration. Radiocesium was not detected in organs collected in August 2012. In September 2011, the authors of the present study visited a local farm to collect samples from pigs who remained inside the radiation alert area. Radiocesium concentration in these pigs was nearly ten times higher than from the rescued pigs.

J. Li (✉) · H. Iitsuka · M. Ikeda · T. Takahashi
Animal Resource Science Center, Graduate School of Agricultural and Life Sciences, The University of Tokyo, Kasama, Japan
e-mail: ajunyou@mail.ecc.u-tokyo.ac.jp

C. Piao
Agricultural Resources and Environment, Faculty of Agricultural, Yanbian University, Jilin, China

N. Kobayashi · A. Hirose · K. Tanoi · T. Nakanishi
Isotope Facility for Agricultural Education and Research, Graduate School of Agricultural and Life Sciences, The University of Tokyo, Tokyo, Japan

M. Kuwahara
Department of Veterinary Medical Sciences, Graduate School of Agricultural and Life Sciences, The University of Tokyo, Tokyo, Japan

T. M. Nakanishi et al. (eds.), *Agricultural Implications of the Fukushima Nuclear Accident (III)*, https://doi.org/10.1007/978-981-13-3218-0_5

Seven of the 16 sows rescued were able to reproduce. The present study showed that the age of sow significantly affected their ability to reproduce. These 7 sows had 15 parturition events and birthed 166 piglets, including two malformed piglets. However, the present study confirmed that body weight did not affect reproductive performance. The average body weight of reproductive and non-reproductive sows was 226.3 versus 230.6 kg, respectively.

Hematology analysis showed that red blood cells (RBC) were lower in rescued pigs than in the non-exposed pigs. The level of HGB, HCT, MCV, and MCH, which are all related to RBC counts, were consistent with the changes in RBC between the two groups. The plasma biochemical indexes that relate to liver and kidney functions also showed differences between the two groups of pigs.

The present study was not scientifically designed and did not contain proper control groups for all tests. As a result, we are not able to conclude the exact effects of the radiation exposure to the pigs' health.

Keywords Radiation · Pig · Reproduction · Malformation · Radiocesium · Fukushima Daiichi Nuclear Power Plant Accident

5.1 Introduction

On March 11, 2011, a magnitude 9.1 earthquake occurred off the Pacific coast of Tohoku, Japan. The earthquake triggered a powerful tsunami wave that destroyed local villages and took a tremendous toll on human life. The earthquake and subsequent tsunami also led to the Fukushima Daiichi Nuclear Power Plant disaster. This accident resulted in a nuclear reactor meltdown followed by the release of radioactive fallout. Residents were evacuated within a 20-km radius of the Fukushima Daiichi Nuclear Power Plant. On June 28, 2011, 26 pigs were rescued from an area 17 km northwest of the Fukushima Daiichi Nuclear Power Plant and transported outside of the alert zone to the ARSC of the University of Tokyo, located 140 km southwest of the Fukushima Daiichi Nuclear Power Plant. The pigs had remained in the alert zone for 107 days with radiation levels measuring approximately 1.9–3.8 μSv/h (https://ramap.jmc.or.jp/map/#lat=37.551093857523306&lon=140.96295470535136&z=15&b=std&t=air&s=25,0,0,0&c=20110429_dr). Details of the rescued pigs are listed in Table 5.1.

After the rescued pigs arrived at ARSC, they were given two restricted feedings per day; this is the same diet fed to the reproductive pigs already housed at the ARSC. Body weights were measured once a month. Tests of groundwater and feed confirmed that neither were contaminated by radioactive material. However, two months after the pigs' arrival at ARSC, the health of some pigs deteriorated, and radiocesium was detected in samples taken from the pigs' organs, testis/ovary, spleen, liver, kidney, psoas major, urine, and blood.

The present study was conducted to evaluate bodyweight changes, lifespan and reproductive performance of the rescued pigs. In addition, an observationally based

Table 5.1 List of the rescued pigs and the details of their breed, sex, and age

	Pig ID	Gender	Date of birth
No.1	D34	♂	2009/9/10
No.2	D554	♂	2004/8/19
No.3	D90	♂	2003/12/12
No.4	L315	♂	2004/4/10
No.5	W851	♂	2009/1/19
No.6	Y281	♂	2006/8/4
No.7	D54	♀	2005/8/1
No.8	D345	♀	2006/9/5
No.9	L301	♀	2006/8/9
No.10	D83	♀	2005/9/2
No.11	Y597	♀	2007/6/1
No.12	Y602	♀	2007/6/1
No.13	Y709	♀	2008/4/10
No.14	B559	♀	2004/8/29
No.15	Y669	♀	2008/2/9
No.16	W156	♀	2006/5/5
No.17	D949	♀	2009/6/22
No.18	D10	♀	2005/7/28
No.19	D424	♀	2007/1/15
No.20	W340	♀	2006/8/29
No.21	D85	♀	2005/9/2
No.22	D801	♀	2005/2/15
No.23	W250	♂	2010/10/22
No.24	W251	♂	2010/10/22
No.25	Y36	♂	2007/11/14
No.26	W179	♂	2010/3/31

Breeds of pigs are identified by the following letters: *D* Duroc, *W* Large White, *L* Landrace, *Y* Middle Yorkshire, *B* Berkshire

health check, hematological tests, and blood plasma biochemical parameters were recorded. Transgenerational effects of radiation exposure were also carried out.

5.2 Methods and Material

1. Animals (Pig)

   ① Twenty-six pigs (16 sows and 10 boars) were rescued. Details of these pigs are shown in Table 5.1.

   ② The control group consisted of six pigs, which were the second generation of the rescued pigs, born from rescued sows and reared at ARSC.

③ Blood samples were taken from both rescued pigs and normal confidential adult sows (NCAS), mentioned above, at 4–5 month intervals.
④ Sick pigs were euthanized and their organs collected. To understand the exposure level of the rescued pigs, local pigs who had been fed in the exclusion zone, were also collected and analyzed.

2. Methods

① The isotopes measurement was carried out by the Isotope Facility for Agricultural Education and Research. The nutrition level is very important for the sows' reproductive performance. To confirm the nutrition level of the rescued sows, in the current study the body weight changes were measured once a month and feed intake was adjusted accordingly. The pigs were fed a commercial pig feed, considered for a sow's diet (Multi Rack, Chubu Shiryo Co. Ltd., Japan), with feed volumes of 1.5–2.0 kg/day/head. The composition of the feed was TDN 75.5%, CP 15.5%, CF 6.0%, and EE 3.5%.

The transgenerational effects of radiation exposure were conducted by producing offspring with the parents that had been irradiated.

② Because exposure to ionizing radiation is known to have a lethal effect (El-Shanshoury et al. 2016) on blood cells, this study evaluated ionizing radiation effect on some blood components in pigs. The blood samples were collected once every four or five months and were drawn consistently at around 15:00 h by jugular venipuncture and transferred to tubes containing EDTA (Terumo Venoject II, Tokyo, Japan). Hematology analyses was conducted immediately with the automated pocH-100iV DIFF hematology analyzer (Sysmex Corporation, Japan). The biochemical index also were analyzed immediately by an automatic dry-chemistry analyzer (DRI-CHEM 3500s; Fujifilm Corporation, Japan).

5.3 Results

5.3.1 Exposure Levels in Pigs

After moving pigs out of the alert area to ARSC, the health of some of the rescued pigs deteriorated. Three pigs died within six months; therefore, the present study was planned to check cesium levels in the pigs' bodies and organs. The organs included ovaries or testes, spleen, liver, kidney, psoas major, urine, and blood. Although radiation tests confirmed that the pigs had taken in only non-contaminated food and water while living in the Fukushima area, the results from organ screens were unexpected (Fig. 5.1). Most of the organs tested showed contamination by radiocesium, even though more than 6–9 months had passed since the Fukushima

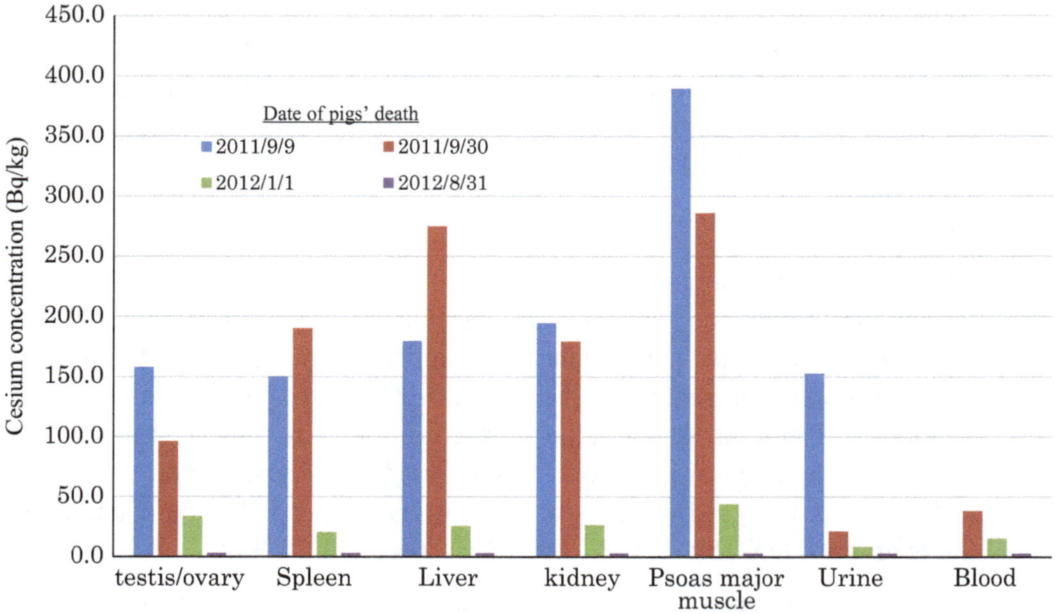

Fig. 5.1 Cesium levels in different organs of rescued pigs (Bq/kg). Different colored bars represent a different pig

Daiichi Nuclear Power Plant disaster, and more than 3–6 months had passed since the pigs had been moved outside of the alert area to ARSC. However, our result confirmed that the rescued pigs, despite consuming only non-contaminated feed concentrate and non-contaminated groundwater, had been exposed to radioactive material. It is likely that the exposure was only due to inhaling contaminated air. Samples from August 1st, 2012 had no radiocesium. These data also indicate that the discharge of radiocesium from the body requires a longer period than previously thought. As shown in Fig. 5.1, the highest levels of radiocesium were present in the psoas major.

To better understand the exposure levels for local pigs who had been feeding in the exclusion zone in September 16th, 2011, in the same area we visited another local pig farm with radiation levels of approximately 1.9–3.8 μSv/h (https://ramap.jmc.or.jp/map/#lat=37.551093857523306&lon=140.96295470535136&z=15&b=std&t=air&s=25,0,0,0&c=20110429_dr). We collected organ samples from these pigs and determined the contamination level of the organs which is shown in Fig. 5.2.

When Figs. 5.1 and 5.2 is compared, the organ radiocesium level of the pigs still feeding in the exclusion zone was nearly ten times higher than the rescued pigs.

Samples were collected from these pigs who fed by grazing, indicating that the pigs most likely acquired the radioactive material from contaminated grass. Again, the highest level of radiocesium was detected in the psoas major muscles, which was consistent with the data collected from the rescued pigs.

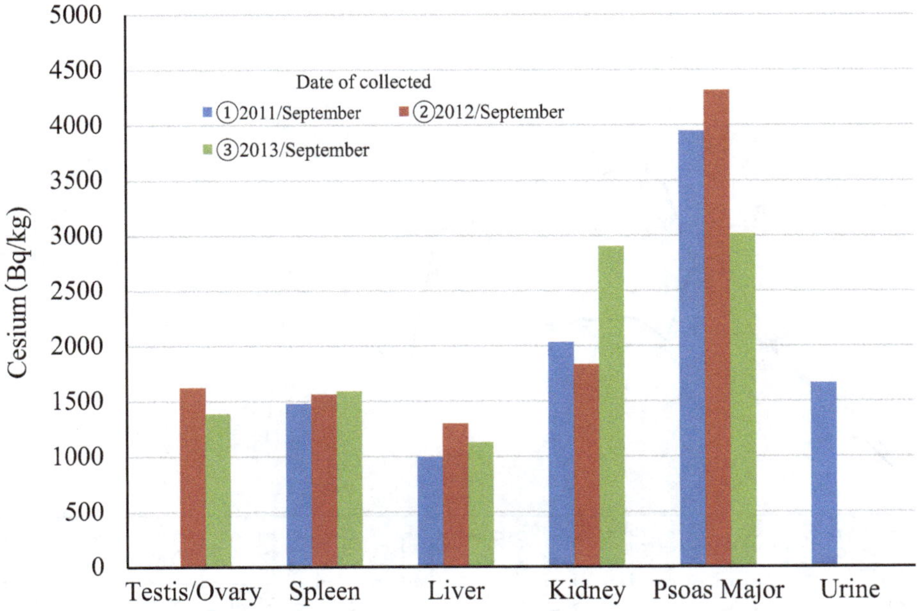

Fig. 5.2 Cesium levels in different organs of local pigs (Bq/kg) who were raised in the alert area after Fukushima Daiichi Nuclear Power Plant disaster. Different colored bars represent a different pig

Table 5.2 Bodyweights of the rescued pigs (kg)

Pigs	No.1	No.2	No.3	No.4	No.5	No.6	No.7	No.8	Average
Duroc(♂)	219	259	220						232.7
Duroc(♀)	221	210	219	218	200	221	187	235	213.9
Large white(♂)	180	119	113	179					147.8
Large white(♀)	213	220							216.5
Yorkshire(♂)	220	200							210.0
Yorkshire(♀)	191	159	159	159					167.0
Berkshire(♀)	203								203.0
Landrace(♂)	203								203.0
Landrace(♀)	165								165.0
Date: 2011/7/6									

5.3.2 *Reproductive Performance*

After being moved to the ARSC, the rescued pigs were allowed some time to recover. The pigs had remained at Fukushima for nearly four months after the accident, dealing with feed shortages, unsanitary stalls, and other environmental stresses. In addition, moving the pigs 140 km to the ARSC induced added stressors, and the animals needed time to become accustom new feeding conditions. When the pigs were moved to the ARSC, the average body weight was determined (Table 5.2).

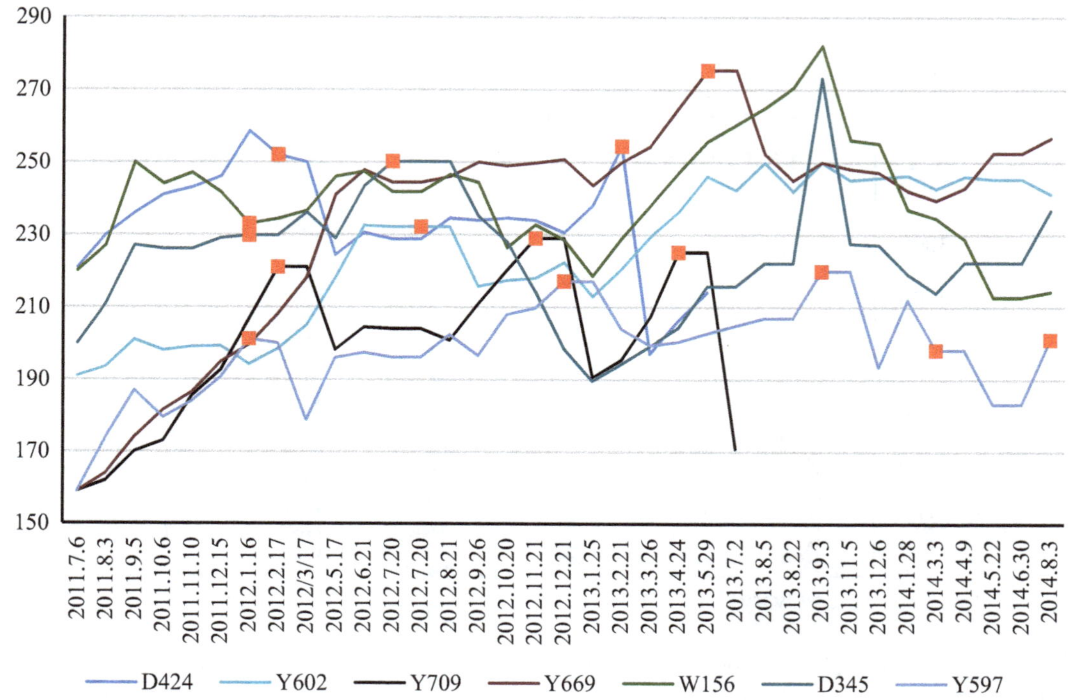

Fig. 5.3 Relationship between bodyweight change and reproduction
■ Parturition

After a two-month recovery period, seven of the sixteen rescued sows gradually showed normal estrous behavior, and mated with boars that also were rescued from Fukushima, and produced offspring. The relationship between body weight and reproductive performance is shown in Fig. 5.3. There was no difference in body weight between reproductive and non-reproductive sows. A total of 7 sows had 15 parturition events and produced 166 piglets, including two malformed piglets (Table 5.3); one piglet had limbs curved disease and the other was intersex. In addition, offspring from the second generation sows who were born from the rescued pigs were produced and reared at ARSC. Six sows had ten parturition events and gave birth to 104 piglets; there were no malformed piglets (Table 5.3).

The present study analyzed sows age on reproductive ability. The age of rescued sows significantly differed between reproductive and non-reproductive sows (Table 5.4).

5.3.3 *Hematology Analyses and Biochemical Indices*

To check the health of the rescued and control pigs, routine blood tests were performed. The following parameters were measured: WBC, RBC, HGB (Hemoglobin), PLT (platelet count), W- SCR (small white cell rate; lymphocyte), W-MCR (middle

Table 5.3 Comparing of reproductive performance between irradiated sows and non-irradiated sows

	Number of sow	Sow of reproduction	Number of parturition	Litter size	Birth weight	Number of piglets	Male	Female	Malformation
Sow of rescued	16	7	15	11.0	1.1	166	85	79	2
		43.80%					51.2%	47.6%	1.2%
Sow of second generation	6	6	10	10.3	1.3	104	49	55	0
		100%					47.1%	52.9%	0%

Sow of second generation: indicating non-irradiated sow

Table 5.4 Relationship between sow age and reproduction success

	Sow	Birthday	Age of rescued	Age of last parturition	
Reproductive sow	Y709	2008/4/10	3.2	2013/5/31	5.1
	Y669	2008/2/9	3.4	2013/6/23	5.4
	Y597	2007/6/1	4.1	2014/9/8	7.3
	Y602	2007/6/1	4.1	2012/8/12	5.2
	D424	2007/1/15	4.5	2013/3/2	6.1
	D345	2006/9/5	4.8	2012/8/17	6.0
	W340	2006/8/29	4.8	2012/2/17	5.5
Non-reproductive sow	L301	2006/5/5	4.9		
	W156	2005/9/2	5.2		
	D85	2005/9/2	5.8		
	D83	2005/8/1	5.8		
	D54	2005/7/28	5.9		
	D10	2005/2/15	5.9		
	D801	2004/8/29	6.4		
	B559	2004/8/29	6.8		
	D949	2006/8/9	2.0	Died soon after been rescued	

white cell rate; monocyte+ Eosinophil granulocyte + basophilic leukocyte), W-LCR (large white cell rate; Neutrophil), W-SCC (small white cell count), W-MCC (middle white cell count), and W-LCC (large white cell count). As shown in Table 5.5, WBC, W- SCC, W-MCR and W-SCC in the irradiated sows was significantly higher than in the NCAS. This indicates that WBC, lymphocyte, monocyte, eosinophil granulocyte, basophilic leukocyte and neutrophil increased. The RBC, HGB and HCT decreased in irradiated sows, while the MCV and MCH increased.

Plasma biochemical indexes, including TP-P (total protein), ALB-P (albumin), TBIL-P (total bilirubin), GOT/AST-P (aspartate amino-transferase), GPT/ALT-P (alanine amino-transferase), ALP-P (alkaline phosphatase), GGT-P (γ-glutamyltransferase), LDH-P (Lactate Dehydrogenase), LAP-P (leucyl amino-peptidase), CPK-P (creatine kinase), AMYL-P (α-amylase), NH3-P (ammonia), TCHO-P (total cholesterol), HDL-C-P (HDL cholesterol), TG-P (triglyceride), UA-P (uric acid), BUN- P (Urea nitrogen), CRE-P (creatinine), GLU-P (glucose), Ca-P (calcium), IP-P (phosphorus), Mg-P (magnesium) were also analyzed (Table 5.6). Blood levels of AST, ALP, GGT, LDH, and TBIL, all related to liver function, were higher in the rescued pigs, indicating some level of liver damage. As shown in Table 5.5, with a decrease in RBC, there was an increase in TBIL. This might be further evidence of liver damage. Furthermore, the high levels of TP and low levels of ALB were indicative of globulin level, like an inflammatory disease, and unstable kidney function. Blood levels of kidney-associated BUN and CRE were also higher in the rescued pigs, further indicating kidney abnormalities, which were confirmed by pathologic anatomy. Low levels of triglycerides and deficiencies in magnesium, potassium, and calcium also were confirmed.

Table 5.5 Comparing of hematology analyses

	WBC	RBC	HGB	HCT	MCV	MCH	MCHC	PLT	W-SCR	W-MCR	W-LCR	W-SCC	W-MCC	W-LCC
Average(C) n = 8	124.7	878.8	17.3	58.8	67.1	19.7	29.4	18.3	55.4	10.0	34.6	69.2	12.4	43.1
Average(I) n = 199	155.0	607.1	12.7	43.1	71.1	20.8	29.4	16.9	54.8	10.3	35.0	80.4	15.9	55.3
SD(C)	10.7	168.0	3.0	10.0	1.8	0.5	0.9	7.7	4.1	1.5	3.0	7.3	2.4	5.4
SD(I)	57.5	132.5	3.0	9.9	5.3	1.3	1.0	9.8	11.1	3.3	10.8	28.2	9.0	35.4
FTEST	0.000107	0.008575	0.000776	0.013811	0.001085	0.002688	0.763371	0.777764	0.004882	0.060379	0.000281	0.000586	0.001978	0.000014
FTEST	**	**	**	**	**	**			**		**	**	**	**
TTEST	0.000012	0.000000	0.000000	0.000014	0.000035	0.000019	0.836276	0.675796	0.302388	0.519669	0.299488	0.009514	0.007178	0.000191
TTEST	**	**	**	**	**	**						**	**	**

C control, *I* irradiated
**$p < 0.01$

Table 5.6 Comparing of biochemical index

	TP-PIII	ALB-P	TBIL-PIII	AST-PIII	ALT-PIII	ALP-PIII	GGT-PIII	LDH-PIII	LAP-P	CPK-PIII	AMYL-PIII
Average(C)n=8	7.4	5.8	0.1	27.0	30.5	111.2	25.2	407.6	32.8	772.6	1816.9
Average(I)n=199	8.4	4.5	0.4	39.5	42.6	76.2	51.1	432.4	52.1	762.3	1865.0
SD(C)	0.6	0.3	0.0	6.1	2.5	18.0	8.1	80.3	4.2	365.6	268.8
SD(I)	0.8	0.6	0.3	25.9	19.0	34.7	32.8	241.0	21.8	578.2	774.4
FTEST	0.261451	0.115884	0.000000	0.000690	0.000013	0.015134	0.000001	0.001503	0.000003	0.338124	0.003008
FTEST			**	**	**	*	**	**	**		**
TTEST	0.000112	0.000000	0.000000	0.000202	0.000000	0.000015	0.000000	0.985323	0.000000	0.913902	0.335975
TTEST	**	**	**	**	**	**	**		**		
	NH3-PII	TCHO-PIII	HDL-C-PIII	TG-PIII	UA-PIII	BUN-PIII	CRE-PIII	GLU-PIII	Ca-PIII	IP-P	Mg-PIII
Average(C)n=8	92.8	76.6	25.7	97.6	0.4	13.5	1.8	89.6	11.5	6.1	2.5
Average(I)n=199	139.5	75.1	23.0	42.8	0.4	17.1	3.2	75.4	8.2	5.2	2.3
SD(C)	17.5	16.5	4.2	38.8	0.0	2.2	0.2	13.8	0.7	1.1	0.2
SD(I)	141.6	22.9	6.8	25.4	0.1	17.6	3.1	18.2	2.3	1.1	0.6
FTEST	0.000006	0.056440	0.167306	0.890043	0.015111	0.000012	0.000000	0.633393	0.003480	0.805897	0.026231
FTEST	**				*	**	**		**		*
TTEST	0.000275	0.350530	0.135123	0.000000	0.002806	0.017142	0.000000	0.049646	0.000000	0.003551	0.037189
TTEST	**			**	**	*	**	*	**	**	*

C control, *I* irradiated
$^*p < 0.05$
$^{**}p < 0.01$

5.4 Discussion and Conclusion

Because of the lack of a scientifically designed control group, we cannot conclude the exact effects of radiation exposure on the rescued pigs. We attempted to provide a reasonable comparison by performing the same tests and analyses on pigs born and raised at the ARSC without radiation exposure. However, we recognize that these "control" pigs were raised at different times and in different environments, making the comparison imperfect. Nevertheless, to our knowledge, this is the first study to observe the effects of radiation exposure on pigs occurring from a nuclear power accident.

The present study confirmed that the pigs that remained in the evacuated area fed on uncontaminated concentrate and drunk uncontaminated groundwater. Inhaling contaminated air may have been the only way the pigs were exposed to radiation. In addition, radiocesium was detected up to a half a year later after the pigs were moved outside the evacuated area. Radiocesium was not the only radioactive material released after the nuclear power plant accident. Thus, we will continue our study, especially to estimate the iodine exposure level and its effect on thyroid function.

To study transgenerational effects of radiation exposure, we made observations of offspring conceived after the parents had been irradiated. The results revealed that only seven of the sixteen sows rescued displayed estrus behavior that resulted in successful births. Two piglets were malformed. Approximately 40~50 days is required for a sperm cell to develop before appearing in the ejaculate (spermatogenesis) (França et al. 2005). Morbeck et al. (1992) indicated that the total time required for a primary follicle containing one layer of granulosa cells to grow to a diameter of 3.13 mm was 98 days. And that a further 19 days were required to reach preovulatory status. The sows described above were mated 7 and 8 months after the nuclear power plant accident. Bille and Nielsen (Bille and Nielsen 1977) reported that in the birth of 29,886 piglets, congenital malformations occurred in 410 piglets (i.e., 1.4% of the piglets born). In the present study, 1.2% of the piglets were malformed, which is consistent with the data from Bille and Nielsen (Bille and Nielsen 1977).

The nine sows considered infertile showed a significant difference in age when rescued. The reproductive sows were approximately 3.2–4.8 years old, while the non-reproductive sows were approximately 4.9–6.8 years old (Table 5.4). While six years of age is not past the reproductive capabilities of a sow, it is, however, an age more easily affected by environmental stressors. Belstra and See (Belstra and See 2004), described that reproductive success generally increases over the first three to four parities, then begins to decline as sows reach the seventh or eighth parity. Sugimoto et al. (1996) showed that profits material increase up to the fifth parity and only show a slight increase from the sixth to tenth parity. Thus, it could be considered that age of 4.9-6.8 year-old-sows was not a non-reproductive age.

It was also confirmed that body weight did not affect reproductive abilities of sows (Fig. 5.3). Rescued sows consisted of five breeds; each breed has a different standard body weight at maturity, so average body weights were difficult to evaluate for this group of pigs; however, body weights for all rescued sows appeared normal.

Table 5.7 List of pig deaths

	Pig	Sex	Date of Birth	Date of death
No.1	D90	M	2003/12/12	2011/9/9
No.2	L315	M	2004/4/10	2013/2/25
No.3	Y281	M	2006/8/4	2013/6/10
No.4	W250	M	2010/10/22	2013/8/7
No.5	Y36	M	2007/11/14	2015/3/4
No.6	D554	M	2004/8/19	2015/4/24
No.7	W851	M	2009/1/19	2016/1/3
No.8	W179	M	2010/3/31	2016/4/24
No.9	W251	M	2010/10/22	2017/11/28
No.10	L301	F	2006/8/9	2011/9/30
No.11	D949	F	2009/6/22	2012/1/1
No.12	D801	F	2005/2/15	2012/8/31
No.13	D10	F	2005/7/28	2012/3/13
No.14	D424	F	2007/1/15	2013/6/10
No.15	D54	F	2005/8/1	2013/7/24
No.16	Y709	F	2008/4/10	2013/7/24
No.17	W340	F	2006/8/29	2014/10/17
No.18	D345	F	2006/9/5	2015/7/31
No.19	Y602	F	2007/6/1	2015/8/3
No.20	W156	F	2006/5/5	2016/2/4
No.21	D85	F	2005/9/2	2016/3/31
No.22	B559	F	2004/8/29	2016/8/9
No.23	Y669	F	2008/2/9	2017/3/22
No.24	D83	F	2005/9/2	2017/7/26
No.25	Y597	F	2007/6/1	2017/7/31

In addition, the sows' reproductive hormones were also confirmed by assaying for plasma steroid hormones, estradiol and progesterone (data not presented). The non-reproductive sows presented low levels of estradiol and progesterone. This could be the main reason why some sows did not show estrous behavior.

High levels of TP and low levels of ALB indicated high levels of globulin. It is consistent with high levels of WBC. Weiss et al. (2009) reported that inflammatory disorders account for a significant percentage of gynecologic disease, particularly in reproductive-age women. Inflammation is a basic method by which humans respond to infection, irritation, or injury. Inflammation is now recognized as a type of nonspecific immune response, either acute or chronic. High levels of TBIL results in low levels of RBC. Meanwhile, the data of AST, ALP, GGT, LDH, NH3 and TBIL and high levels of CRE and BUN indicate liver and kidney functional disorder.

As of April 27th, 2018, 25 of the rescued pigs have died (Table 5.7), with only one pig still alive. Altogether, the rescued pigs showed high mortality and low immunity, so a more detailed study of the effects of radiation exposures is needed. Also, the study of radiocesium transfer coefficients for the body and organs is an on-going project that resulted from the rescue of this group of pigs.

Recent events occurring in the Fukushima Daiichi nuclear disaster area have focused attention on studies evaluating the levels of radioactive material in animals and in agricultural products, but less concern has arisen regarding animal health. The radiation exposure levels were most acute during the first 30 days because of the large quantities of short-lived radionuclides present in the exclusion zone. The pigs used in this study were rescued from their pigsties, which were located only about 17 km from the center of the meltdown area, and remained in this area during the most highly radiative time after the meltdown. In medical laboratory animals, the pig is the closest species to human in evolutionary terms, with the exception of primates. As an animal model, the pig is highly regarded by many scientific fields, including comparative biology, developmental biology and medical genetics (Guo and Shi-ming 2015).

Pigs share many physiological similarities with humans, and offers breeding and handling advantages (when compared to non-human primates), making it an optimal species for preclinical experimentation. The adduced examples are taken from the following fields of investigation: (a) the physiology of reproduction, where pig oocytes are being used to study chromosomal abnormalities (aneuploidy) in the adult human oocyte; (b) the generation of suitable organs for xenotransplantation using transgene expression in pig tissues; (c) the skin physiology and the treatment of skin defects using cell therapy-based approaches that take advantage of similarities between pig and human epidermis; and (d) neurotransplantation using porcine neural stem cells grafted into inbred miniature pigs as an alternative model to non-human primates xenografted with human cells (Vodicka et al. 2005).

Thus, the results of the present study also could provide critical information about the health effects of radiation exposure in humans.

We made observations of hair loss over time as shown in Fig. 5.4. This symptom only occurred once, and if the depilation was the result of radiation exposure, the symptoms were less severe than expected. Sieber et al. (1993) indicated that hair loss is dose-dependent for exposures between 1.0 and 15.0 Gy (1.0–15.0 Sv), and occurs in a linear relationship. No further increase in hair loss was observed for doses ≥15.0 Gy, as 20–30% of the hair remained.

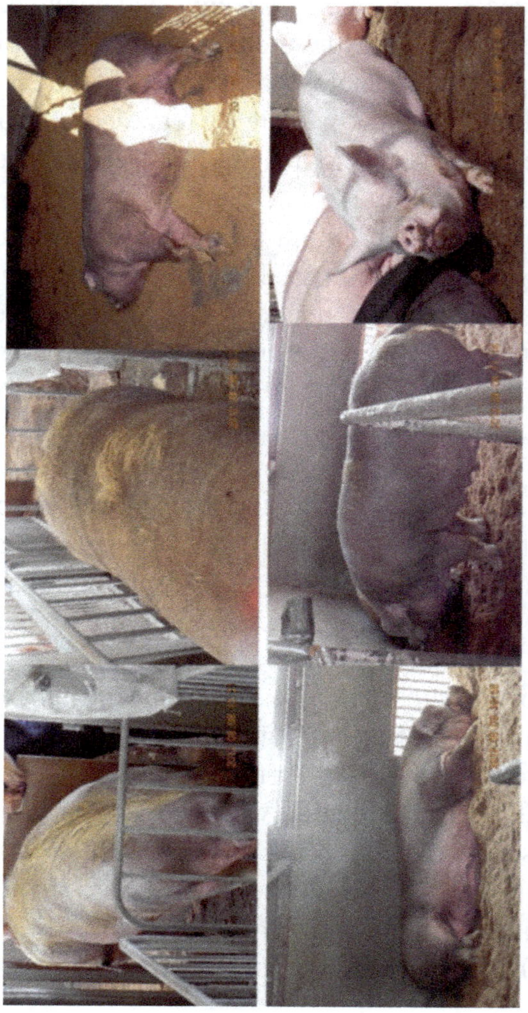

Fig. 5.4 Depilation of the pigs presented after approximately 18 months after the nuclear power plant accident

Acknowledgment Funding by JSPS KAKENHI (Grant-in-Aid for Scientific Research (C)) Grant Number JP25517004, (Grant-in-Aid for Scientific Research (B)) Grant Number 15TK0025, (Grant-in-Aid for Scientific Research (A)) Grant Number JP00291956 and Livestock Promotion Agency of Japan Racing Association.

References

Belstra B, See T (2004) Age, parity impact breeding Traits. Nat Hog Farmer, April 15. http://www.nationalhogfarmer.com/mag/farming_age_parity_impact

Bille N, Nielsen NC (1977) Congenital malformations in pigs in a post mortem material. Nord Vet Med 29(3):128–136

El-Shanshoury H, El-Shanshoury G, Abaza A (2016) Evaluation of low dose ionizing radiation effect on some blood components in animal model. J Radiat Res Appl Sci 9:282–293. https://doi.org/10.1016/j.jrras.2016.01.001

França LR, Avelar GF, Almeida FF (2005) Spermatogenesis and sperm transit through the epididymis in mammals with emphasis on pigs. Theriogenology 63:300–318. https://doi.org/10.1016/j.theriogenology.2004.09.014

Guo W, Shi-ming Y (2015) Advantages of a miniature pig model in research on human hereditary hearing loss. J Otol 10(3):105–107. https://doi.org/10.1016/j.joto.2015.11.001

Morbeck DE, Esbenshade KL, Flowers WL, Britt JH (1992) Kinetics of follicle growth in the prepubertal gilt. Biol Reprod 47:485–491. https://doi.org/10.1095/biolreprod47.3.485

Press Release of the Ministry of the Environment. http://www.env.go.jp/press/press.php?serial=16323

Sieber VK, Wilkinson J, Aluri GR, Bywaters T (1993) Quantification of radiation-induced Epilation in the pig. Int J Radiat Biol 63(3):355–360

Sugimoto T, Nibe A, Takahashi H, Onozato M (1996) A case study on the optimum breeding age of a sow keeping. (Japanese). Jpn J Swine Sci 32(2):41–46

This Extension Site of the Distribution Map for Radiation Dose. https://ramap.jmc.or.jp/map/#lat=37.551093857523306&lon=140.96295470535136&z=15&b=std&t=air&s=25,0,0,0&c=20110429_dr

Vodicka P, Smetana K Jr, Dvoránková B, Emerick T, Xu YZ, Ourednik J, Ourednik V, Motlík J (2005) The miniature pig as an animal model in biomedical research. Ann N Y Acad Sci 1049:161–171. https://doi.org/10.1196/annals.1334.015

Weiss G, Goldsmith LT, Taylor RN, Bellet D, Taylor HS (2009) Inflammation in reproductive disorders. Reprod Sci 16(2):216–229. https://doi.org/10.1177/1933719108330087

Chapter 6
A Composting System to Decompose Radiocesium Contaminated Baled Grass Silage

Takahiro Yoshii, Tairo Oshima, Saburo Matsui, and Noboru Manabe

Abstract Due to the Fukushima Daiichi nuclear power plant accident, a tremendous amount of organic waste (e.g., baled grass silage) contaminated with radioactivity was generated in Tohoku region, northeastern Japan. To establish a safe and efficient way to treat cesium contaminated silage, we investigated the use of aerobic, high temperature composting. Radiocesium (^{137}Cs and ^{134}Cs) contaminated silage (2000 kg, approximately 2700 Bq/kg), water (4000 kg) and matured compost soil (as inoculum, 16,000 kg) were mixed by a wheel loader, and then the mixture was piled up. Air was supplied from the bottom of a compost pile continuously, and the fermentation continued for 7 weeks. The temperature at 100 cm below the surface reached approximately 100 °C. The water content decreased to less than 30% after 7 weeks. The level of radioactive cesium in the final product (18,000 kg) was 265 Bq/kg, which was below the tolerance value for fertilizer (400 Bq/kg) suggested by the Japanese government. The radioactive cesium within silage remained in the final products. We cultivated tomato (fruit), soybean (seed), carrot (root), Italian ryegrass (leaf feed for livestock), Swiss chard (leaf), cosmos (flower) and field mustard (seed) in an experimental farm fertilized with the matured compost made from the radiocesium contaminated silage, for 3 months. Radiocesium levels of edible parts and non-edible parts of each crop were lower than 20 Bq/kg, which was less than one-fifth of the Japanese government value for food (100 Bq/kg). This

T. Yoshii
Institute of Environmental Microbiology, Kyowa Kako Co., Ltd, Tokyo, Japan

T. Oshima
Institute of Environmental Microbiology, Kyowa Kako Co., Ltd, Tokyo, Japan

Faculty of Bioscience and Biotechnology, Tokyo Institute of Technology, Tokyo, Japan

S. Matsui
Graduate School of Global Environmental Studies, Kyoto University, Kyoto, Japan

Research and Development Initiative, Chuo University, Tokyo, Japan

N. Manabe (✉)
Graduate School of Agricultural and Life Sciences, The University of Tokyo, Tokyo, Japan

Faculty of Human Sciences, Osaka International University, Osaka, Japan
e-mail: n-manabe@oiu.jp

T. M. Nakanishi et al. (eds.), *Agricultural Implications of the Fukushima Nuclear Accident (III)*, https://doi.org/10.1007/978-981-13-3218-0_6

research demonstrated that the final product can be used safely as an organic fertilizer.

Keywords Aerobic · High temperature composting · Compost · Cultivation of crops · Fertilizer · Radiocesium · Silage · Waste treatment

6.1 Composting Organic Waste Contaminated with Radioactive Cesium

Due to radiation fallout at the Fukushima Daiichi nuclear power plant accident caused by the giant tsunami associated with the Tohoku earthquake on 11th March 2011, vegetation (e.g., baled grass silage) and carcasses of animals (e.g., livestock, wild boars) were contaminated with radioactive cesium. This contaminated organic matter has mostly remained untouched since 2011. Although the final treatment of organic waste from farms has yet to be been decided, it is essential to investigate the most optimum and safest way to do this to accelerate the recovery process.

Since radioactive cesium-contaminated waste is too large to remove and store, it is essential to reduce its physical weight and volume. Incineration is one solution, but this requires expensive equipment and is undesirable from the standpoint of the environment.

Instead, we have proposed to treat organic waste with microorganisms, that is, to decompose by composting, especially using aerobic, high temperature composting devised by Sanyu limited company (Kagoshima, Japan) (Oshima and Moriya 2008). We have already succeeded to treat contaminated carcasses of cows, pigs and wild boars in towns in Fukushima prefecture. In short, we could demonstrate the conversion of dead animal bodies into a smaller amount of compost soil.

In this article, we converted radioactive contaminated baled grass silage into compost soil which is less bulky. We also showed that the final product is usable as a fertilizer, with the resulting vegetables having a radioactivity level below 100 Bg/kg.

6.2 Reduction in the Volume and Weight of Silage Contaminated with Radiocesium by an Aerobic, High-Temperature Composting System

For over 5 years after the Tohoku earthquake, approximately 2,600,000 kg of contaminated silage with low levels of radiocesium (mainly ^{137}Cs and ^{134}Cs) was left untouched on farms in Kurihara city, Miyagi prefecture, which is located about 150 km north of Fukushima Daiichi nuclear power plant (Fig. 6.1a). From May 2016, we began to compost contaminated silage after receiving a request from Kurihara city local government. The basic information on an aerobic, high-temperature compost system was reported by Manabe et al. (2016). Briefly, we set up two fermenters (4 m width × 3 m height × 6 m depth) in a temporary warehouse (165 m^2; 10 m × 16.5 m) covered with plastic sheets at Kurihara city (Fig. 6.1b, c). Two pipes for aeration were buried in the floor (Fig. 6.1c). The composting process is described in Fig. 6.1. We determined the quantities of radioactive cesium in each silage bale by using in-vehicle sodium iodide scintillator (Mirion Technologies-Canberra Japan Co. Ltd.; with ±10% accuracy). The mean concentration of radioactive cesium in silage was 2723 Bq/kg. Approximately 2000 kg of contaminated silage, 16,000 kg of mature compost (as inoculum) and 4000 kg of water were mixed in a fermenter in order to adjust water content to 45%. The temperature at 100 cm below the top of the compost pile reached 100 °C within 3 days. The compost pile was transferred to another fermenter using a wheel loader once a week and the mixing was repeated seven times. Although the dust and air moisture in the temporary warehouse were collected by circular sprinkling dust collector throughout the fermentation, radioactive cesium in the dust collector was not detected (<2 Bq/kg). The atmospheric radioactivity within a 200-m radius around the examination facility did not fluctuate during the composting process. Thus, the environmental effects of composting low levels of radiocesium contaminated silage appear not to be an issue. The end product of the silage composting became a powder-like substance and the water content was approximately 30%. The total weight of the end product was 18,000 kg. Assuming that the compost added as inoculum was not decomposed during the fermentation, it can be concluded that we can decomposed more than 90% of the contaminated silage by aerobic, high-temperature composting. We detected the presence of *Calditerricola satsumensis*, the genus *Geobacillus* and *Planifilum* as the dominant thermophiles in the composting process by using bacterial 16S rRNA gene targeting denatured gradient gel electrophoresis analyses. The radioactive cesium level of the end product was 265 Bq/kg, which indicated 100 Bq/kg less than the Japanese government tolerance value (400 Bq/kg).

Furthermore, we performed the dissolution test of radiocesium in contaminated silage and the compost. The elution rate of radiocesium from contaminated silage with water or 2% citrate were approximately 60% or 80%, respectively (Fig. 6.2) However, the radiocesium elution rate of the compost made from contaminated silage with water and 2% citrate were 30 and 40%, respectively. Therefore, it was conceivable that a part of radiocesium in the contaminated silage might be changed into an insoluble form by the composting process.

6.3 Dynamics of Radiocesium in Crops Grown with Radioactive Contaminated Silage Compost

In our earlier report, we cultivated seven different crops including soybean (seed), sweet corn (seed), eggplant (fruit), bitter gourd (fruit), potato (rhizome), cabbage (leaf) and ginger (root) on cubic holes filled with the radiocesium contaminated compost soil in a field of the experimental ranch of the University of Tokyo (Manabe

Fig. 6.1 Aerobic, high temperature composting process of radiocesium contaminated silage. Kurihara city is located about 150 km north of Fukushima Daiichi nuclear power plant (**a**).

et al. 2014a, b, c, 2016). The radiocesium levels in the roots, stems, leaves, and fruits of each crop were less than 20 Bq/kg (the Japanese government radiocesium limit in food is 100 Bq/kg).

The compost produced by composting of radiocesium contaminated silage in Kurihara city was 265 Bq/kg and can sell as commercial organic fertilizer. The results of the chemical composition of radiocesium contaminated silage compost indicated that amounts per weight of nitrogen (N) and phosphate (P_2O_5) were 2.6 and 4.6% respectively, which is higher than the compost of non-contaminated cattle feces in Kurihara city. However, the amount of potassium (K_2O) in the silage compost was 0.5%, which is one-quarter the amount of the compost made from non-contaminated cattle feces in Kurihara city (2.2%). Surplus potassium in soil is considered to compete against radiocesium crop absorption. In other words, radioactive cesium will be more easily transferred to plants from potassium deficient soil. We examined the movement of radiocesium into crops from soil when the contaminated silage compost was applied.

We built a greenhouse (5.4 m × 20.0 m) close to the composting facility in Kurihara city. We divided the greenhouse into four sections (4.0 m × 2.6 m) (Fig. 6.3a) before tilling the soil 20 cm deep (Fig. 6.3b). To fertilize the soil in each experimental section, a chemical fertilizer (as a control), 3 kg/m^2 of contaminated silage compost, 20 kg/m^2 of contaminated silage compost, and 2 kg/m^2 of non-contaminated cattle faces fermented compost (as an organic fertilizer control) were applied. Radiocesium concentrations in each soil identified by a germanium semiconductor detector was 39.8 Bq/kg in the chemical fertilizer applied-soil, 75 Bq/kg in the 3 kg/m^2 of contaminated-silage compost applied-soil, 87.6 Bq/kg in the 20 kg/m^2 of contaminated silage compost applied soil, and 76 Bq/kg in the 2 kg/m^2 of non-contaminated cattle feces fermented compost soil. We planted seven crops [tomato (fruit), soybean (seed), carrot (root), Italian ryegrass (leaf feed for livestock), Swiss chard (leaf), cosmos (flower) and field mustard (seed)] on each soil and conducted the cultivation for about 3 months. The effect of contaminated silage compost on the growth of tomato and Swiss chard were similar to that of chemical fertilizer treatment, and it was better than cattle feces-fermented compost. Radioactive cesium levels in the edible and non-edible parts of all harvesting crops were less than 20 Bq/kg. These results indicated that the radiocesium transfer rate to crops from soil applied with contaminated silage compost was very low.

Fig. 6.1 (continued) The outside appearance of the temporary warehouse and the fermenter (**b** and **c**). The circular sprinkling dust collection system (**d**). The contaminated baled grass silage with low levels of radiocesium contamination was left untouched since 2011 (**e**). Measurement of radiocesium level of each bale using an in-vehicle sodium iodide scintillator (**f**). Appearance of the compost after 2 weeks of fermentation (**g**). Mixing the compost pile using a shovel loader (**h**). The end product produced after 7 weeks of composting (**i**)

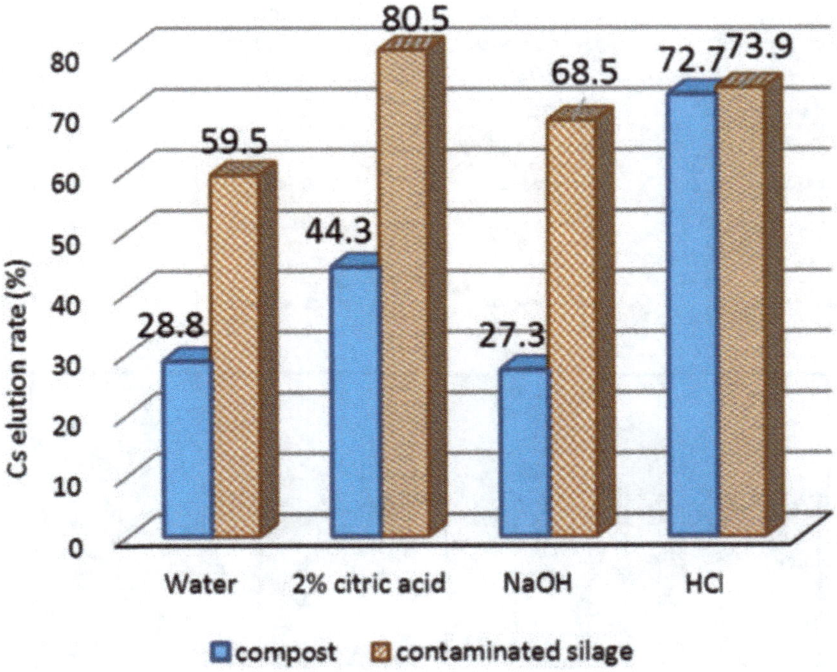

Fig. 6.2 The radioactive cesium elution rate of the contaminated silage and the compost. The silage or compost soil were mixed with ten times volume of water, 2% citrate, 1 N HCl or 1 N NaOH and then agitated for 6 h at room temperature. After centrifugation, the supernatant was filtered by suction through a 0.45-μm Millipore filter. The radiocesium elution rates were determined by measuring radioactivity in each fraction

6.4 Conclusion

Low levels of radioactive cesium contaminated silage were decomposed by aerobic, high-temperature composting system. The final product contained radioactivity of 265 Bq/kg, which was lower than the Japanese government maximum tolerable level in fertilizers (400 Bq/kg). Radiocesium in the silage compost might have changed into a more insoluble form during the composting process because radiocesium in the compost was less soluble in water and citrate solution than before. We cultivated crops fertilized with radiocesium-contaminated silage compost. Radiocesium levels in the edible and non-edible parts of each crop were less than 20 Bq/kg, which is below the Japanese government maximum tolerable level in food (i.e., 100 Bq/kg). These results strongly suggested that aerobic, high-temperature composting system was an extremely useful and safe way to treat low levels of radiocesium-contaminated organic waste. In addition, because the contaminated silage compost can be reused as organic fertilizer or soil conditioner safely, a final landfill site is not required. For reconstruction of sustainable agriculture in radiocesium-contaminated areas of Japan, it can be helpful to treat the organic waste contaminated with radioactive cesium by aerobic, high-temperature composting.

Fig. 6.3 The examination of cultivation of crops in soil fertilized radiocesium contaminated silage compost. Overview of the experimental sections in the greenhouse (**a**). We fertilized the soil with a chemical fertilizer (control treatment in the 1st section). In the 2nd and 3rd sections, the contaminated silage compost (265 Bq/kg) was applied to the soil (3 kg/m^2 and 20 kg/m^2, respectively). And cow feces compost (2 kg/m^2; non-contaminated) was applied to the 4th section. Mixing of compost in the 3rd section (**b**). Cultivation of crops in a greenhouse (**c**). Swiss chard (leaf), tomato (fruit), carrot (root), and cosmos (flower) were cultivated in soil fertilized with the contaminated silage compost 20 kg/m^2 (**d**, **e**, **f,** and **g**)

References

Manabe N, Takahashi T, Li JY (2014a) Future of livestock waste processing with an emphasis on safety of livestock products. Sustain Livest Prod Hum Welf 68:447–451

Manabe N, Takahashi T, Li JY, Tanaka T, Tanoi K, Nakanishi T (2014b) Effect of radionuclides due to the accident of the Fukushima Daiichi nuclear power plant on the recycling agriculture including livestock. New Food Ind 56:45–50

Manabe N, Takahashi T, Li JY, Tanaka T, Tanoi K, Nakanishi T (2014c) Farm animals and livestock products contamination with radioactive cesium due to the Fukushima Daiichi nuclear power plant accident. Sustain Livest Prod Hum Welf 68:1085–1090

Manabe N, Takahashi T, Piao C, Li J, Tanoi K, Nakanishi T (2016) Adverse effects of radiocesium on the promotion of sustainable circular agriculture including livestock due to the Fukushima Daiichi Nuclear Power Plant accident. In: Nakanishi TM, Tanoi K (eds) Agricultural implications of the Fukushima nuclear accident. Springer-Verlag GmbH, Berlin, pp 91–98. https://doi.org/10.1007/978-4-431-55828-6_8

Oshima T, Moriya T (2008) A preliminary analysis of microbial and biochemical properties of high temperature compost. Ann N Y Acad Sci 1125:338–344

Chapter 7
Weathered Biotite: A Key Material of Radioactive Contamination in Fukushima

Toshihiro Kogure, Hiroki Mukai, and Ryosuke Kikuchi

Abstract The eastern area of Fukushima Prefecture, where the Fukushima Daiichi nuclear power plant is located, is covered mainly with weathered granitic soil originated from the geology of this area. Weathered biotite (WB), or partially to almost vermiculitized biotite, is abundant in the soil. WB has frequently been found as radioactive soil particles sorbing radiocesium and has been identified as "bright spots" by autoradiography. Laboratory experiments using the ^{137}Cs radioisotope indicated that WB collected from Fukushima sorbed ^{137}Cs far more efficiently than other clay minerals from ^{137}Cs solutions whose concentration was comparable to that expected for the radioactive contamination in Fukushima. This supports the abundance of radioactive WB particles in the actual contaminated soil. The Cs-desorption property of WB was also different from those of other minerals. If the period of immersion in the Cs solution was more than a few weeks, the sorbed Cs in the WB were hardly desorbed by ion-exchange with any electrolyte solutions. These results imply that decontamination of the radioactive soils is difficult if using "mild" chemical treatments and that most radioactive Cs are now fixed stably (dare one say "safely") by WB in the soil of the Fukushima area.

Keywords Weathered biotite · Radioactivity · Cesium · Sorption · Contaminated soil · Desorption · Frayed edge sites · Ion-exchange

7.1 Introduction

The accident at the Fukushima Daiichi nuclear power plant (FDNPP) in March 2011 released a significant amount of radiocesium (Cs), which caused serious and long-term radioactive contamination of the land around the power plant. It is essential to understand the state and dynamics of radiocesium in the environment to consider its

T. Kogure (✉) · H. Mukai · R. Kikuchi
Department of Earth and Planetary Science, Graduate School of Science, The University of Tokyo, Tokyo, Japan
e-mail: kogure@eps.s.u-tokyo.ac.jp; h-mukai@aist.go.jp; rkikuchi@eps.s.u-tokyo.ac.jp

T. M. Nakanishi et al. (eds.), *Agricultural Implications of the Fukushima Nuclear Accident (III)*, https://doi.org/10.1007/978-981-13-3218-0_7

influence on life, agriculture, the decontamination processes, etc. in Fukushima at present and in the future. More than 6 years after the accident, our knowledge of radiation and radiocesium in Fukushima has been increased owing to the work of a number of research groups in Japan. For instance, measurements of the depth profile of radiation in the soil revealed that most radiocesium remains at a shallow depth and it hardly moves downward with time (e.g., Honda et al. 2015), which suggests that radiocesium is being trapped rigidly in specific materials such as clay minerals in the soil. Many researchers have suggested, mainly based on laboratory experiments, that micaceous minerals such as illite and vermiculite are important for the sorption and retention of Cs in the soil (Comans et al. 1991; Cornell 1993; Evans et al. 1983; Francis and Brinkley 1976; Poinssot et al. 1999; Zachara et al. 2002). For instance, Cornell (1993) summarized the potential adsorption sites for Cs in the micaceous minerals as follows: (1) cation exchange sites on the surface, (2) layer edge sites, (3) frayed edge sites (FES), and (4) internal interlayer sites. Among them, it has been suggested that FES, which are formed around the edges of platy micaceous crystals by weathering, strongly and selectively adsorb Cs (Brouwer et al. 1983; McKinley et al. 2004; Nakao et al. 2008). However, it was not certain whether such micaceous minerals really retain radiocesium in the soil around Fukushima. By analyzing the contaminated soils in Fukushima, we first reported that WB or partially vermiculitized biotite, originating from granitic rocks which constitute the geology of this area, is a dominant sorbent of radiocesium (Mukai et al. 2014). Next, we demonstrated by laboratory experiments that WB sorbed Cs more efficiently than other clay minerals from solutions if the Cs concentration in the mineral was dilute, as in the actual soil in Fukushima (Mukai et al. 2016a).

This paper reviews our recent research with respect to WB in Fukushima, including its structure and Cs-sorption/desorption properties, and discusses the role of WB in the radioactive contamination in Fukushima. Details of the experiments to obtain the results presented here are described in the original papers already published (Mukai et al. 2014, 2016a, b; Kikuchi et al. 2015; Motai et al. 2016).

7.2 Speciation of the Radioactive Particles in the Soil of Fukushima

Although the air dose rate in the areas around FDNPP is quite high due to radiocesium in the soil, the actual concentration of radiocesium in the soil is generally too low to specify its location, even if using recent advanced micro-analytical techniques such as X-ray microanalysis with synchrotron radiation. At present, only autoradiography, which is capable of detecting radiation with a certain (but not enough) spatial resolution, is a practical method for finding radioactive particles in the soil. Autoradiography using imaging plates (IPs), which are reusable and detect radioactive rays efficiently and proportionally to the intensity of radiation, has been applied frequently to find the distribution of radiation or radiocesium in various samples including soils, plant tissues, feathers of birds, etc. (e.g., Nakanishi 2016). IP autoradiography of the samples from Fukushima often showed an inhomogeneous distribution of radiation in the samples, represented by a number of "bright spots" in the IP images obtained by placing the IPs in contact with the samples for a certain time. However, results or reports that identified the materials forming the bright spots were few.

Mukai et al. (2014) collected soil particles of around 50 μm in size by sieving litter soil collected from Iitate village. The soil radioactivity was ~ 10^6 Bq/Kg, and the researchers dispersed the soil particles directly on a special IP which had a fine grid pattern formed by a laser marker (Fig. 7.1a). Using the grid pattern, the soil particles which formed bright spots were easily located under an optical microscope (Fig. 7.1b–d). Then the radioactive particles were picked up by a vacuum tweezer on a micro-manipulator (Fig. 7.2), and moved onto a substrate with double-stick tape for electron microscopy. About 50 radioactive particles were characterized by their morphologies and chemical compositions by a scanning electron microscope (SEM) with an energy-dispersive X-ray spectrometer (EDS), and they were roughly divided into three types: an aggregate of fine mineral particulates, particles rich in organic matter, and "weathered biotite" (Fig. 7.3) (Mukai et al. 2014). The particle

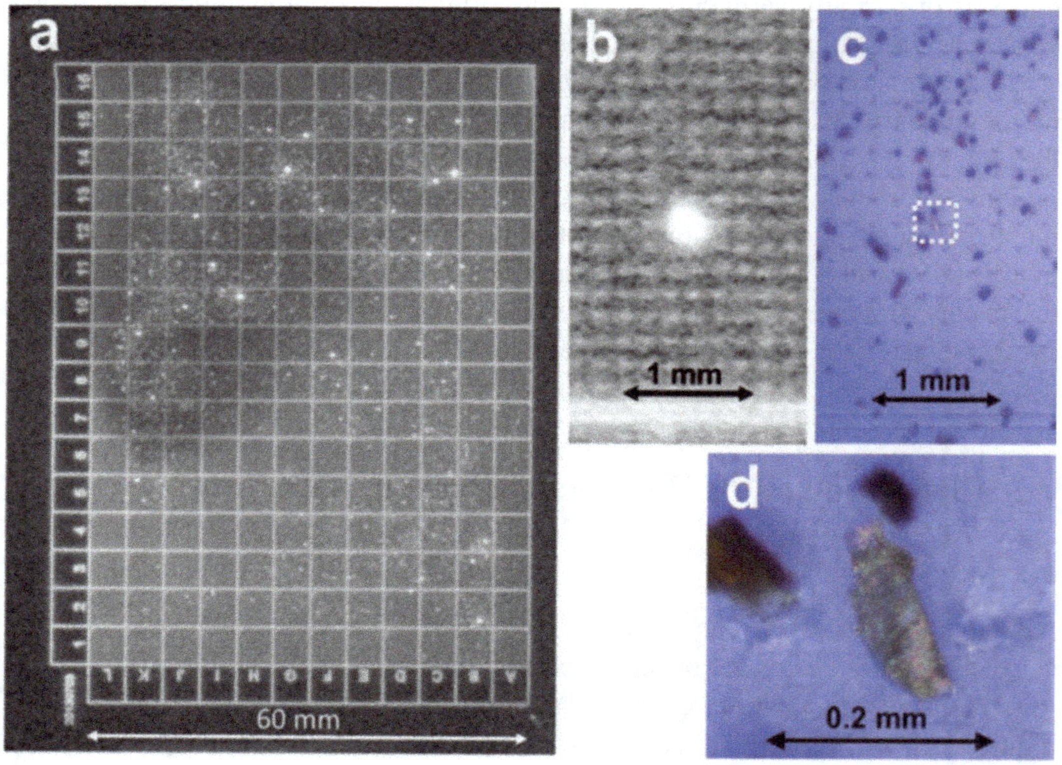

Fig. 7.1 (**a**) Readout image from an imaging plate (IP, Fuji Film FDL-UR-V) with a grid pattern formed using a laser marker, and with radioactive soil particles dispersed on the IP and exposed for around 1 week in a dark box. (**b**) Magnified readout image including a "bright spot" with the image of the grid pattern. (**c**) The position of the IP corresponding to the image in (**b**). (**d**) The soil particle corresponding to the bright spot in (**b**). Probably the particle is weathered biotite, considering its platy morphology. (Mukai et al. 2014)

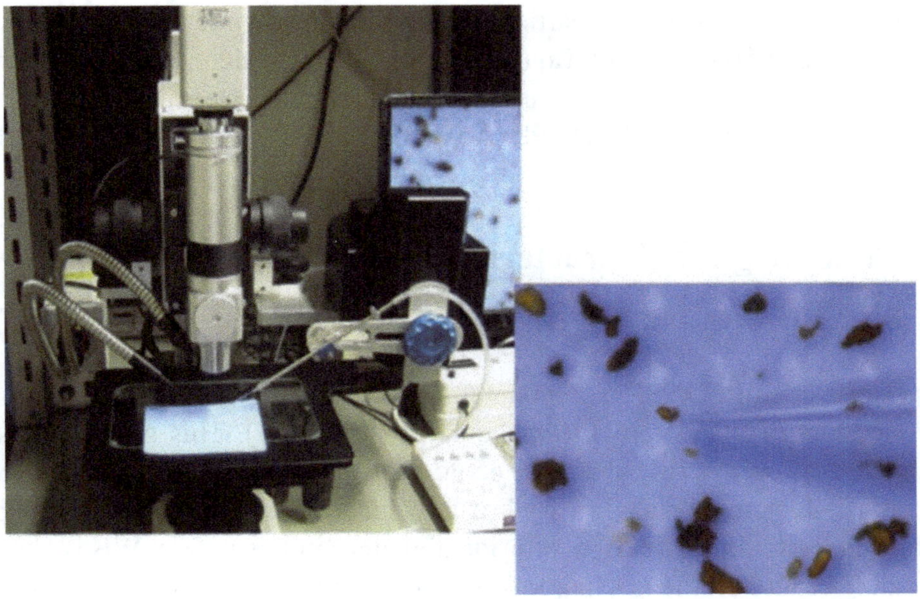

Fig. 7.2 (Left) High-magnification stereomicroscope with a micro-manipulator, used to pick up the soil particles dispersed on IPs. Vacuum tweezers (an evacuated capillary tube) are attached to the micro-manipulator. (Right) A soil particle picked up by the capillary tube

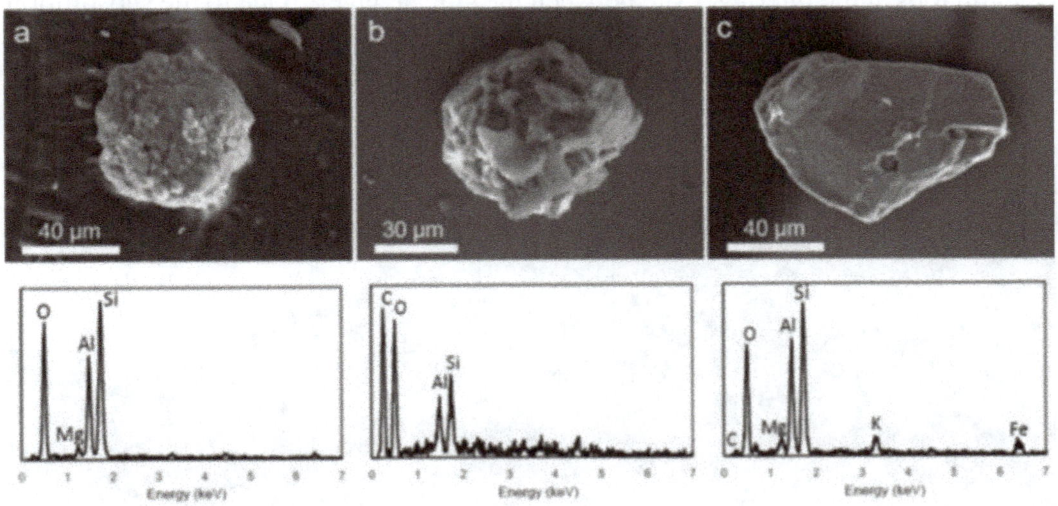

Fig. 7.3 SEM images (upper) and EDS spectra (lower) from the whole particle for typical radioactive soil particles found by IP autoradiography. (**a**) Aggregate of fine mineral particulates. (**b**) Particle consisting of organic matter and a small amount of minerals. (**c**) Weathered biotite with a platy shape. (Mukai et al. 2014)

shown in Fig. 7.3b had a definite platy morphology indicating a mono-crystalline phyllosilicate mineral and had a composition similar to biotite but with less potassium than normal biotite. It is well known that biotite partially changes gradually into vermiculite by the oxidation of iron and by losing potassium through weathering. Here we call such partially vermiculitized biotite “weathered biotite (WB)”. The population ratios of the three types of radioactive particles were almost the same. The radioactivity of each particle estimated from the intensity of the IP signal was in the range of 0.005 ~ 0.05 Bq (Motai et al. 2016), which is far lower than that (a few Bq per particle) of radiocesium-bearing microparticles emitted directly from the broken pressure vessel of FDNPP (Adachi et al. 2013; Yamaguchi et al. 2016).

7.3 Mineralogical Characterization of Weathered Biotite (WB)

The western side of FDNPP is a mountainous area named the Abukuma highland, which consists of granitic rock (Abukuma granitic-body). This granitic rock has been altered by weathering during a geological time to a thick sandy soil called "Masado" in Japanese (Fig. 7.4). By erosion and sedimentation, Masado is spread over forests, agricultural fields, and residential areas in this region. WB is abundant in Masado because biotite is a major constituent mineral of granite. Biotite is structurally classified into a trioctahedral 2:1 type phyllosilicate, and its composition can be roughly expressed as $K(Mg, Fe^{2+}, Fe^{3+})_3Si_3AlO_{10}(OH, F)_2$. Due to weathering on the terrestrial surface, ferrous irons in biotite were oxidized to ferric, which resulted in the leaching of potassium (K) ions and substitution by hydrated magnesium or calcium ions at the interlayer site between the silicate layers. Due to the substitution or hydration, the interlayer space is expanded by ca. 0.4 nm, increasing the thickness of the unit layer from 1.0 nm to 1.4 nm. If this substitution is complete, the resultant

Fig. 7.4 Weathered granitic soils or "Masado" consisting of hills in Fukushima (Ono-Town, Tamura-county, Fukushima Prefecture). The rounded rocks are original granite buried in the soil

new mineral is called "vermiculite". However, WB is generally at an intermediate stage between biotite and vermiculite, forming interstratification of the K-occupied (biotite) and hydrated (vermiculite) interlayers. Such an interstratified structure can be identified directly using recent high-resolution transmission electron microscopy (HRTEM). Figure 7.5 shows a HRTEM image of biotite-vermiculite interstratification in WB collected from Fukushima. The arrowed interlayer regions where the contrast of K is missing were originally occupied by hydrated Ca^{2+} or Mg^{2+} forming a larger space but collapsed due to dehydration in the vacuum in TEM (Kogure et al. 2012). X-ray diffraction (XRD) of WB in Fukushima shows various complicated patterns, depending on the ratio of biotite and vermiculite interlayers, and their mixing feature (Fig. 7.6) (Kikuchi et al. 2015). If WB is immersed in a concentrated Cs solution experimentally, Cs selectively substitutes the hydrated interlayer sites, as revealed by the distinct bright contrast of Cs in the high-angle-annular-dark-field (HAADF) scanning TEM (STEM) image (Fig. 7.7) (Okumura et al. 2014; Kikuchi et al. 2015). However, considering its radioactivity (~0.1 Bq per particle), the actual concentration of radiocesium in the radioactive WB in Fukushima is extremely low compared to the concentration of the artificially Cs-sorbed WB like that shown in

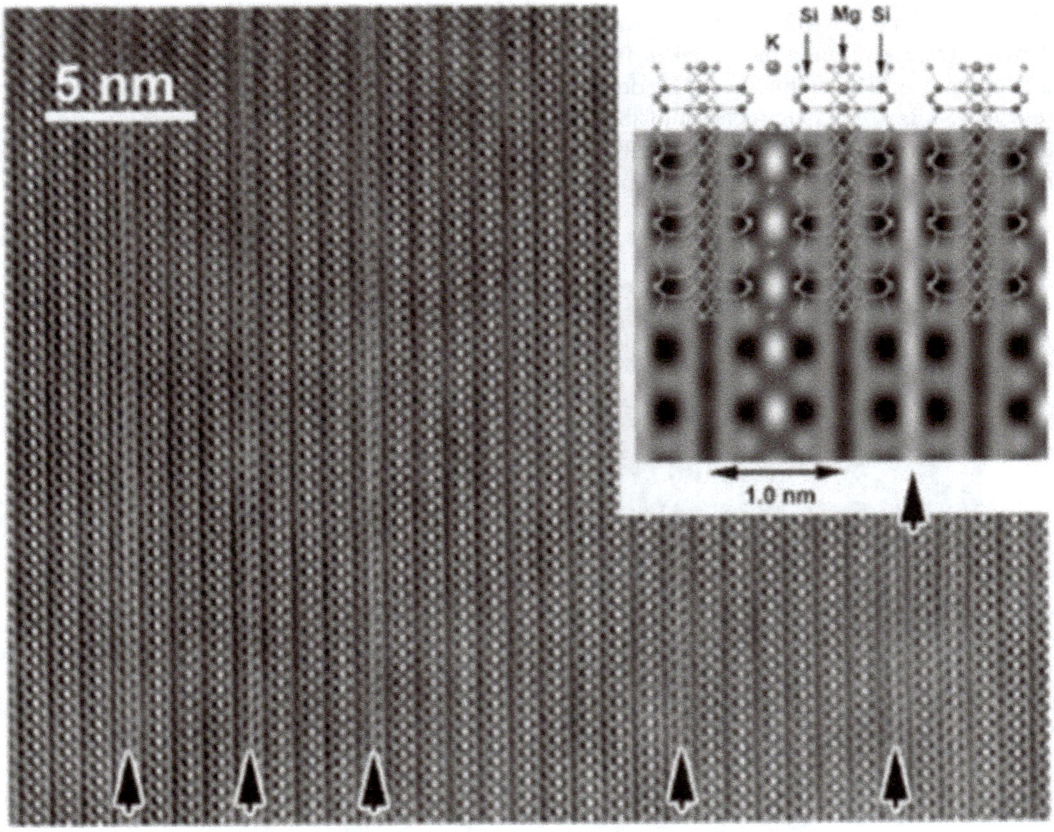

Fig. 7.5 High-resolution transmission electron microscope (HRTEM) image of biotite-vermiculite interstratification in the weathered biotite collected from Fukushima. The arrowed interlayer regions where the contrast of potassium is missing were originally hydrated but collapsed by dehydration in the vacuum in TEM. The inset at the top-right is the simulated contrast for the biotite structure with the potassium (K)-occupied (left) and K-missing (right) interlayers

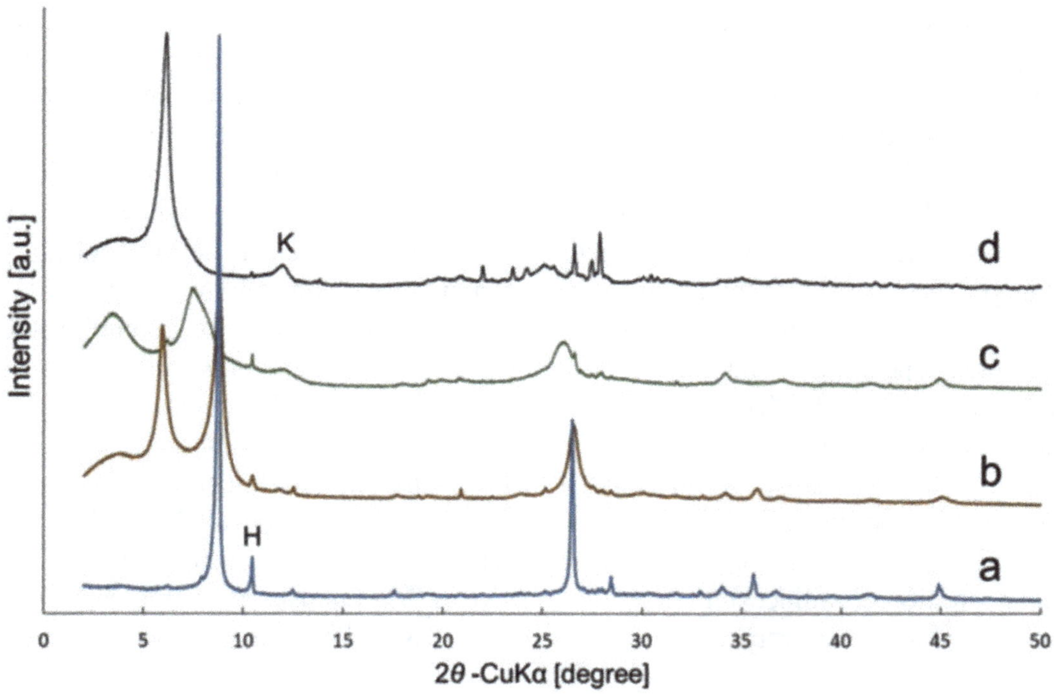

Fig. 7.6 Oriented XRD patterns of various "weathered biotite" specimens in Fukushima. (**a**) Almost fresh or original biotite taken from granite. (**b**) Interstratification of the biotite and vermiculite layers. The biotite layers are dominant and the two types of layers are rather segregated. (**c**) Finely-mixed interstratification of the biotite and vermiculite layers. (**d**) Interstratification of the two types of layers and the vermiculite layers are dominant. The broad peak marked with "K" is kaolinite or halloysite, and the sharp peak marked with "H" is hornblende. (Kikuchi et al. 2015)

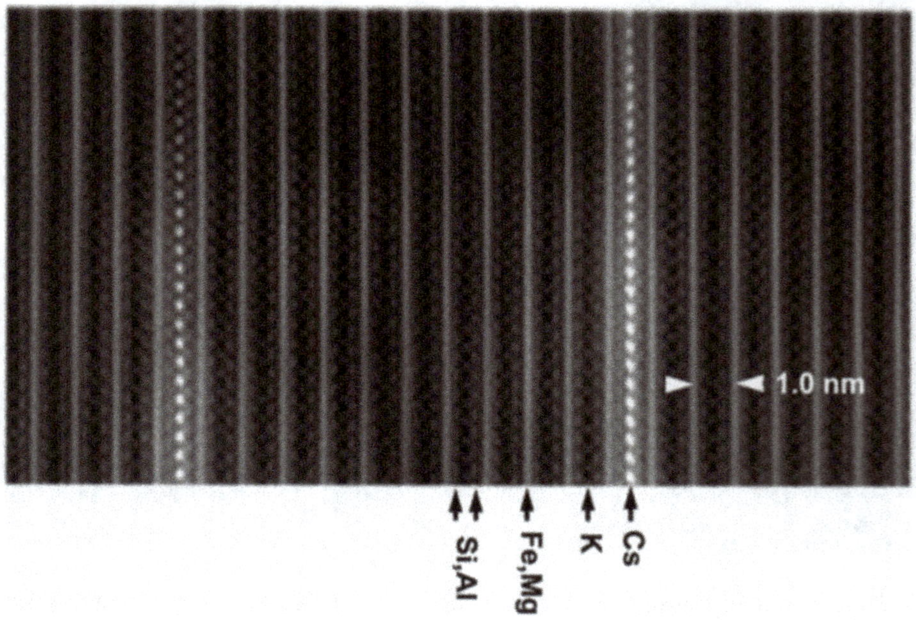

Fig. 7.7 High-angle annular dark-field (HAADF) image of weathered biotite to which cesium was sorbed experimentally, recorded using a scanning transmission electron microscope (STEM). (Kikuchi et al. 2015)

Fig. 7.7. Hence, direct localization of radiocesium sorbed in the weathered biotite by high-resolution electron microscopy or any micro-analytical techniques is not possible. We need to apply other techniques to determine (or at least form a conjecture on) the sorption site for the radiocesium in WB.

As stated above, several researchers have demonstrated experimentally that micaceous minerals such as illite and vermiculite have a high affinity with Cs, and have suggested that this property is due to the FES located around the edge of the platy crystals of the minerals, where Cs with a larger ionic radius than other alkali ions are thought to be selectively and strongly fixed owing to the tapered interlayer spaces (Mckinley et al. 2004; Poinssot et al. 1999; Zaunbrecher et al. 2015). To consider its correctness and the actual sorption sites of radiocesium in WB, the following experiment was conducted (Mukai et al. 2016b). In general, the special resolution of IP autoradiography is not good enough to distinguish the distribution of radiation in a radioactive particle of even a few hundred microns. Instead, we cut platy radioactive WB crystals using a focused-ion-beam (FIB) micro-processing instrument into several fragments and separated them from each other enough that they could be resolved in the IP autoradiography using the micro-manipulator (Fig. 7.2). As shown in Fig. 7.8, all the fragments had similar radiation, regardless of the location in the grain, or with/without the original edge of the platy crystal, indicating that radioactivity was not concentrated around the edge but rather homogeneously distributed in the crystal. However, it is unlikely that radiocesium is diffused deeply inside the crystal structure of WB in Fukushima. The actual particles of WB are not uniformly dense but have a laminated structure with many open spaces caused by cleavages of biotite (Fig. 7.9). Probably, solutions containing radiocesium easily filled the spaces by capillary action and Cs was sorbed to appropriate sites on the cleaved surfaces of WB. This issue will be discussed further in the next section.

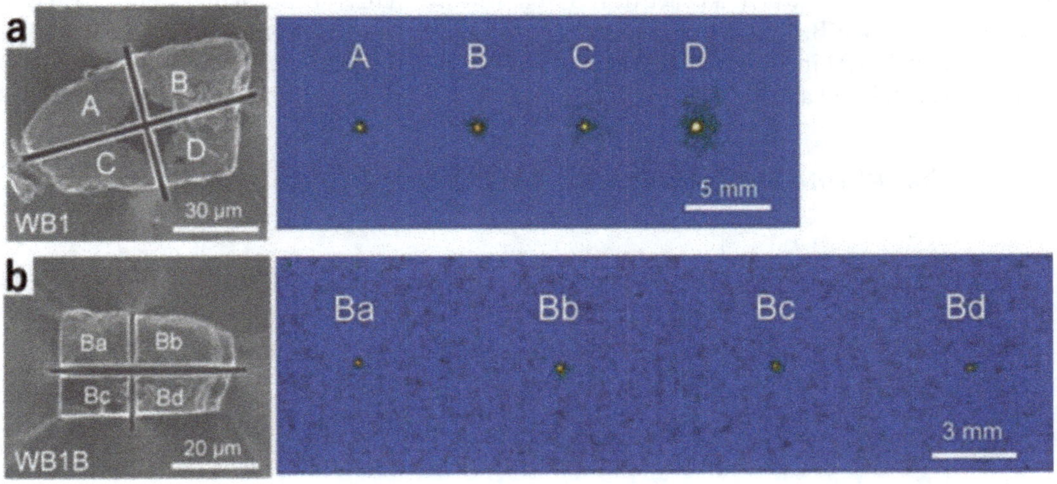

Fig. 7.8 (**a**) SEM image of radioactive WB cut into four fragments by FIB, and a readout IP image from the four fragments, after separating them from each other. (**b**) SEM image of the fragment "B", cut into four fragments, and readout IP image of the four fragments, after separating them from each other. (Mukai et al. 2016b)

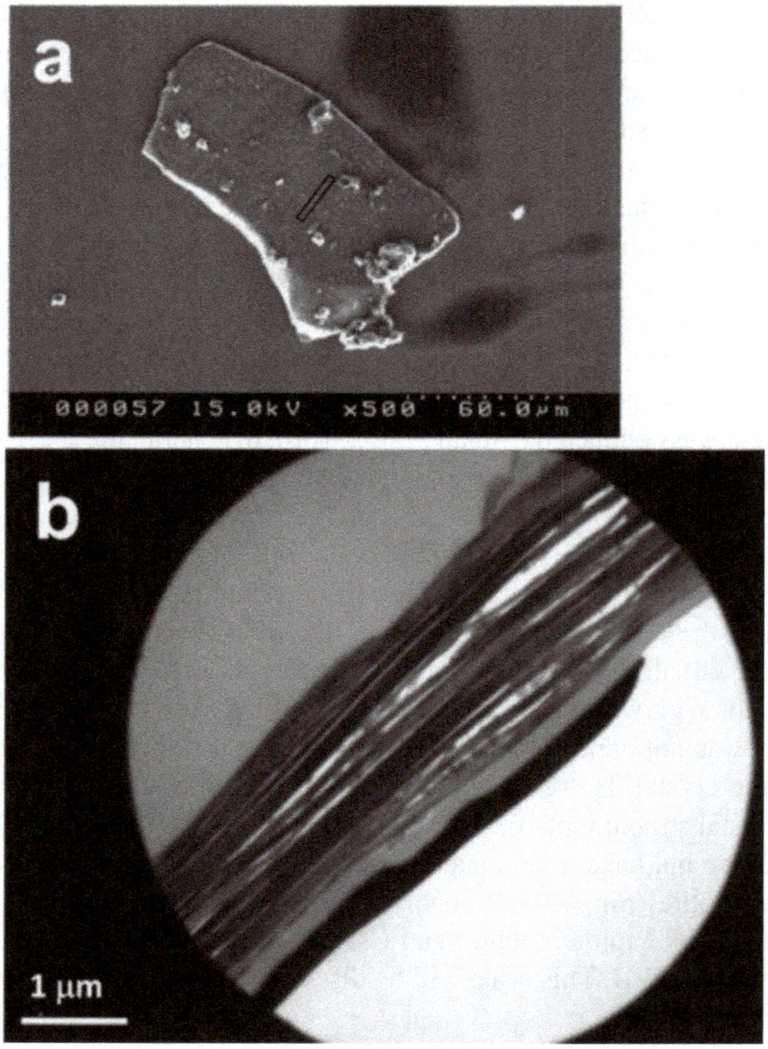

Fig. 7.9 (**a**) SEM image of a radioactive WB from which the cross-sectional thin section for TEM was fabricated by FIB, from the area indicated with the elongated square around the center. (**b**) Cross-sectional TEM image of the WB in (**a**), showing a laminated structure with many cleavages and spaces. (Mukai et al. 2014)

7.3.1 Sorption and Desorption Behavior of Cs to WB

Based on the analyses of the radioactive soil particles in the field as described above, it is suggested that WB is an important material sorbing radiocesium in the soil, influencing the dynamics and fate of radiation in the soil around Fukushima. Hence, we investigated the Cs-sorption/desorption properties of this mineral in the laboratory. For instance, a question to be answered is why WB was frequently found as radioactive soil particles in Fukushima? Was it owing to the abundance of WB in Fukushima or its superior Cs-sorption ability? A study by NIMS (the URL is provided in the Reference section) which surveyed the Cs-sorption ability of various clay minerals after the accident indicated that WB (vermiculite in the study) does not predominantly sorb Cs, in contrast to other micaceous clay minerals like illite and smectite.

However, these data were obtained using a solution with a cesium concentration as low as sub-ppm (~10^{-5} mol L^{-1}). It is natural to expect that the Cs-sorption ability of minerals is dependent on the concentration of cesium in the solution and/or the solid-solution ratio because there are various sorption sites in the minerals. The actual concentration of radiocesium in the rain which caused the radioactive contamination of the soil in Fukushima is considered to be very low. For instance, the amount of rainfall in Iitate village, a seriously contaminated area in Fukushima, over a few weeks after the nuclear accident was around 10 mm according to the records of the Japan Meteorological Agency (the URL is provided in the Reference section). On the other hand, the amount of ^{137}Cs per unit area deposited on Iitate village was ~10^6 Bq/m^2, according to a report by JAEA (the URL is provided in the Reference section). From these values, the concentration of radiocesium in the rain should have been of the order of 10 ppt (10^{-10} mol L^{-1}). To discuss the contamination event affecting the Fukushima soil, sorption experiments should be conducted with such a low concentration. However, this is close to or below the detection limit of the most sensitive analytical instruments. This problem is solved if radiocesium itself is used as the cesium source and the sorption/desorption amount is estimated by measuring the radiation of the sorbed radioisotope, as used in previous studies (Poinssot et al. 1999; Ohnuki and Kozai 2013). Moreover, we evaluated the sorption amounts of ^{137}Cs in the minerals by measuring the radiation in the individual mineral particles quantitatively using IP autoradiography, instead of counting gamma-rays. An advantage of using IP autoradiography is that we can investigate a reaction between a solution and *multi*-minerals. Such reaction can reproduce more practically the event in which radiocesium in the rain was sorbed to the soil composed of several mineral species.

The following RI experiment was conducted: Several mineral species including WB, with four or five particles ~50 μm in size for each species, were arranged within an area of ~6 mm × 6 mm on an acrylic substrate with double-stick tape. Then, 50 μL solutions containing 3.7, 37 and 370 Bq/mL (0.185, 1.85 and 18.5 Bq in the solutions) of ^{137}Cs were dropped to cover all the particles on the substrates. After immersion in the solutions for a certain time, the substrates were washed away with running water and dried, then placed in contact with an IP. The readout images of the IPs are presented in Fig. 7.10. Under all conditions with different ^{137}Cs concentrations and immersion periods, the amount of ^{137}Cs sorbed by WB collected from Fukushima was much higher than that by other minerals. In the case of lower concentrations of ^{137}Cs and/or shorter reaction periods, radiation was detected only from WB. At the concentration of 18.5 Bq/50 μL and with a reaction length of 1 day, the amount of ^{137}Cs sorbed by WB was about two orders of magnitude higher than that sorbed by the other clay minerals. These results definitely indicate that WB has an ability to sorb radiocesium very efficiently in the soil of Fukushima, and our observation above, that WB was frequently found as the radioactive soil particles in Fukushima, is reasonable. Besides its high Cs-sorption ability, the Cs-sorption mechanism of WB at a low concentration is probably different from that of other minerals such as smectite. Figure 7.11 shows the amount of radiocesium sorbed to WB and ferruginous smectite (SWa-1, a reference sample of Clay Mineral Society, USA) as a function of immersion time in the Cs solution. For smectite, the sorption

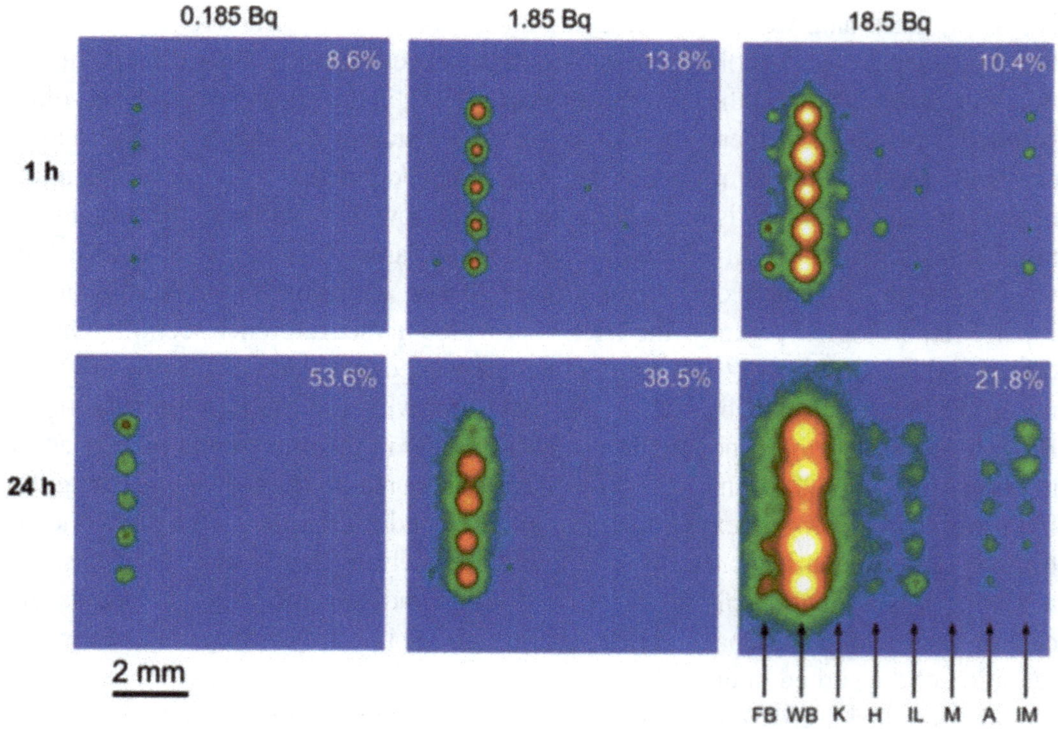

Fig. 7.10 A matrix of the readout images of IPs exposed by the substrates with various mineral particles (five particles for each species) sorbed radiocesium from the solutions. The radioactivity input to the solution and reaction time are at the top and left, respectively. The figure at the top-right of each image is the percentage of radioactivity (or ^{137}Cs) sorbed to the whole mineral particles, estimated from the IP signal. The abbreviations at the bottom-right mean FB: fresh biotite, WB: weathered biotite, K: kaolinite, H: halloysite. IL: illite, M: montmorillonite, A: allophan, IM: imogolite. (Mukai et al. 2016a)

was completed very quickly (probably in less than 1 h) but for biotite, the progress of the sorption was slow, continuing probably more than 1 day.

Next, we investigated the Cs-desorption property of WB, using the Cs-sorbed WB particles prepared by the experiments in Fig. 7.10. In the experiments, substrates with the ^{137}Cs-sorbed WB particles were immersed in various electrolyte solutions and the amount of leached radioactive Cs was estimated also by IP autoradiography before and after the immersion (Fig. 7.12a). The results of the experiments are summarized in Fig. 7.12b. The solutions of NH_4NO_3, KNO_3, and $CsNO_3$ (1 M) desorbed only small amounts of ^{137}Cs from the WB. In addition, the desorption ratios of ^{137}Cs of these solutions remained almost unchanged with time. In contrast, ^{137}Cs was considerably desorbed by $LiNO_3$ and $NaNO_3$, whereas $Mg(NO_3)_2$ and $Ca(NO_3)_2$ were not so effective at desorbing ^{137}Cs. The acidic solutions (HCl and HNO_3, 0.1 M) also desorbed ^{137}Cs effectively from WB. In this experiment, the immersion time of WB in the ^{137}Cs solution to sorb radioactive Cs was 24 h. We extended this to 168 h, 336 h, and 672 h. Radioactive particles of WB collected from Fukushima, or WB which sorbed radioactive Cs in the field ("natural radioactive WB"), were also subjected to the same desorption experiment (Fig. 7.13). $LiNO_3$ and $NaNO_3$ solutions were effective in desorbing ^{137}Cs from WB when the sorption

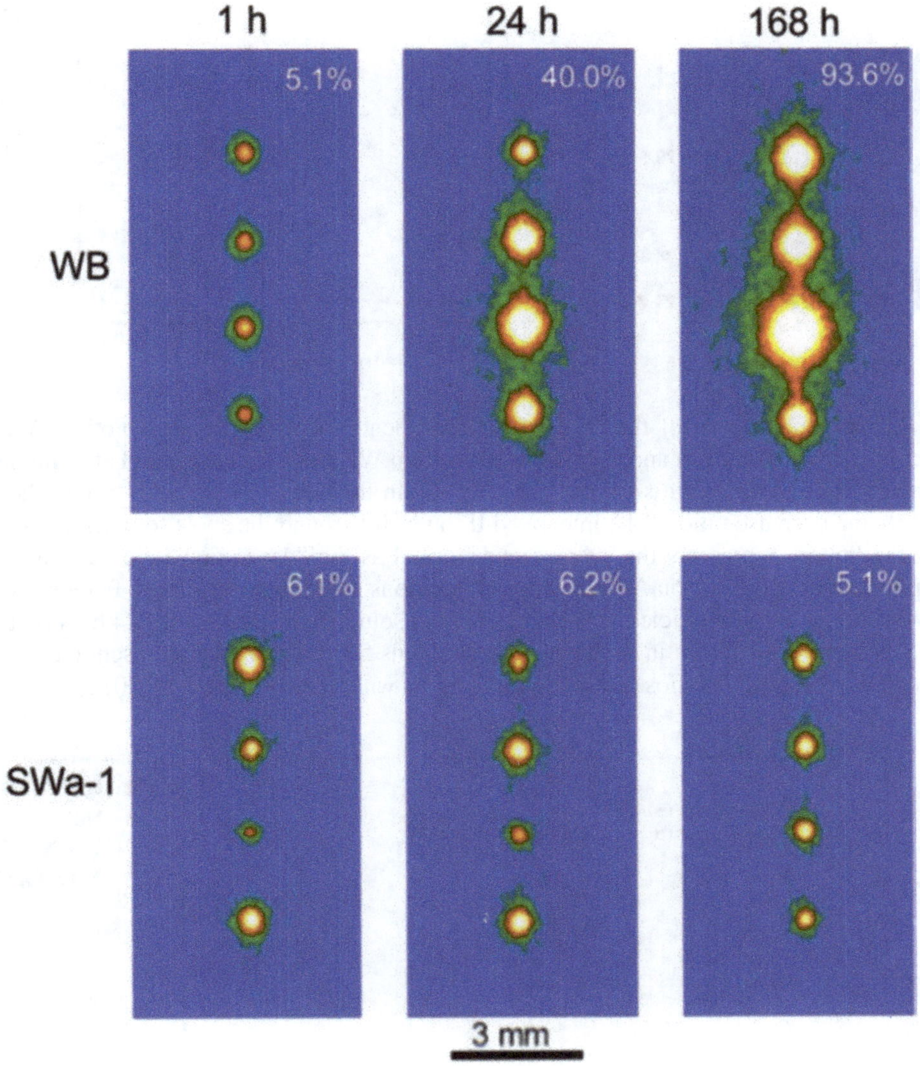

Fig. 7.11 A matrix of the readout images of IPs exposed by the substrates with four mineral particles of WB (top) and SWa-1 (bottom), reacted with 1.85 Bq ^{137}Cs solution for various immersion times which are shown at the top. Notice that only one mineral species was placed on the substrates and the mineral particles were different for each run. (Mukai et al. 2016a)

time was 24 h and 168 h. However, with a longer time, the desorption ratio was significantly decreased, becoming similar to NH_4NO_3. In addition, radioactive Cs was hardly desorbed by these solutions from the natural radioactive WB. On the contrary, in the case of HCl, almost no decrease of the desorption ratios was observed with an extension in the sorption time. Furthermore, nearly half of the radioactivity of the natural radioactive WB was removed after treatment with HCl. In the treatment with the NH_4NO_3 solution, small amounts of the radioactive Cs were desorbed, irrespective of the sorption time. Probably these results in Fig. 7.13 indicate a kind of "aging effect", namely as the sorption period is lengthened, Cs probably migrates along the interlayer regions of WB, to reside in more stable sites, and becomes almost impossible to remove by the normal ion-exchange process. Because natural radioactive WB particles collected from Fukushima have been in the field for more

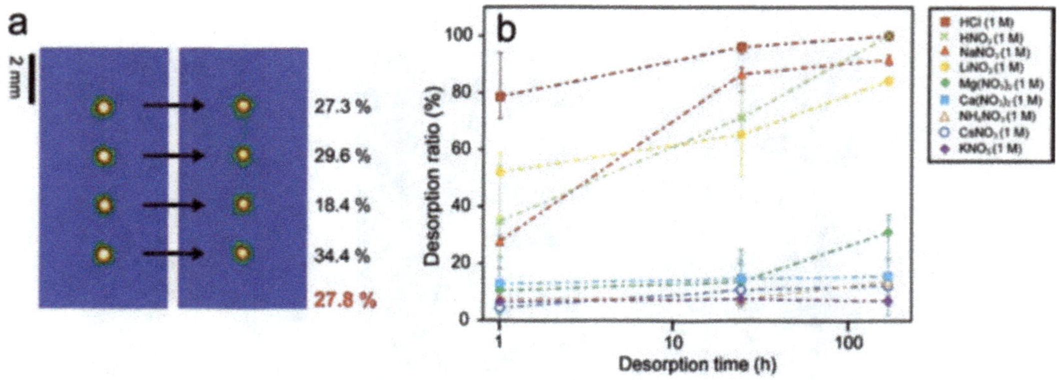

Fig. 7.12 (**a**) An example of the desorption experiments for the weathered biotite (WB). The readout IP images on the left and right were taken from WB particles before and after immersion, respectively, in an electrolyte solution (1 M $NaNO_3$ in this case) for a certain time. The figure represents the decrease ratio of the integrated IP intensity around the spots for each particle, and that at the bottom represents the average and is used as the "desorption ratio (%)" in (**b**). (b) Desorption ratios of ^{137}Cs from WB for various solutions and immersion times. For each sample, ^{137}Cs was sorbed to four particles of WB from a 50 μl solution with 2.5 Bq for 24 h. WB particles sorbing ^{137}Cs were immersed in the electrolyte solutions 50 μl. Error bars represent the minimum and maximum desorption ratios in four particles as shown in (a). (Mukai et al. 2018)

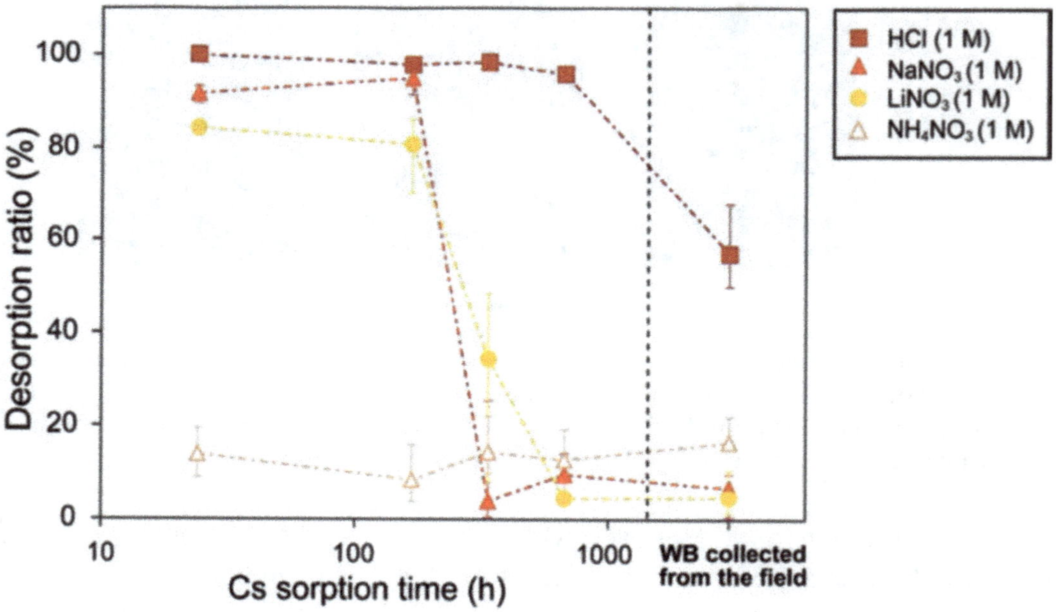

Fig. 7.13 Desorption ratios of WB as a function of ^{137}Cs-sorption time (24 h, 168 h, 336 h, and 672 h), and the desorption ratio of radioactive WB collected from the field at Fukushima. Desorption time was 168 h. (Mukai et al. 2018)

than 2 years, radiocesium has already been fixed at such stable sites. The effectiveness of HCl in removing a certain amount of Cs regardless of "aging" is probably due to the different leaching mechanism from ion-exchange; HCl partially dissolved the WB structure, particularly around the surface, and radiocesium fixed around the surface was leached to solutions with the other constituent elements.

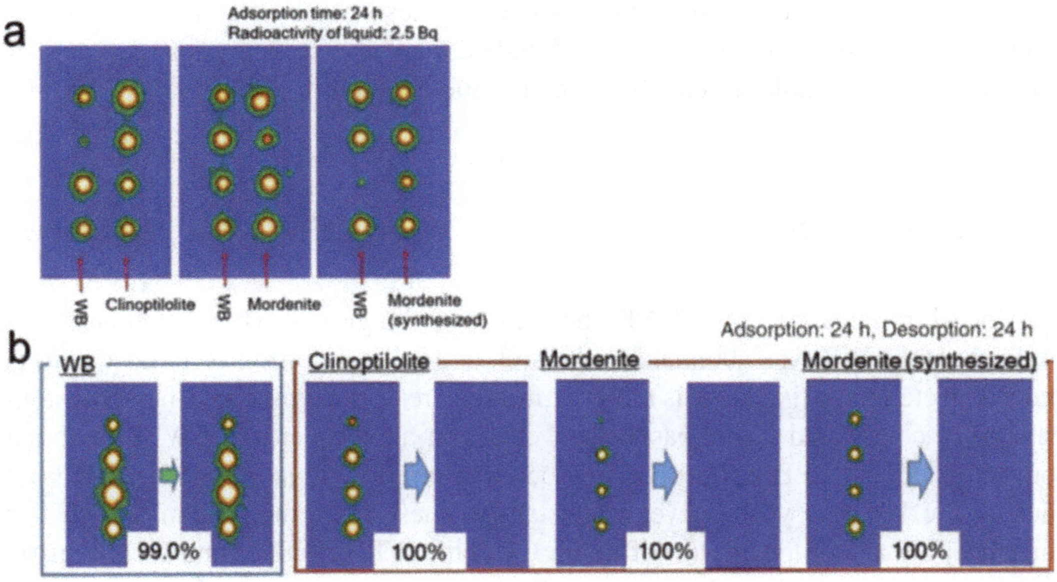

Fig. 7.14 (**a**) Comparison of the Cs-sorption abilities between WB and several zeolites. The experimental procedure was the same as those in Fig. 7.10. (**b**) Desorption properties of WB and several zeolites. The solution for desorption was CsCl (1 M)

These distinct properties of WB can explain several aspects of the radioactive contamination in Fukushima. For instance, most of the radiation that fell on the ground was trapped in a very shallow depth of the soil and has hardly moved into deeper soils (Honda et al. 2015). The distribution coefficients of radiocesium between sediments and river water are far larger (rich in sediment) in Fukushima than in Chernobyl (Konoplev et al. 2016). The transfer factors of radiation (radiocesium) from the soil to plants have decreased rapidly since the accident (e.g., Takeda et al. 2013). Probably these observations in the field in Fukushima are related to the high sorption and fixation ability of WB.

Cs in WB is hardly desorbed by ion-exchange with other alkali or alkali-earth cations in normal electrolyte solutions, particularly for WB that has remained for a long period in the field. Six years after the accident is probably long enough to change radiocesium in WB in such a passive state. Hence, decontamination of the radioactive soil by a process based on conventional ion-exchange is probably unlikely. Partial dissolution of WB in soils by strong acidic or basic solutions may be an answer but such solutions are not mild to the environment. Thermal decomposition of WB by calcination with additives is an alternative but the cost-performance ratio should be considered. On the other hand, WB is a promising candidate as an immobilizer of radiocesium, applicable to soil-improvement materials for agriculture or barriers at storage facilities for radioactive waste. For instance, Fig. 7.14 indicates the difference of sorption/desorption properties between WB and zeolites, typical sorbents used to remove radiocesium from contaminated water. The Cs-sorption ability from dilute Cs solutions is almost the same between WB and zeolites, but the sorbed Cs in zeolites was easily desorbed by electrolyte solutions whereas that in WB was hardly removed, indicating that WB is superior as a retainer of radiocesium, if radiocesium should be immobilized permanently.

7.4 Conclusions

Radiocesium released from FDNPP has fallen on the ground where, fortunately or unfortunately, WB was abundant in the soil, and a large portion of radiocesium is now expected to be fixed in this mineral. In such areas, the dynamics of radiocesium should be considered on the basis of the character and properties of WB. Although our knowledge of WB has been increased considerably by the research carried out during the last few years, several substantial questions remain unanswered. For instance, the actual location(s) of dilute Cs in the WB structure, and the atomistic mechanism of "aging" as shown in Fig. 7.13, etc. are not clear yet. For future studies, these unknowns should be elucidated using fundamental research. Such research is still necessary because the radiation of ^{137}Cs will not rapidly decay for several decades and because similar disasters may occur in the future.

Acknowledgements The authors are grateful to Dr. S. Motai and Ms. E. Fujii for their collaboration in the research, Dr. T. Hatta and Dr. H. Yamada for donating the radioactively contaminated soils from Fukushima, Prof. Y. Watanabe for donating the zeolite specimens, Dr. A. Hirose, Prof. K. Tanoi, and Prof. TM. Nakanishi for the assistance with the sorption/desorption experiments using RI. This study was financially assisted by a Grant-in-Aid for Science Research (15H04222, 15H02149 and 24340133) by JSPS, Japan. This study was also supported through contracted research with the Japan Atomic Energy Agency (JAEA) for Fukushima environment recovery, entitled "Study on Cs adsorption and desorption process on clay minerals".

References

Adachi K, Kajino M, Zaizen Y, Igarashi Y (2013) Emission of spherical cesium-bearing particles from an early stage of the Fukushima nuclear accident. Sci Rep 3:2554

Brouwer E, Baeyens B, Maes A, Cremers A (1983) Cesium and rubidium ion equilibria in illite clay. J Phys Chem 87:1213–1219

Comans RNJ, Haller M, Depreter P (1991) Sorption of cesium on Illite – nonequilibrium behavior and reversibility. Geochim Cosmochim Acta 55:433–440

Cornell RM (1993) Adsorption of cesium on minerals – a review. J Radioanal Nucl Chem 171:483–500

Evans DW, Alberts JJ, Clark RA (1983) Reversible ion-exchange fixation of cesium-137 leading to mobilization from reservoir sediments. Geochim Cosmochim Acta 47:1041–1049

Francis CW, Brinkley FS (1976) Preferential adsorption of Cs-137 to micaceous minerals in contaminated freshwater sediment. Nature 260:511–513

Honda M, Matsuzaki H, Miyake Y, Maejima Y, Yamagata T, Nagai H (2015) Depth profile and mobility of ^{129}I and ^{137}Cs in soil originating from the Fukushima Dai-ichi Nuclear Power Plant accident. J Environ Radioact 146:35–43

Japan Atomic Energy Agency (JAEA). Extension site of distribution map of radiation dose, etc. Available at: http://ramap.jmc.or.jp/map/eng/. Accessed 28 Dec 2015

Japan Meteorological Agency. Archives of the past meteorological data (in Japanese). Available at: http://www.data.jma.go.jp/obd/stats/etrn/index.php. Accessed 28 Dec 2015

Kikuchi R, Mukai H, Kuramata C, Kogure T (2015) Cs-sorption in weathered biotite from Fukushima granitic soil. J Mineral Petrol Sci 110:126–134

Kogure T, Morimoto K, Tamura K, Sato H, Yamagishi A (2012) XRD and HRTEM evidences for fixation of cesium ions in vermiculite clay. Chem Lett 41:380–382

Konoplev A, Golosov V, Laptev G, Nanba K, Onda Y, Takase T, Wakiyama Y, Yoshimura K (2016) Behavior of accidentally released radiocesium in soil-water environment: looking at Fukushima from a Chernobyl perspective. J Environ Radioact 151:568–578

McKinley JP, Zachara JM, Heald SM, Dohnalkova A, Newville MG, Sutton SR (2004) Microscale distribution of cesium sorbed to biotite and muscovite. Environ Sci Technol 38:1017–1023

Motai S, Mukai H, Watanuki T, Ohwada K, Fukuda T, Machida A, Kuramata C, Kikuchi R, Yaita T, Kogure T (2016) Mineralogical characterization of radioactive particles from Fukushima soil using μ–XRD with synchrotron radiation. J Mineral Petrol Sci 111:305–312

Mukai H, Hatta T, Kitazawa H, Yamada H, Yaita T, Kogure T (2014) Speciation of radioactive soil particles in the Fukushima contaminated area by IP autoradiography and microanalyses. Environ Sci Technol 48:13053–13059

Mukai H, Hirose A, Motai S, Kikuchi R, Tanoi K, Nakanishi TM, Yaita T, Kogure T (2016a) Cesium adsorption/desorption behavior of clay minerals considering actual contamination conditions in Fukushima. Sci Rep 6:21543

Mukai H, Motai S, Yaita T, Kogure T (2016b) Identification of the actual cesium-adsorbing materials in the contaminated Fukushima soil. Appl Clay Sci 121–122:188–193

Mukai H, Tamura K, Kikuchi R, Takahashi S, Yaita T, Kogure T (2018) Cesium desorption behavior of weathered biotite in Fukushima considering the actual radioactive contamination level of soils. J Environ Rdioact 190–191:81–88

Nakanishi TM (2016) An overview of our research. In: Nakanishi, Tanoi (eds) Agricultural implications of the Fukushima nuclear accident. The first three years. Springer Open, London

Nakao A, Thiry Y, Funakawa S, Kosaki T (2008) Characterization of the frayed edge site of micaceous minerals in soil clays influenced by different pedogenetic conditions in Japan and northern Thailand. Soil Sci Plant Nutr 54:479–489

National Institute of Material Science (NIMS). Database of promising adsorbents for decontamination of radioactive substances after Fukushima Daiichi Nuclear Power Plants accident. Available at: http://reads.nims.go.jp/index_en.html. Accessed 23 Nov 2017

Ohnuki T, Kozai N (2013) Adsorption behavior of radioactive cesium by non-mica minerals. J Nucl Sci Technol 50:369–375

Okumura T, Tamura K, Fujii E, Yamada H, Kogure T (2014) Direct observation of cesium at the interlayer region in phlogopite mica. Microscopy 63:65–72

Poinssot C, Baeyens B, Bradbury MH (1999) Experimental and modelling studies of caesium sorption on illite. Geochim Cosmochim Acta 63:3217–3227

Takeda A, Tsukada H, Nakao A, Takaku Y, Hisamatsu S (2013) Time-dependent changes of phytoavailability of Cs added to allophanic andosols in laboratory cultivations and extraction tests. J Environ Radioact 122:29–36

Yamaguchi N, Mitome M, Akiyama-Hasegawa K, Asano M, Adachi K, Kogure T (2016) Internal structure of cesium-bearing radioactive microparticles released from Fukushima nuclear power plant. Sci Rep 6:20548

Zachara JM, Smith SC, Liu CX, McKinley JP, Serne RJ, Gassman PL (2002) Sorption of Cs^+ to micaceous subsurface sediments from the Hanford site, USA. Geochim Cosmochim Acta 66:193–211

Zaunbrecher LK, Cygan RT, Elliott WC (2015) Molecular models of cesium and rubidium adsorption on weathered micaceous minerals. J Phys Chem A119:5691–5700

Chapter 8
Radiocesium Accumulation in Koshiabura (*Eleutherococcus sciadophylloides*) and Other Wild Vegetables in Fukushima Prefecture

Naoto Nihei and Keisuke Nemoto

Abstract Wild vegetables naturally grow in the mountains, and their new buds and leaves are routinely eaten by local residents. In Fukushima, wild vegetables are more contaminated than agricultural products because most forests have not been decontaminated and radiocesium still remains in the forest soil. Radiocesium concentrations in wild vegetables can vary depending on the species, and in the case of koshiabura (*Eleutherococcus sciadophylloides*), it was found to have the highest concentration among wild vegetables. To acquire basic knowledge about radiocesium accumulation in koshiabura, we collected young trees which had been grown in the forest of Date City, Fukushima and investigated the radiocesium concentration in each part and its seasonal transition.

Keywords Koshiabura · Radiocesium · Wild vegetable

8.1 Monitoring and Examination of Agricultural Products

A large amount of radioactive substances, especially radiocesium (^{134}Cs and ^{137}Cs), were spread throughout the environment by the Fukushima Daiichi Nuclear Power Plant Accident, and forests, croplands, and residential areas were contaminated. Agriculture is an important industry in Fukushima Prefecture, and it produces various agricultural products in its warm climate and rich natural environment. Since the accident, several measures in the production process such as deep plowing, stripping, and applying potassium fertilizer have been performed to prevent agricultural products from absorbing radiocesium.

N. Nihei (✉) · K. Nemoto
Graduate School of Agricultural and Life Sciences, The University of Tokyo, Bunkyo-ku, Tokyo, Japan
e-mail: anaoto@mail.ecc.u-tokyo.ac.jp

T. M. Nakanishi et al. (eds.), *Agricultural Implications of the Fukushima Nuclear Accident (III)*, https://doi.org/10.1007/978-981-13-3218-0_8

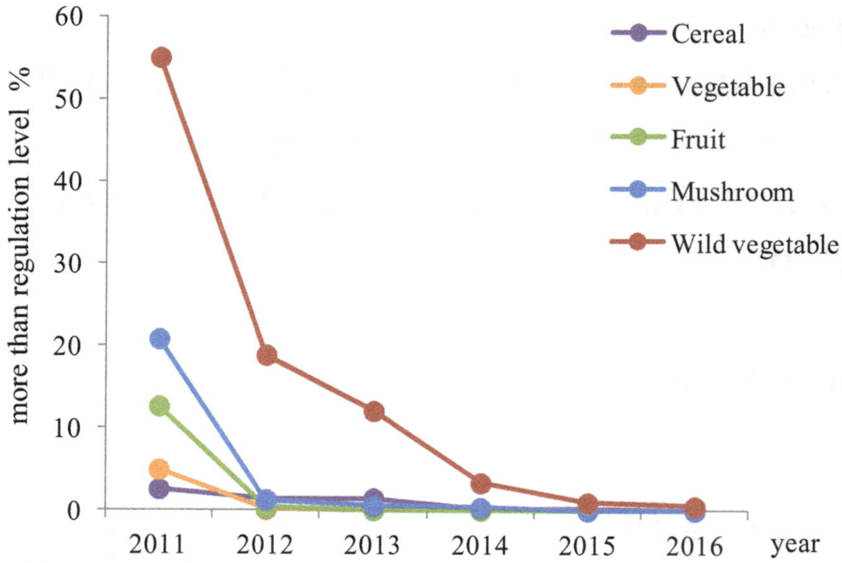

Fig. 8.1 The percentage of agricultural products whose radiocesium concentration exceeded regulation levels in the monitoring investigation

In addition, agricultural products are examined before being sent to markets and confirmed that their radiocesium concentrations are lower than the standard value of 100 Bq/kg (Nihei et al. 2016). Approximately 500 kinds or 100,000 agricultural products (excluding rice and grass) were examined up until March 2016. The results of these investigations describe how the circumstance of contamination of agricultural products in Fukushima Prefecture had changed since the accident. Figure 8.1 shows the percentage of agricultural products whose radiocesium concentration exceeded regulation levels (i.e., 500 Bq/kg in 2011 and 100 Bq/kg after 2012), as expressed within the categories of cereals (e.g., wheat and soybeans, excluding rice), vegetables (e.g., tomatoes, eggplants, and spinach), fruits (e.g., peaches, apples, and grapes), mushrooms, and wild vegetables. Although the percentage of products that exceeded regulation levels in 2011, just after the accident, was 3% for cereals, 5% for vegetables, 13% for fruits, 21% for mushrooms, and 55% for wild vegetables, the percentage of all categories decreased considerably from 2012. It is estimated that this is a result of the measures to reduce radiocesium uptake as well as the physical decrease of radiocesium. Mushrooms and wild vegetables were more contaminated than the other categories. Though there are several studies about the high concentration of radiocesium in mushrooms (Yoshida et al. 1994), few studies have focused on radiocesium in wild vegetables which are an important part of the Japanese diet (Kiyono and Akama 2013).

8.2 Wild Vegetables and Local People

Wild vegetables naturally grow in the mountains of Japan, and it is the new buds and leaves of these plants that are commonly eaten. These plants have been an important part of peoples' lives in the mountain villages, not just providing nutrition but also allowing the local people to experience more intimately the changing of the seasons. These edible plants are sometimes preserved by drying or salting. Collecting wild vegetables is a great pleasure for local people. The locations of rare wild vegetables are so precious that they are sometimes kept a secret from other family members. Wild vegetables have been closely related to Japanese people's lives for a long time, and have no doubt contributed to enriching Japanese culture. In addition, they have supported the local traditions in communities by creating a unique food culture in each region as well as being a food source for the local people. In Fukushima Prefecture, 70% of the land area is forested, and the Fukushima citizens have a strong connection with the forests. Unfortunately, most forests near Fukushima Daiichi Nuclear Power Plant were contaminated by the nuclear accident, and high concentrations of radiocesium have been found in wild vegetables in many regions.

8.3 Reasons for High Radiocesium Concentration in Wild Vegetables

The forests where wild vegetables grow are too expansive to decontaminate – cleaning up contaminated fallen leaves and other debris on the ground was performed only within 20 m from residential areas. Radiocesium still remains in forest soil, and wild vegetables growing there are thought to absorb radiocesium from the soil. In addition, the application of potassium fertilizer which is used as a countermeasure to prevent crops from absorbing radiocesium has not been performed in forests, as wild vegetables are not cultivated but grow naturally. Also, because the ratio of organic matter is high and clay mineral is low in the forest soil, the sorption of radiocesium to the soil is mild and thus transfer of radiocesium to wild vegetables will be higher than in cultivated soil. Finally, because wild vegetables are perennial plants or trees, it is considered likely that these plants absorbed and accumulated radiocesium which fell directly on their leaves and barks at the time of the accident in 2011.

8.4 Radiocesium Concentration of Each Category of Wild Vegetables

There are many kinds of wild vegetables and their edible parts can vary depending on the species. In the case of koshiabura (*Eleutherococcus sciadophylloides*) and fatsia (*Aralia elata*), new buds on the top of branches are eaten, and for bamboo (*Phyllostachys bambusoides*), butterbur (*Petasites japonicas*) and Udo (*Aralia cordata*), new sprouts growing from rhizomes are eaten; the petioles of butterbur can also be consumed when mature. In addition, new fronds of ferns, such as royal fern (*Osmunda japonica*), bracken (*Pteridium aquilinum*), and ostrich fern (*Matteuccia struthiopteris*), are considered to be wild vegetables. Figure 8.2 shows the results of

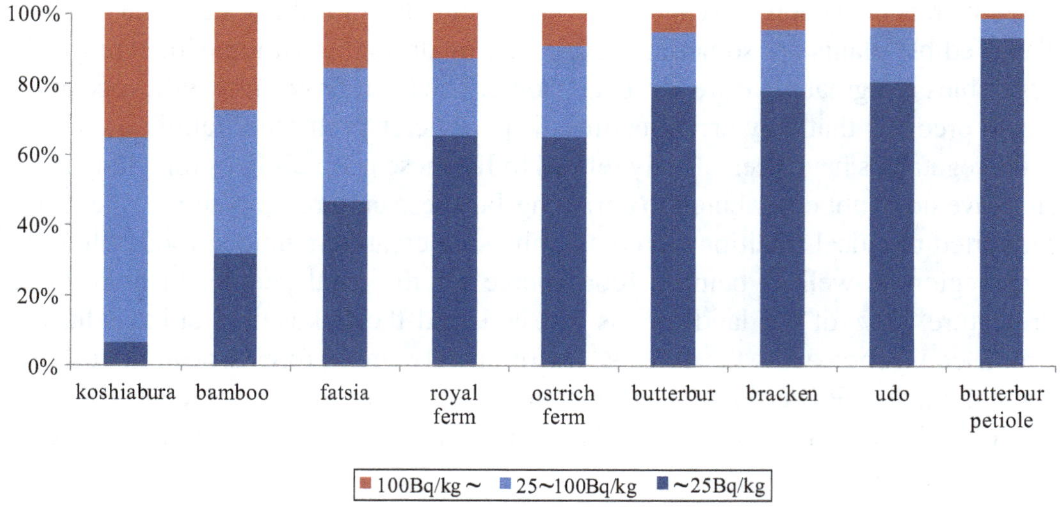

Fig. 8.2 The radiocesium concentration of nine types of wild vegetables (expressed as the average values measured during the monitoring survey from 2011 to 2015)

a monitoring survey of contaminated wild vegetables over a 6-year period (2011–2016). More than 10% of koshiabura, bamboo, fatsia, and royal fern have had radiocesium concentrations of over 100 Bq/kg, but less than 10% of ostrich fern, butterbur (both sprout and petiole), bracken, and Udo exceeded the standard value. Therefore, not all wild vegetables had high radiocesium concentrations but variation in radiocesium concentrations did exists, with koshiabura having the highest concentration.

Koshiabura is a deciduous tree belonging to the Araliaceae family and found in every part of Japan from Hokkaido to Kyushu region. Although it is a popular wild vegetable with tasty and nutritious new buds, its sale continuous to be regulated in many regions of Fukushima Prefecture 6 years after the accident. Therefore, to acquire the basic knowledge about radiocesium accumulation in koshiabura, we collected young trees which had been grown in the forest of Date City, Fukushima and investigated the radiocesium concentration in each part and its seasonal transition.

8.5 The Seasonal Transition of Radiocesium Concentration in Koshiabura

To investigate the seasonal transition of radiocesium concentration in koshiabura, leaves were collected from one individual plant in early May, late June, and late September in 2015 and 2016 (Fig. 8.3). The radiocesium concentration was found to be highest in early May, and decreased to less than half after late June and late September. When leaves of oak trees (*Quercus serrata*) were also investigated in the same site in late September, they showed almost similar radiocesium concentrations to that of koshiabura. Also, there was no difference in the radiocesium concentration

Fig. 8.3 Koshiabura (upper: early May, lower: late June)

in koshiabura between the same season of 2015 and 2016. Secondly, as koshiabura has several new leaves on its apical and lateral buds, each growing leaf was investigated for its radiocesium concentration in early May. Radiocesium concentration was highest in new leaves which had just started foliation, and decreased as the leaves finished foliation and aged. All leaves had finished foliation by late June, and there was no difference between the concentrations of each leaf at that time.

Factors such as the symbiosis with mycorrhizal fungi of koshiabura have been implicated as the reason for its high radiocesium accumulation potential (Sugiura et al. 2015), further research is needed considering that the concentration was the same for another species (i.e., *Q. serrata*) in the current investigation in late September. On the other hand, newly grown leaves in early May showed the highest radiocesium concentration. The monitoring investigation was performed on the edible parts of each wild plant when its taste was optimum. Therefore, one reason why radiocesium concentration in koshiabura was so high was because the monitoring investigation coincided with the stage of newly grown leaves in early spring. Though further research is necessary to clarify why high radiocesium concentration occurs in newly grown leaves in early spring, one of the possible factors is that the behavior of radiocesium is similar to that of potassium. In the investigation in Date City, the concentration of potassium in leaves showed a similar seasonal transition as potassium. Several studies reported that Japanese cedar (*Cryptomeria japonica*) (Tubouchi et al. 1996) and butterbur (Kiyono and Akama 2015) also showed a high concentration of potassium in leaves in spring and the gradual decrease after spring. Potassium is an essential element for growth and photosynthesis, and it moves into growing cells quickly at the time of foliation. Cesium is also a group-1 alkali metal and its movement is considered to be similar to that of potassium and thus it is easily accumulated in new leaves. Although high concentrations of manganese have been reported to accumulate in the leaves of koshiabura (Mizuno 2008), the relationship between cesium and manganese is still unclear and the potential effect of manganese accumulation on cesium accumulation needs to be further investigated.

8.6 Conclusion

Seven years have passed since the accident, and agriculture is gradually resuming in the regions where the evacuation order has been cancelled. To help the agricultural industry recover while ensuring the production of safe agricultural products, we hope to continue the monitoring investigations as outlined in this chapter.

Wild vegetables are an irreplaceable pleasure for people residing in mountain villages. However, consuming wild vegetables is not completely without risk because high concentrations of radiocesium in spring corresponds to the season of collecting wild vegetables. Although we have performed some countermeasures such as the application of potassium fertilizer and the removal of the soil surface, it is assumed that it will take several years to find out whether these measures will suppress the absorption of radiocesium in koshiabura. The movement of radiocesium in wild vegetables is still unclear, and further research and continuous monitoring are necessary.

Acknowledgement The authors would like to thank Nobuhito Ohte (Kyoto University) and Riona Kobayashi (The University of Tokyo) for their technical advice, and to Junnichiro Tada (Radiation Safety Forum) and Takashi Kurosawa (Date City) for koshiabura experiment at Date City.

References

Kiyono Y, Akama A (2013) Radioactive cesium contamination of edible wild plants after the accident at the Fukushima Daiichi Nuclear Power Plants. Jpn J Environ 55:113–118

Kiyono Y, Akama A (2015) Seasonal variations of radioactive cesium contamination in cultivated Petasites japonicas. J Jpn For Soc 97:158–164

Mizuno T (2008) Investigations on the metal hyperaccumulator plants and their application for plant nutrition study. Bull Grad Sch Bioresour Mie Univ 35:15–25

Nihei N, Tanoi K, Nakanishi TM (2016) Monitoring inspection for radiocesium in agricultural, livestock, forestry and fishery products in Fukushima prefecture. J Radioanal Nucl Chem 307:2217–2220

Sugiura Y, Takenaka C, Kanesasi T, Deguchi Y, Matsuda Y, Ozawa S (2015) Radiocesium accumulation characteristics in woody plants koshiabura. Abstract of The Ecological Society of Japan, http://www.esj.ne.jp/meeting/abst/62/PA1-177.html (in Japanese)

Tubouchi A, Maekawa T, Hiyoshi S, Ueyama Y, Hisajima T (1996) Seasonal concentration changes of various compositions in plant leaf. Ann Rep Environ Res Centre Fukui Prefecture 23:53–62

Yoshida S, Muramatsu Y, Ogawa M (1994) Radiocesium concentration in mushrooms collected in Japan. J Environ Radioact 22:141–154

Chapter 9
The Transition of Radiocesium in Peach Trees After the Fukushima Nuclear Accident

Daisuke Takata

Abstract In this chapter, we introduce the effects of radiocesium released by the Fukushima Daiichi nuclear accident on fruit trees, especially the change of radiocesium in fruit during the past 6 years. We investigated radiocesium and ^{40}K in peach during the maturity of its fruit chronologically for 6 years. In the investigation during one crop period, the concentration of radiocesium in young fruit 15 days after the full bloom was the highest, and this result was common in all the investigated years. After that, the concentration of radiocesium decreased as the fruit became bigger; the decrease until 60 days after the full bloom was considerable. This tendency was also common among all investigations conducted until 2016. Though the concentration of ^{40}K during the same period also decreased in the same way as radiocesium, the rate of the decrease from 15 to 30 days after the full bloom was different. When looking at the chronological transition, the concentration of radiocesium in harvested fruit decreased by one third every year from 2011 to 2013. However, such decrease could not be seen from 2014 to 2016. While the concentration in the harvested fruit tended to stop decreasing, the concentration in fruit 15 days after the full bloom tended to decrease over the years from 2012 to 2016. During the past 6 years, there was no year-over-year decrease in the concentration of ^{40}K in fruit. The reason why the transition of radiocesium in fruit varied according to their stage of maturity was because the difference in timing to use the tree's nutrient reserves.

To understand the year-over-year transition of radiocesium in peach, the amount of ^{137}Cs in every part of the tree was measured. When comparing the distribution of ^{137}Cs and ^{40}K in the peach trees, it was found that ^{137}Cs was existing in the body of the tree, which was contaminated by fallout, while ^{40}K was distributed more in the leaves and fruit. Also, while the weight of the trees and the amount of ^{40}K in the tree body increased with time, the amount of ^{137}Cs decreased over the years. It is considered that radiocesium stored in the mature woody parts such as stem and branches had been transferred to fruit, leaves, or young branches.

D. Takata (✉)
Faculty of Food and Agricultural Sciences, Fukushima University, Fukushima, Japan
e-mail: r841@ipc.fukushima-u.ac.jp

T. M. Nakanishi et al. (eds.), *Agricultural Implications of the Fukushima Nuclear Accident (III)*, https://doi.org/10.1007/978-981-13-3218-0_9

Keywords Peach · Cs · Fruit tree · Fukushima Daiichi Nuclear Power Plant accident

9.1 Introduction

Fukushima Prefecture has 6820 ha of orchards in 2016 and it is one of the top fruit producing areas in Japan (National Statistics Center 2016). When looking at each kind of fruit, the area of the land under cultivation with peaches is the second largest in Japan (1810 ha), and this means that the prefecture is one of the major producers of fruit in Japan even after the Fukushima Daiichi nuclear accident. With regards to changes to the retail price of fruit, Komatsu (2014) and the author's previous study (Takata 2016), show that most fruits have returned to their pre-accident prices, yet the situation is not the same as before.

Six years after the accident, the concentration of ^{134}Cs is thought to be in low in fruit because its half-life is about 2 years. On the other hand, ^{137}Cs, whose half-life is longer than ^{134}Cs can still be detected. However, previous studies found that radiocesium in fruit decreased more quickly than its actual half-life; for example, the concentration of ^{137}Cs decreased by one third in the second year, and by another one third in the third year after the accident (Takata et al. 2014, 2016). These results correspond to what was found in several types of fruit after the Chernobyl nuclear power plant accident (Antonopoulos-Domis et al. 1990; Madoz-Escande et al. 2004). However, there are few reports which traced and investigated products in the same region to examine a long-term transition of the concentration of radiocesium in fruit trees, and the future transition is uncertain. To understand this point, it is necessary to investigate the concentration of radiocesium not only in fruit continuously but also in trees which are the source of Cs in fruit. The author's previous study found that radiocesium in soil has almost no relationship with Cs in fruit in most cases. This is because (1) the radiocesium transfer factor (TF) ratio for fruit is lower than annual crops in Japan, (2) trees are no longer affected by fallout which fell directly on bark, and largely contributed to the concentration of radiocesium in the fruit (Takata et al. 2012b, 2013a), (3) since radiocesium in the trees remain *in situ* year after year, this has become a source of radiocesium in fruit (Takata et al. 2012d, 2013c), and (4) most kinds of fruit in Fukushima Prefecture (except a few kinds such as blueberries) do not form their rooting zones within 5 cm from surface horizon and thus they do not absorb radiocesium, which was distributed only on the surface horizon unevenly (Takata et al. 2013b). The fourth reason is also related to the fact that potassium fertilizer has no effect in most cases. Therefore, it is necessary to investigate trees continuously to understand the source of radiocesium in fruit.

In this chapter, new research findings, after the author's previous two reports stated above, are reported. Especially, understanding the dynamics of radiocesium in fruit and redistribution of ^{137}Cs in trees 6 years after the accident is the focus.

9.2 The Year-Over-Year Transition of Radiocesium in Fruit

After the nuclear power plant accident, one variety of peach "Akatsuki" was investigated continuously regarding the concentration of radiocesium in the fruit in Shimo-oguni district, Ryozen-machi, Date City from harvest in 2001 to 2016. In 2011, the concentration of radiocesium in the harvested fruit was measured. From 2012 to 2016, the concentration of radiocesium and ^{40}K was measured throughout fruit development: from Day 15 after the full bloom to harvest (approximately Day 103).

Figure 9.1 shows the transition of radiocesium (calculated using dry mass) in peach fruit. The red arrows indicate the concentration in the harvested fruit. As shown in the author's previous research, the concentration decreased approximately by one third from 2011 to 2012. It also decreased by one third from 2012 to 2013. Though it was expected that the year-over-year decrease would continue in the same way, the concentration in the fruits in 2014 showed a slightly higher value than the level in the prior year. In 2015 and 2016, the concentration of radiocesium in the harvested fruit decreased slightly, but this decrease was not significant, and the transition after 2013 tended to be a gradual decrease or almost no change. This fact has both good and concerning aspects. The positive aspect is that it is possible to continue to produce fruit without concerns when the concentration of radiocesium is lower than the current standard value of 100 Bq/FW. It is expected that fruit will not require dedicated countermeasures, whereas some other crops such as rice needs continuous countermeasures (e.g., application of potassium fertilizer) to lower the radiocesium concentration. Applying potassium fertilizer and absorption of radiocesium from soil will not be elaborated on in this chapter because it has already been covered in the author's previous articles. On the other hand, the concerning points are regional problems and issues related to sales methods. Regional problems refer to the fruit production in regions showing a high concentration of radionuclides, such as Hama-dori region. Hama-dori region represented by Namie-machi is famous

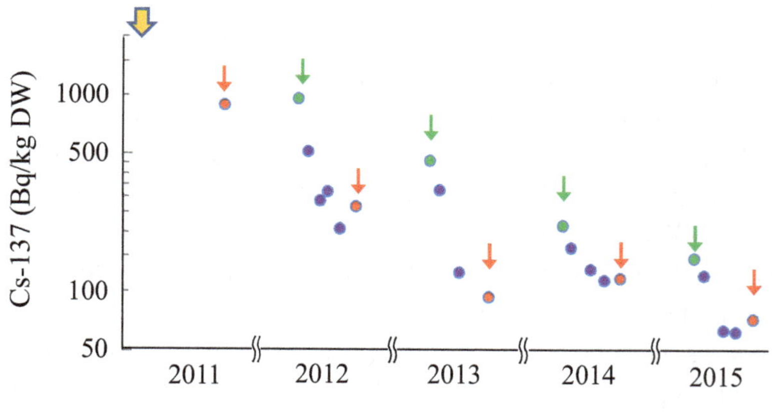

Fig. 9.1 Seasonal change of ^{137}Cs concentration in peach fruit from 2011 to 2016. The pink bars indicate the period of peach fruit growth from full bloom to harvest. The yellow arrow indicates the time of the accident (i.e., March 2011). The green dots indicate the young peach fruits 15 days after full bloom. The red dots indicate the peach fruits at harvest time. The purple dots are the peach fruits between youngest fruits and harvested fruits

for Asian pear production, and the findings presented above does not instill confidence to resume production there. Yet, it is unrealistic to reuse the same trees to resume fruit production in these regions because the trees have gone without pruning and pest control, and the trellis are likely to have deteriorated. Also, the soils were contaminated with a large amount of radionuclides and it would be necessary to replace the soil and to plant new trees. Therefore, the data shown in this study has no negative effect on fruit production, and what should be considered first is not issues of cultivation but financial aspects such as compensation for farmers during the time without income, the replacement and planting of new trees, and decontamination of surface soil. On the other hand, Japanese persimmon produced mainly in Date City can carry with it a severe problem because of how this fruit is sold. In Date City, persimmons are processed and sold as Anpo-gaki. It is a kind of Hoshigaki (dried persimmon) and the fruits are semi-dried. In Japan, because a standard value of radiocesium is determined in products at the point of sale, the radiocesium concentration is measured in the dried persimmon to compare it with the standard value (100 Bq/kg). Therefore, we need to be careful in producing raw materials for dried persimmon because radiocesium in persimmon is concentrated after drying. Although, Anpo-gaki started to be sold again in 2013, it is necessary for producers whose fruits show radiocesium concentrations close to the standard value to carry out drastic measures such as replanting to ensure the concentration in the fruit is below the standard value. The green arrows in Fig. 9.1 show the concentration of radiocesium in fruit 15 days after the full bloom. It has been decreasing year-over-year until 2016, while the concentration at the time of harvest stopped decreasing. The concentration of radiocesium in trees may contribute to the different trends between 15-day-old fruit and harvested fruit. Young organs in peach trees grow using their nutrient reserves for 3–4 weeks after flowering, and inadvertently this results in the transfer of radiocesium to young organs. Young fruit show high a concentration of radiocesium because they rely on the tree's nutrient reserves. As the fruit matures and nutrients in the soil contribute more, the radiocesium concentration is believed to decrease. The decrease of the concentration following maturity of fruits is common with the transition of ^{40}K in fruits shown in Fig. 9.2. Though the ratio of the transition is different between these two elements, they show a similar tendency. On the other hand, the year-over-year transition of ^{40}K could not be seen. This may be because the source of ^{40}K and ^{137}Cs transferred into fruit is different.

9.3 The Year-Over-Year Transition of Radiocesium in Trees

Radiocesium was detected mainly in organs exposed to the atmosphere such as bark, because fruits trees are perennial plants and grew in the fields even at the time of the accident (in winter) (Takata et al. 2012a, d). Of course, fruit also absorb radiocesium from the soil through their roots, but it has already been confirmed that when considering the source of Cs in fruit, the ratio of radiocesium due to the fallout that adhered to the bark and was transferred to fruits is much higher than absorption through the

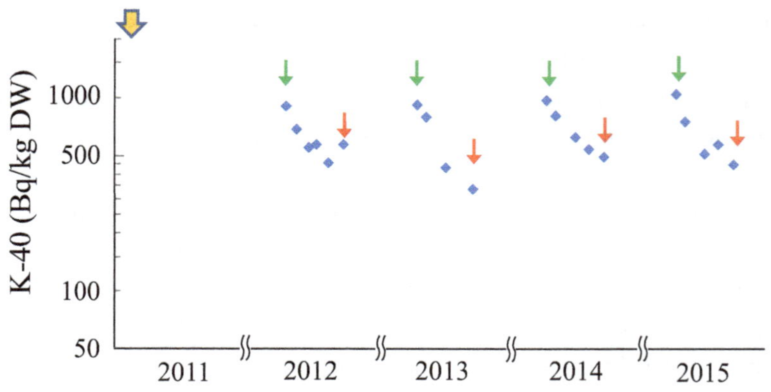

Fig. 9.2 Seasonal change of ^{40}K concentration in peach fruit from 2011 to 2016. The pink bar indicates the period of peach fruit growth from full bloom to harvest. The yellow arrow indicates the time of accident (i.e., March 2011. The green arrows indicate the young peach fruits 15 days after full bloom. The red arrows indicate the peach fruits at harvest time. The blue dots indicate the peach fruits between the youngest fruits and harvested fruits

roots. Therefore, it is important to grasp the transition of the concentration of radiocesium not only in the soil but also in the trees to understand the concentration in fruits. As stated in an earlier section, the concentration of radiocesium in peach fruit decreased as they matured and decreased over 3 years after the accident. It is important to understand the amount of radiocesium in trees to consider the cause of rapid decrease of the concentration over the 3 years after the accident; there is no report which investigated the amount of radiocesium in trees year-over-year. We introduce a study (Takata et al. 2016) which dug up trees after leaf fall to investigate the distribution of ^{137}Cs and compare the dynamic of ^{137}Cs in trees with that of ^{40}K.

We used another variety of peach "Yuzora" planted in the fruit research center of Fukushima Agricultural Technology Center. The individual parts of peach trees of a similar age and planted in the same year were separated for the investigation from 2012 to 2014. Stems and branches were collected, and the branches were sorted according to their age. The roots were collected and sorted according to their thickness. The concentration of ^{137}Cs and ^{40}K in each part were measured and calculated using the dry mass of each part to convert the measured values into the quantities (expressed by Bq) of each organ.

Figure 9.3 shows that the weight of the trees increased year-over-year. The concentration of ^{137}Cs in the soil was the highest in the surface horizon (0–3 cm) and decreased in the deeper horizons. This result corresponded to the investigation in orchards in Fukushima after the nuclear power plant accident (Takata et al. 2012c). On the other hand, the concentration of ^{40}K in the soil was almost the same across all horizons depths. Figures 9.4a and 9.5a shows that the amount of ^{137}Cs and ^{40}K, respectively, in each organ calculated by multiplying the weight of each organ and the concentration of the radionuclides together; Figs. 9.4b and 9.5b shows the distribution which converts the whole amount of the radionuclides during each crop period into 100. The distribution of ^{137}Cs in trees after 10 months, which is equivalent to one crop period was almost similar with what was reported for peach trees

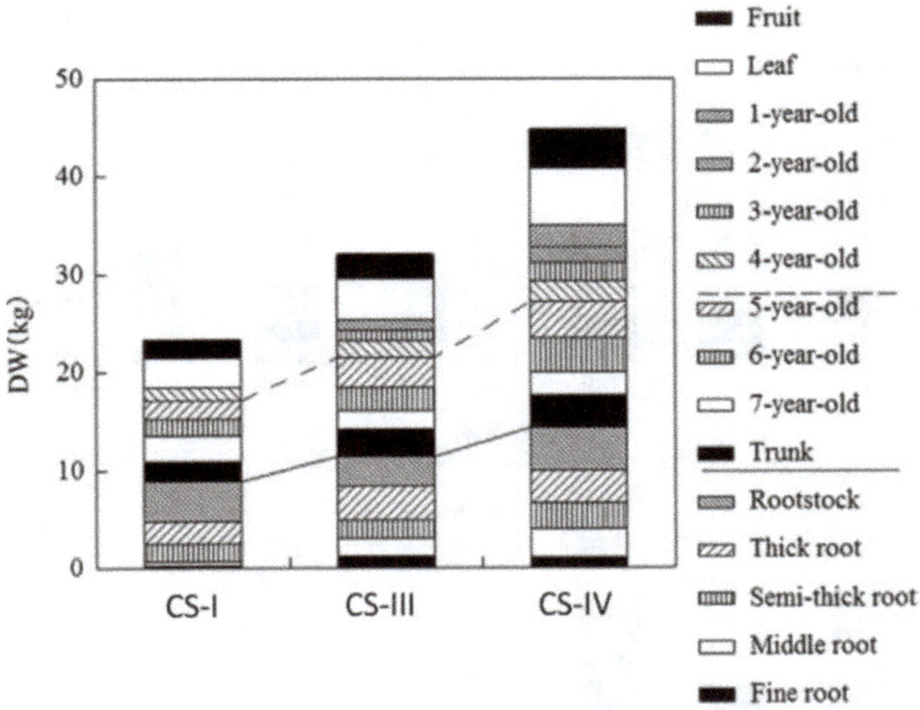

Fig. 9.3 Annual changes in dry-weight of peach trees after the Fukushima Daiichi Nuclear Power Plant Accident (Takata 2016. CS: Cropping season after FDNPP accident, CS-I: 2011 summer, CS-III: 2013 summer, CS-IV: 2014 summer. Category of roots was divided by their thickness. Thick root: 10–20 mm in diameter, Semi-thick root: 5–10 mm, middle root: 2–5 mm, fine root: <2 mm

from Tokyo 5 months after the accident (Takata et al. 2012d). The percentage of total radiocesium in fruits and leaves in peach trees from Tokyo 5 months after the accident was about 20% (Takata et al. 2012d), while that value was 13% in peach trees from Fukushima 10 months after the accident in the present study (Fig. 9.4b). Though the sites of those two investigations were inconsistent with the amount of radiocesium fallout and the difference in the amount of fallout should have affected the distribution manner within the tree, the difference of the amount of ^{137}Cs in fruits and leaves between the fifth and tenth month may have been due to the effect of translocation of Cs. Tagami and Uchida (2016) suggested that ^{137}Cs in leaves is translocated to the tree body just before leaf fall in a study on cherry trees, which belongs to the same plant family as peach trees. Therefore, it is considered that cesium in leaves is translocated to the tree body just before leaf fall also in peach trees.

Figure 9.4a shows that the total amount of ^{137}Cs in trees was highest after 1 year and decreased in subsequent years. When looking at the year-over-year transition of each organ, ^{137}Cs in five and seven-year-old branches and stems decreased. The five-year-old branches were the youngest branches which received fallout, and the seven-year-old branches were the main branches at the time of the accident. There was no year-over-year transition of the distribution of ^{137}Cs in organs which did not

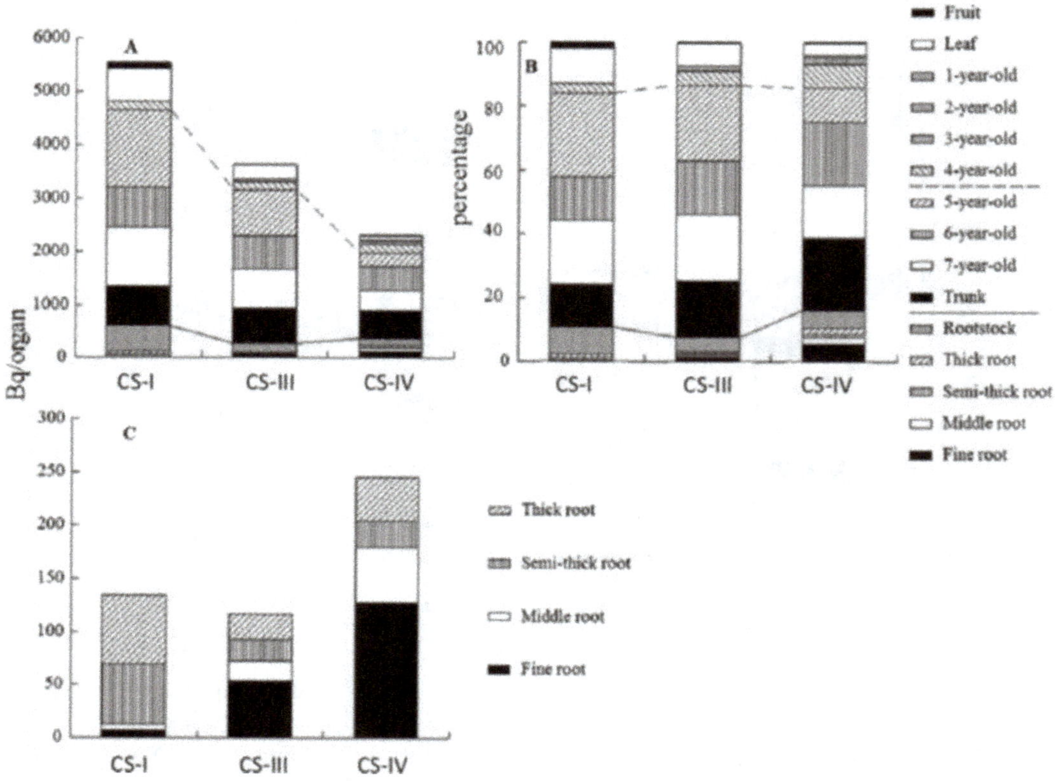

Fig. 9.4 Annual changes in ^{137}Cs content (A) and its distribution (percentage) (B) in peach trees after the Fukushima Daiichi Nuclear Power Plant Accident. Fig. 9.4C is an enlarged view of root ^{137}Cs content (Takata 2016. Organs above the dashed line grew after the accident. Branch age conforms to cropping season 4 (2014). CS: Cropping season after FDNPP accident, CS-I: 2011 summer, CS-III: 2013 summer, CS-IV: 2014 summer. Category of roots was divided by their thickness. Thick root: 10–20 mm in diameter, Semi-thick root: 5–10 mm, middle root: 2–5 mm, fine root: <2 mm

receive the fallout: that includes branches younger than 4 years old, fruit and leaves (Fig. 9.4b), though the percentage of ^{137}Cs which was transferred to fruit and leaves was 13.1% in the first year and decreased to 7.5% and 3.9% over the subsequent years. On the other hand, the distribution of ^{137}Cs increased year-over-year in the four-year-old branches, which developed in the year of the accident. This could be because branches accumulated radiocesium as they grew, while new leaves and fruits were produced every year. Also, it is considered that the source of Cs in new-grown organs were the stems, five-year-old and seven-year-old branches which decreased the amount of ^{137}Cs in them. However, five-year-old branches were pruned during winter and the cause of the decrease could be either the pruning or redistribution. Therefore, it is estimated that the source of ^{137}Cs in new-grown organs were the stems and seven-year-old branches, which were not pruned.

Although there was no general tendency in the transition of the amount of ^{137}Cs in the roots; the roots thinner than 5 mm and rootlets had higher amounts of ^{137}Cs year- over-year (Fig. 9.4c). Two reasons can be considered to explain this phenom-

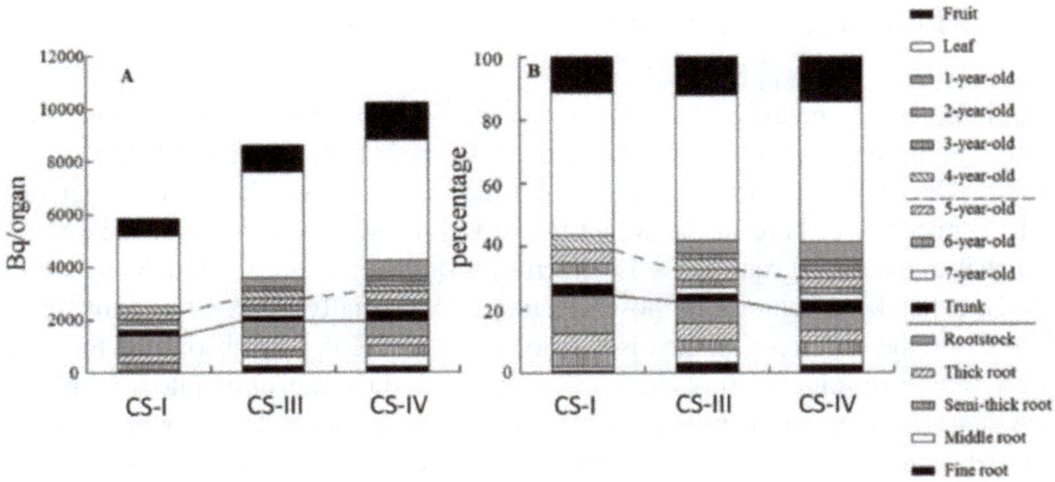

Fig. 9.5 Annual changes in ^{40}K content (A) and its distribution (percentage) (B) in peach trees after the Fukushima Daiichi Nuclear Power plant Accident (Takata 2016. Organs above the dashed line grew after the accident. Branch age conforms to cropping season 4 (2014). CS: Cropping season after FDNPP accident, CS-I: 2011 summer, CS-III: 2013 summer, CS-IV: 2014 summer. Category of roots was divided by their thickness. Thick root: 10–20 mm in diameter, Semi-thick root: 5–10 mm, middle root: 2–5 mm, fine root: <2 mm

enon: (1) the absorption from the soil and (2) translocation from thick roots and terrestrial organs. However, as explained in an earlier section, radiocesium existed mainly in the surface horizon, especially to a depth of 5 cm rather than the lower horizons. Because the main root area of peach trees is deeper than the horizon, it is unrealistic that ^{137}Cs can be absorbed by the roots. Therefore, it is considered that translocation from thicker roots contributes to the increase in the amount of ^{137}Cs in medium and fine roots. The amount of ^{137}Cs in the trees decreased more than the estimated quantity removed by fruit, leaves and pruned branches (Fig. 9.3a). It is necessary to consider other possibilities to explain this result such as ^{137}Cs exudation from the surface of branches or roots and decay of rootlets. As mentioned earlier, a proportion of ^{137}Cs is translocated to newly grown fine roots, and these new roots often decay in a single season. Therefore, ^{137}Cs transferred to the soil by falling out or decay is not possibly negligible.

On the other hand, the total ^{40}K in trees was smallest in the first year and increased with years; the increase in the fruit and leaves were considerable (Fig. 9.5). This result is closely related to the increase of weight of the trees (Fig. 9.3), and completely opposite to the year-over-year transition of ^{137}Cs which was mentioned earlier.

During the 4 years of the investigation after the accident, it was found that the absorption of radiocesium from soil was much smaller than derived from the tree. The source of ^{137}Cs in the fruits could be the stems and main branches, and the results show that ^{137}Cs transferred from the stems or branches to fruits was not supplemented by the soil. On the other hand, ^{40}K is transported from the soil through the stems and main branches. It suggests that the dynamics of Cs and K in fruits trees is different.

9.4 The Current Investigation

It is important to investigate the movement of ^{137}Cs in trees, but its concentration will become too low to investigate like ^{134}Cs when it reaches its half-life in the future. The dynamics of Cs need to be clarified in case a similar nuclear accident was to occur elsewhere in the world in the future. Although the dynamics of K, which is in the same group as Cs, is sometimes compared to understand the dynamic of Cs, it has been found that those dynamics are partially different in fruit trees. Therefore, the difference in the distribution of Cs and K in soil, absorption through roots, the transition from roots to terrestrial parts, and the transition rate from old to new organs can be obstacles to understanding the dynamics of Cs. Though comparing stable isotope ^{133}Cs is another method, there are several problems to its use as an alternative to ^{137}Cs. The analysis using high precision equipment is necessary to measure ^{133}Cs in the natural environment though it is more abundant than ^{137}Cs. Even if ^{133}Cs is added to the environment to make the measurement easier, its dynamics and concentration may be different than ^{137}Cs. Also, even if ^{133}Cs was not added, the natural ^{133}Cs may show a different dynamic from ^{137}Cs because the natural ^{133}Cs concentration is much higher than ^{137}Cs. A simultaneous measurement of ^{133}Cs and ^{137}Cs is essential to solve these problems and it should be performed before the concentration of ^{137}Cs becomes too low to measure. Now we found several results on the similarity and difference of dynamics and abundance between ^{133}Cs and ^{137}Cs. For example, the transfer factor ratio of ^{133}Cs and ^{137}Cs to fruit is likely to be similar when they are applied at the same time to blueberries. Other experiments such as spraying with ^{133}Cs have been performed together with investigations using ^{137}Cs. Also, the difference of dynamics of Cs which stored in terrestrial and subterranean tree parts has been analyzed.

Acknowledgments This article is based on the collaborative research with Dr. Eriko Yasunaga, Dr. Haruto Sasaki, Dr. Keitaro Tanoi, Dr. Natsuko I. Kobayashi, Dr. Seiichi Oshita (the University of Tokyo) and Mamoru Sato, Kazuhiro Abe (Fukushima Prefecture Agricultural Technology Centre), with the assistance of Mr. Kengo Izumi, Mr. Kenichiro Ichikawa (the University of Tokyo). This study was partly supported by of Ministry of agriculture, forestry and fisheries (MAFF, Japan).

References

Antonopoulos-Domis M, Clouvas A, Gagianas A (1990) Compartment model for long-term contamination prediction in deciduous fruit trees after a nuclear accident. Health Phys 58:737–741

Komatsu T (2014) Actual condition of sale and the trend of direct selling by fruit farm management after a nuclear hazard. Jpn J Farm Manage 52:47–52

Madoz-Escande C, Henner P, Bonhomme T (2004) Foliar contamination of Pharsalus vulgaris with aerosols of Cs-137, Sr-85, Ba-133 and Te-123m: influence of plant development stage upon contamination and rain. J Environ Radioact 73:49–71

National Statistics Center (2016) e-stat – FRUIT tree production and shipping statistics -. http://www.e-stat.go.jp/SG1/estat/List.do?lid=000001160923

Tagami K, Uchida S (2016) Seasonal change of radiocesium concentration in deciduous tree leaves. KEK Proc 17:72–76

Takata D (2016) Translocation of radiocerium in fruit trees. In: Nakanishi TM, Tanoi K (eds) Agricultural implications of the Fukushima nuclear accident -the first three years. Springer, Tokyo, pp 119–144

Takata D, Yasunaga E, Tanoi K, Nakanishi T, Sasaki H, Oshita S (2012a) Radioactivity distribution of the fruit trees ascribable to radioactive fall out: a study on stone fruits cultivated in low level radioactivity region. Radioisotopes 61:321–326 (in Japanese with English abstract and tables)

Takata D, Yasunaga E, Tanoi K, Nakanishi T, Sasaki H, Oshita S (2012b) Radioactivity distribution of the fruit trees ascribable to radioactive fall out (II): transfer of raiocaesium from soil in 2011 when Fukushima Daiichi nuclear power plant accident happened. Radioisotopes 61:517–521 (in Japanese with English abstract and tables)

Takata D, Yasunaga E, Tanoi K, Kobayashi N, Nakanishi T, Sasaki H, Oshita S (2012c) Radioactivity distribution of the fruit trees ascribable to radioactive fall out (III): a study on peach and grape cultivated in South Fukushima. Radioisotopes 61:601–606 (in Japanese with English abstract and tables)

Takata D, Yasunaga E, Tanoi K, Nakanishi T, Sasaki H, Oshita S (2012d) Radioactivity distribution of the fruit trees ascribable to radioactive fall out (IV): cesium content and its distribution in peach trees. Radioisotopes 61:607–612 (in Japanese with English abstract and tables)

Takata D, Sato M, Abe K, Yasunaga E, Tanoi K (2013a) Radioactivity distribution of the fruit trees ascribable to radioactive fall out (V): shifts of caesium-137 from scion to other organs in 'Kyoho' grapes. Radioisotopes 61:455–459 (in Japanese with English abstract and tables)

Takata D, Sato M, Abe K, Yasunaga E, Tanoi K (2013b) Radioactivity distribution of the fruit trees ascribable to radioactive fall out (VI): effect of heterogeneity of caesium-137 concentration in soil on transferability to grapes and fig trees. Radioisotopes 62:533–538 (in Japanese with English abstract and tables)

Takata D, Yasunaga E, Tanoi K, Nakanishi T, Sasaki H, Oshita S (2013c) Radioactivity distribution of the fruit trees ascribable to radioactive fall out (VII): seasonal changes in radioceasium of leaf, fruit and lateral branch in peach trees. Radioisotopes 62:539–544 (in Japanese with English abstract and tables)

Takata D, Sato M, Abe K, Tanoi K, Kobayashi N, Yasunaga E (2014) Shift of radiocaesium derived from Fukushima Daiichi Nuclear Power Plant accident in the Following Year in peach trees. In: 29th international horticultural congress, impact of Asia-Pacific horticulture 117, p 215

Takata D, Ichikawa K, Sato M, Abe K, Kobayashi N, Tanoi K, Yasunaga E (2016) Seasonal change of radiocesium concentration in deciduous tree leaves. The 17th workshop

Chapter 10
Application of the Artificial Annual Environmental Cycle and Dormancy-Induced Suppression of Cesium Uptake in Poplar

Yusaku Noda, Tsutomu Aohara, Shinobu Satoh, and Jun Furukawa

Abstract In perennial woody plants, dormancy-induced alteration of potassium (K) localization is assumed one of the mechanisms for adapting and surviving the severe winter environment. To establish if radio-cesium (^{137}Cs) localization is also affected by dormancy initiation, the artificial annual environmental cycle was applied to the model tree poplar. Under the short day-length condition, the amount of ^{137}Cs in shoots absorbed through the roots was drastically suppressed, but the amount of ^{42}K was unchanged. Potassium uptake from the rhizosphere is mainly mediated by KUP/HAK/KT and CNGC transporters. However, in poplar, these genes were constantly expressed under the short-day condition and there were no up- or down-regulation. These results indicated the suppression of ^{137}Cs uptake was triggered by the short-day length, however, the key transporter and the mechanism remains unclear. We hypothesized that Cs and K transport systems are separately regulated in poplar.

Keywords Artificial annual environmental cycle · Poplar · Dormancy · Uptake · Transporter

---

Y. Noda
Graduate School of Life and Environmental Sciences, University of Tsukuba, Tsukuba, Japan

T. Aohara · S. Satoh
Faculty of Life and Environmental Sciences, University of Tsukuba, Tsukuba, Japan

J. Furukawa (✉)
Faculty of Life and Environmental Sciences, University of Tsukuba, Tsukuba, Japan

Center for Research in Isotopes and Environmental Dynamics, University of Tsukuba, Tsukuba, Japan
e-mail: furukawa.jun.fn@u.tsukuba.ac.jp

T. M. Nakanishi et al. (eds.), *Agricultural Implications of the Fukushima Nuclear Accident (III)*, https://doi.org/10.1007/978-981-13-3218-0_10

10.1 Introduction

In 2011, radionuclides were released into the environment due to the Fukushima Daiichi Nuclear Power Plant accident. Among the released radionuclides, radiocesium (^{137}Cs) has been considered the main environmental contaminant. A large part of the contaminated land was forested and, after the accident, the ^{137}Cs deposition to tree canopies, uptake through leaves and/or bark, and the translocation to growing branch were actively investigated (Kato et al. 2012; Takata 2013; Kanasashi et al. 2015). However, ^{137}Cs uptake from the contaminated soil through the root was not well investigated because of methodological challenges. Therefore, the physiological knowledge of Cs transfer from the soil and its distribution among the tree organs is limited.

Cesium has a chemical property similar to potassium (K), but it is not an essential nutrient for plants. Generally, the rhizosphere Cs concentration (almost all is stable ^{133}Cs) is less than approx. 200 μM and not toxic to plant growth (White and Broadley 2000). Cesium uptake and translocation within the plant body are considered to be mediated by K^+ transport systems (White and Broadley 2000). Arabidopsis HAK5 (AtHAK5) is one of the best investigated root K^+ uptake transporters, and its mutant, *athak5*, showed a tolerance against 300 μM Cs treatment under low K conditions (Ahn et al. 2004; Gierth et al. 2005; Qi et al. 2008). *AtCNGC*s are Voltage-Independent Cation Channels (VICCs) type K^+ permeable channels and *AtCNGC1* is the candidate gene for Cs uptake identified by quantitative trait locus analysis in *Arabidopsis* (Leng et al. 1999; Kanter et al. 2010).

To understand ^{137}Cs behavior within woody plants, Cs content in each organ should be revealed under each seasonal environment, because K distribution among tree organs is changed by the seasonal transition. To combine the seasonal change of Cs distribution and the genetic information related to Cs transport, we used poplar as an experimental plant, although it is not the major tree species in the Fukushima region. Poplar is a perennial deciduous tree and the seasonal cycle of growth and dormancy is distinct. The genome of *Populus trichocarpa* is available (Tuskan et al. 2006) and the findings related to the Cs and K transport mechanisms obtained from crop research are applicable. Moreover, the high transfer rate of Cs from soil to poplar leaves has been observed under ^{137}Cs contaminated regions in Europe (International Atomic Energy Agency (IAEA) 2010).

In trees, the phase shift from growth to dormancy is a basic winter adaptation mechanism (Jansson et al. 2010) and the transition of meristems into dormant buds is crucial for cold adaptation to protect the meristems against hazardous frosts. By perceiving the change in photoperiod and temperature, woody plants can shift their growth stage (Welling et al. 2002), and the initiation of cold acclimation under the short-day length increases endogenous abscisic acid levels and reduces endogenous gibberellic acid levels (Olsen et al. 1997; Welling et al. 1997; Mølmann et al. 2005). In beech (*Fagus sylvatica* L.), leaf K content is rapidly decreased before shedding and the retrieved K is deposited in the cortex and pith tissues of the stem (Eschrich et al. 1988). Japanese native poplar (*Populus maximowiczii*) can decrease leaf K concentration after dormant bud formation (Furukawa et al. 2012). In addition, the increase of K concentration in xylem sap was observed during the winter period in field grown *Populus nigra* (Furukawa et al. 2011). These K behaviors implied the existence of a re-translocation mechanism and it is assumed that K and Cs are transported to K required organs for growth regulation and/or survival.

In this chapter, we will outline the experimental method of the artificial annual environmental cycle for cultivating small scale sterilized poplar and the investigation of ^{137}Cs and ^{42}K uptake through their roots under long and short photoperiod conditions.

10.2 Application of the Artificial Annual Environmental Cycle to Poplar

In poplar, dormancy is primarily initiated in response to short-day lengths (Sylven 1940; Nitsch 1957), and the recent genetic and physiological understandings of dormancy initiation and break make poplar a highly suitable model tree for investigating growth rhythms. And the *Populus trichocarpa* genome was sequenced and expressed sequence tags of *Populus* were also identified (Kohler et al. 2003; Sterky et al. 2004). To utilize these advantages, we used poplar Hybrid aspen T89 (*Populus tremula* x *tremuloides*) as our experimental plants for investigating the shift of K and Cs distribution.

Hybrid aspen T89 were cultured in half-strength Murashige & Skoog (MS) medium in a sterilized container (height 14 cm, diameter 11 cm) (Fig. 10.1a) under light- and temperature-controlled conditions; 16 h light (light intensity

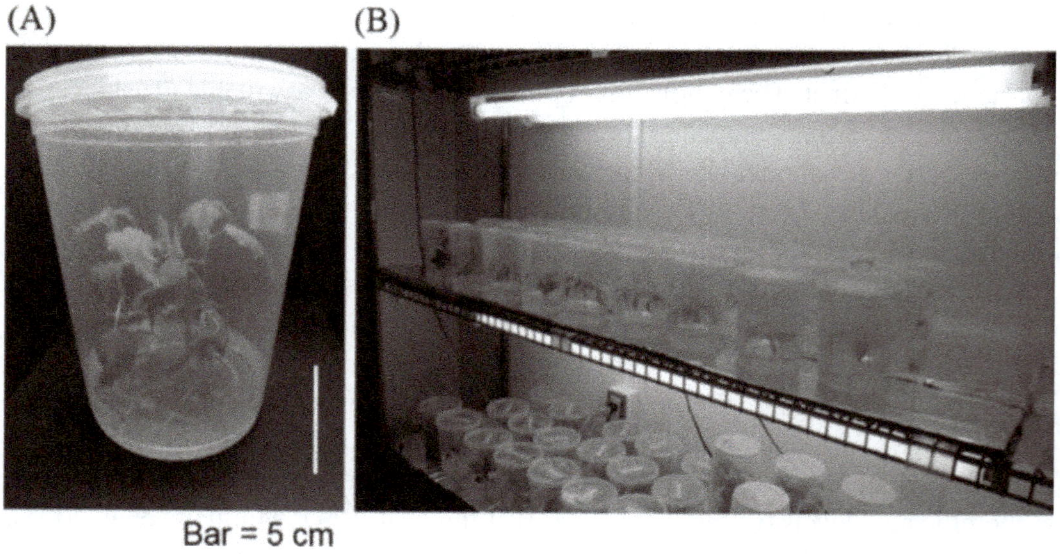

Fig. 10.1 Growth condition of poplar. (**a**) The container for sterilized culture. The lid of the container has a microscopic pore air inlet. Bar = 5 cm. (**b**) Cultivation condition of artificial annual environmental cycle

Table 10.1 Culture conditions in artificial annual environmental cycle

	Mimicked season	Kurita et al. 2014			Our condition		
		Temperature (°C)	Light/ dark (h)	Cultivation period (week)	Temperature (°C)	Light/ dark (h)	Cultivation period (week)
Stage-1	Spring/ summer	25	14/10	4	23	16/8	3
Stage-2	Autumn	15	8/16	4	23	8/16	10
Stage-3	Winter	5	8/16	8–12	4	8/16	4

37.5 μmol m^{-1} s^{-1}) and 8 h dark cycle, 23 °C (Fig. 10.1b). Once a month, each plant was cut approx. 5 cm below the shoot apex and replanted in a container containing new MS medium under sterile conditions.

The annual cycle of poplar growth is re-created in the artificial annual environmental cycle, which is a modification of a similar method by Kurita et al. (2014). The poplar was treated with three stages: Stage-1 (Spring and Summer) 23 °C, 16 h light /8 h dark, Stage-2 (Autumn) 23 °C, 8 h light/16 h dark and Stage-3 (Winter) 4 °C, 8 h light/16 h dark (Table 10.1). Poplars can break its dormant bud after the return to Stage-1 within 2 weeks. In our artificial annual environmental cycle, in contrast to Kurita et al. (2014), the temperature in Stage-2 was set to 23 °C. Our preliminary experiments revealed that the dormant bud formation can be triggered by 4 weeks of short day-length treatment when the temperature is less than 23 °C. To minimize the environmental factors related to the dormancy initiation, the temperature was kept constant in Stage-1 and -2.

10.3 Measurement of ^{137}Cs and ^{42}K Distributions in Poplar

Under the artificial annual environmental cycle using *Populus alba* L., the shift of growth condition from Long Day (LD) to Short Day (SD) condition decreased the phosphate contents in lower leaves (Kurita et al. 2014). This change of phosphate content suggested the existence of a mechanism for the re-translocation of phosphate from lower leaves to younger upper leaves with seasonal changes. Regarding calcium (Ca) translocation, Furukawa et al. (2012) indicated the Ca transport from root to shoot is also regulated by the shift from LD to SD in *Populus maximowiczii*. To investigate if a similar shift occurs for ^{137}Cs, Cs uptake through roots and its behavior within the plant body was compared in LD and SD conditions (Noda et al. 2016).

Poplars were grown under Stage-1 (LD) condition for 3 or 9 weeks (LD3 and LD9) in a light- and temperature-controlled environment. For investigating the effect of seasonal transition, a part of the LD3 plants were grown under Stage-2

(SD) conditions for an additional 2, 4 and 6 weeks (SD2, SD4 and SD6). After those cultivations, ^{137}CsCl (25 kBq, with 0.1 μM ^{133}CsCl) or ^{42}K (8 kBq, with 0.1 μM ^{39}KCl) (Aramaki et al. 2015; Kobayashi et al. 2015) solution was added to the growth medium to observe root absorption. Cesium-137 distribution was investigated with the autoradiography technique in LD3, LD9 and SD6 plants and the harvested plants were cut into four parts, apex (shoot apex and top three leaves), leaf (remaining leaves and petioles), stem and root.

Figure 10.2a displays the localization of ^{137}Cs through root absorption under LD3, LD9 and SD6 conditions. In LD3 plants, ^{137}Cs was localized entirely and the radiation intensity around the apex was higher than other organs. Similarly, LD9 which was the same age as SD6 plants and grown under LD condition also showed the same ^{137}Cs behavior as LD3 plants. In SD6 plants, ^{137}Cs was mainly localized in the stem and root, furthermore, the total ^{137}Cs quantity seemed to have decreased. Comparing ^{137}Cs in the shoot (all shoot organs) and root under LD3, LD9 and SD6, the quantity of ^{137}Cs in shoots under SD6 condition was the lowest (Fig. 10.2b). However, ^{137}Cs quantity in roots was similar under each condition.

In respect to the amount of ^{137}Cs in each organ, ^{137}Cs mainly accumulated in leaves under LD3 condition (48.8% of applied ^{137}Cs) (Fig. 10.3a); in the LD9 condition, accumulation pattern was similar to LD3. However, under SD6 condition, leaf ^{137}Cs content was the second highest (32.1%) and ^{137}Cs mostly accumulated in the stem (39.7%). As for the ^{137}Cs concentration, the effect of SD transition on ^{137}Cs distribution mainly resulted in suppressing Cs transport into shoot apices and leaves (Fig. 10.3b).

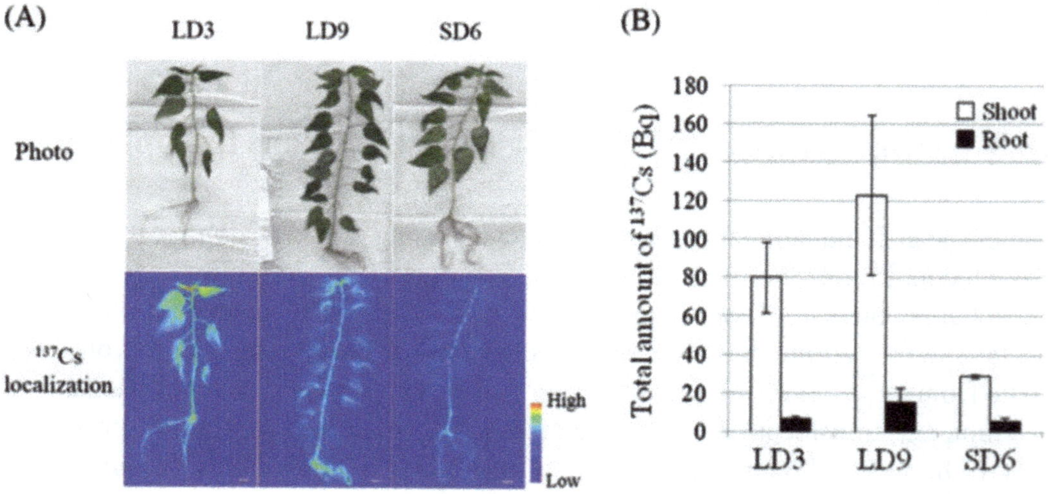

Fig. 10.2 Effect of short-day (SD) transition for ^{137}Cs uptake activity in poplar. (**a**) Cesium-137 localization through root application under long-day (LD) conditions (LD3, LD9 and SD6). Upper images are photo and lower images indicate autoradiograph. Poplars were treated with ^{137}Cs for 48 h. The color change from blue to red indicates ^{137}Cs accumulation in autoradiography imaging. Bar = 1 cm. (**b**) Alteration of total amounts of ^{137}Cs in poplar with SD transition. Poplars in each condition were treated with ^{137}Cs for 48 h. Error bars represent standard deviation (n = 3). [Modified from Noda et al. (2016)]

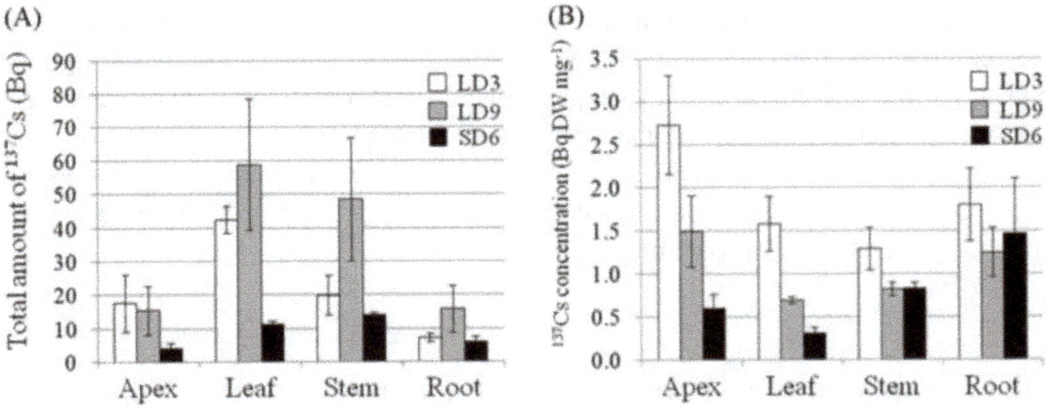

Fig. 10.3 Detailed ^{137}Cs accumulation patterns under different growth condition. Poplars were separated into apices (including upper three leaves), leaves, stem and roots. (**a**) Detailed ^{137}Cs accumulation in each organ after 48 h treatment under long-day (LD) and short-day conditions (SD) (LD3, LD9 and SD6). (**b**) Cesium-137 concentrations in each organ after 48 h treatment under LD3, LD9 and SD6 conditions. Three plants were tested for each condition. Error bars represent standard deviation (n = 3). [Modified from Noda et al. (2016)]

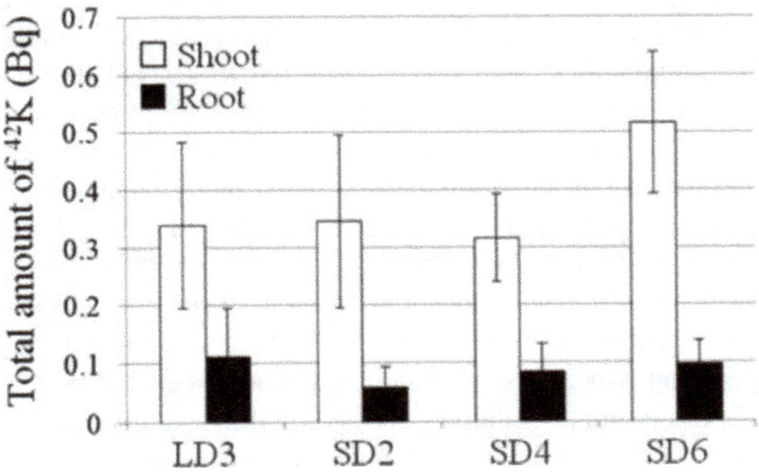

Fig. 10.4 Effect of short-day (SD) transition for ^{42}K uptake activity in poplar. Total amount of ^{42}K in poplar shoot and root with SD transition. ^{42}K under each condition was treated for 24 h. Three plants were tested for each condition. Error bars represent standard deviation (n = 3). (Modified from Noda et al. (2016))

From the result of ^{137}Cs uptake assays, it was expected that the amount of ^{42}K absorbed through the root was also down-regulated by SD transition. To explain the K uptake activity with seasonal change, poplar was treated with exogenous ^{42}K application to the root and the quantity of K in shoot and root under LD3, SD2, SD4 and SD6 conditions were measured after 24 h incubation (Fig. 10.4). Contrary to our expectations, there was no difference in the amount of ^{42}K in either root or shoot among the four conditions. These data suggest that the K demand for the rhizosphere was not changed by SD transition in poplar up to 6 weeks.

Comparing the data presented in Figs. 10.2 and 10.4, ^{137}Cs accumulation decreased under SD6 condition but ^{42}K accumulation remained constant through the SD transition. This inconsistency suggests that Cs accumulation might be separately controlled, and not part of the major K uptake systems in poplar. In addition, it is also implied that the responsible transporter for poplar Cs uptake might be poorly involved in K uptake because no decrease of K accumulation was observed under the low Cs accumulation condition.

10.4 Expression of Potassium Influx Transporters in Poplar Root

To identify the Cs uptake responsible transporter, we investigated the expression patterns of some candidate genes under SD condition. Among various K^+ uptake transporters, we focused on high-affinity K^+ transporters, KUP/HAK/KT family (Rubio et al. 2000; Mäser et al. 2001; Gupta et al. 2008), and low-affinity channel, cyclic-nucleotide-gated channel (CNGC) (Hua et al. 2003; Harada and Leigh 2006; Ahmad and Maathuis 2014). As for one of the KUP/HAK/KT family genes, we focused on *Populus tremula K⁺ uptake transporter 1*, *PtKUP1* (POPTR_0003s13370), identified from hybrid aspen. *PtKUP1* was used for the complementation test using K uptake-deficient *E. coli* mutant. The addition of toxic level Cs to the culture media inhibited the growth of *E. coli* expressing PtKUP1 strongly, suggesting PtKUP1 can transport both K and Cs (Langer et al. 2002). In addition to *PtKUP1*, there are eight *AtHAK5* orthologue genes in *Populus trichocarpa*. Arabidopsis HAK5 (AtHAK5) is a well-known root K^+ uptake transporter and it has been reported that the induction of AtHAK5 is enhanced by K^+ deficiency (Ahn et al. 2004; Gierth et al. 2005) or by Cs^+ applications when there is sufficient K^+ (Adams et al. 2013). Based on the similarity of those nine putative KUP/HAK/KT family K^+ transporters, POPTR_0010s10450 and POPTR_0001s00580 were chosen as highly homologous genes with AtHAK5. We identified POPTR_0010s10450 and POPTR_0001s00580 homologues in hybrid aspen T89 and named those *Populus tremula* x *tremuloides HAK-like1* (*PttHAK-like1*) and *PttHAK-like2*, respectively. Similar to *HAK-like* genes, two *CNGC-like* genes were selected from nine *AtCNGC1* homologue genes in *P. trichocarpa*. POPTR_0012s01690 and POPTR_0015s02090 had a higher similarity to ATCNGC1 and those homologues in hybrid aspen T89 were named *PttCNGC1-like1* and *PttCNGC1-like2*, respectively.

The expression patterns of these genes were measured during the transition to the SD condition (Fig. 10.5). There was no significant change in *PtKUP1* expression. *PttHAK-like1* showed steady expression until SD4 condition and was slightly up-regulated by about 1.5-fold under SD6 condition. *PttHAK-like2* exhibited the decreasing tendency in SD2 and SD4 plants, but remained statistically constant through the SD transition. Two *PttCNGC1-like* genes also expressed relatively constantly under SD condition.

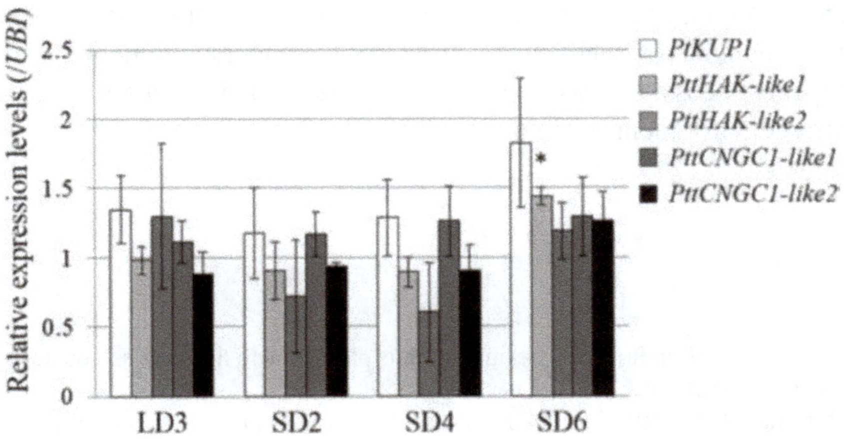

Fig. 10.5 Effect of short-day (SD) transition on transcriptional expression of *HAK* and *CNGC* orthologue genes in poplar root. Total RNA was isolated from root and these gene transcript levels were analyzed by qRT-PCR. *UBIQUTIN* was used as the reference gene. All gene expression levels were normalized by expression level of LD3 condition. Error bars represent standard deviation. * indicated significant difference to the level of LD3 expression level (* <0.01). (Modified from Noda et al. (2016))

In this experimental condition, dormant buds were formed up to 4 weeks after starting the short-day treatment and, therefore, the re-translocation of K should have already commenced at SD6. However, the results showed that the ^{42}K accumulation through root uptake and gene expression related to the root K uptake were almost constant. Steady-state of K requirement under SD condition seems to indicate abundant K was stored in the plant body and the change of K demand might be covered by the internal re-translocation.

10.5 Perspectives in Cs⁺ Transporter Research

Despite the constant K accumulation patterns under SD conditions, Cs accumulation was drastically decreased in SD6 plant (Figs. 10.2 and 10.4). Cesium uptake and translocation is considered to be regulated by plant K transport system, however, no down-regulation in candidate genes was observed at SD6 (Fig. 10.5). However, it is not known if elemental transport is only regulated by gene expression and not by protein activity (White and Broadley 2000). But the inconsistence of $^{42}K^+$ and $^{137}Cs^+$ localization patterns indicates a possible existence of an unknown Cs uptake system which preferentially transports Cs rather than K.

Based on our experimental design to identify permeable Cs transporters, it would appear that the mechanisms regulating the specificity of these transporters are very complex and it might be difficult to identify the responsible genes through reverse genetics. Our studies have also demonstrated that Cs uptake in poplar is regulated by the photoperiod, therefore, the mechanisms of the dormancy initiation might be involved in the suppression of Cs uptake. Using our data and the increasing knowledge of dormancy, the mechanisms of Cs uptake from contaminated soil to forest trees should be revealed and we hope that the understanding of Cs circulation within the forest ecosystem will be utilized in the prediction of Cs transfer among the terrestrial environment in the near future.

References

Adams E, Abdollahi P, Shin R (2013) Cesium inhibits plant growth through jasmonate signaling in *Arabidopsis thaliana*. Int J Mol Sci 14:4545–4559

Ahmad I, Maathuis FJM (2014) Cellular and tissue distribution of potassium: physiological relevance, mechanisms and regulation. J Plant Physiol 171:708–714

Ahn SJ, Shin R, Schachtman DP (2004) Expression of *KT/KUP* genes in Arabidopsis and the role of root hairs in K^+ uptake. Plant Physiol 134:1135–1145

Aramaki T, Sugita R, Hirose A, Kobayashi NI, Tanoi K, Nakanishi TM (2015) Application of ^{42}K to Arabidopsis tissues using real-time radioisotope imaging system (RRIS). Radioisotopes 64:169–176

Eschrich W, Fromm J, Essiamah S (1988) Mineral partitioning in the phloem during autumn senescence of beech leaves. Trees 2:73–83

Furukawa J, Abe Y, Mizuno H, Matsuki K, Sagawa K, Kojima M, Sakakibara H, Iwai H, Satoh S (2011) Seasonal fluctuation of organic and inorganic components in xylem sap of *Populus nigra*. Plant Root 5:56–62

Furukawa J, Kanazawa M, Satoh S (2012) Dormancy-induced temporal up-regulation of root activity in calcium translocation to shoot in *Populus maximowiczii*. Plant Root 6:10–18

Gierth M, Mäser P, Schroeder JI (2005) The potassium transporter *AtHAK5* functions in K^+ deprivation-induced high-affinity K^+ uptake and AKT1 K^+ channel contribution to K^+ uptake kinetics in Arabidopsis roots. Plant Physiol 137:1105–1114

Gupta M, Qiu X, Wang L, Xie W, Zhang C, Xiong L, Lian X, Zhang Q (2008) KT/HAK/KUP potassium transporters gene family and their whole life cycle expression profile in rice (*Oryza sativa*). Mol Gen Genomics 280:437–452

Harada H, Leigh RA (2006) Genetic mapping of natural variation in potassium concentrations in shoots of *Arabidopsis thaliana*. J Exp Bot 57:953–960

Hua BG, Mercier RW, Leng Q, Berkowitz GA (2003) Plants do it differently. A new basis for potassium/sodium selectivity in the pore of an ion channel. Plant Physiol 132:1353–1361

Jansson S, Bhalerao RP, Groover AT (2010) Genetics and genomics of populus. Springer, New York

Kanasashi T, Sugiura Y, Takenaka C, Hijii N, Umemura M (2015) Radiocesium distribution in sugi (*Cryptomeria japonica*) in eastern Japan: translocation from needles to pollen. J Environ Radioact 139:398–406

Kanter U, Hauser A, Michalke B, Dräxl S, Schäffner AR (2010) Cesium and strontium accumulation in shoots of *Arabidopsis thaliana*: genetic and physiological aspects. J Exp Bot 61:3995–4009

Kato H, Onda Y, Gomi T (2012) Interception of the Fukushima reactor accident-derived ^{137}Cs, ^{134}Cs, and ^{131}I by coniferous forest canopies. Geophys Res Lett 39. https://doi.org/10.1029/2012GL052928.

Kobayashi NI, Sugita R, Nobori T, Tanoi K, Nakanishi TM (2015) Tracer experiment using $^{42}K^+$ and $^{137}Cs^+$ revealed the different transport rates of potassium and caesium within rice roots. Funct Plant Biol 43:151–160

Kohler A, Delaruelle C, Martin D, Encelot N, Martin F (2003) The poplar root transcriptome: analysis of 7000 expressed sequence tags. FEBS Lett 542:37–41

Kurita Y, Baba K, Ohnishi M, Anegawa A, Shichijo C, Kosuge K, Fukaki H, Mimura T (2014) Establishment of a shortened annual cycle system; a tool for the analysis of annual re-translocation of phosphorus in the deciduous woody plant (*Populus alba L.*). J Plant Res 127:545–551

Langer K, Ache P, Geiger D, Stinzing A, Arend M, Wind C, Regan S, Fromm J, Hedrich R (2002) Poplar potassium transporters capable of controlling K^+ homeostasis and K^+-dependent xylogenesis. Plant J 32:997–1009

Leng Q, Mercier RW, Yao W, Berkowitz GA (1999) Cloning and first functional characterization of a plant cyclic nucleotide-gated cation channel. Plant Physiol 121:753–761

Mäser P, Thomine S, Schroeder JI, Ward JM, Hirschi K, Sze H, Talke IN, Amtmann A, Maathuis FJ, Sanders D, Harper JF, Tchieu J, Gribskov M, Persans MW, Salt DE, Kim SA, Guerinot ML (2001) Phylogenetic relationships within cation transporter families of Arabidopsis. Plant Physiol 126:1646–1667

Mølmann JA, Asante DKA, Jensen JB, Krane MN, Ernstsen A, Junttila O, Olsen JE (2005) Low night temperature and inhibition of gibberellin biosynthesis override phytochrome action and induce bud set and cold acclimation, but not dormancy, in *PHYA* overexpressors and wild-type of hybrid aspen. Plant Cell Environ 28:1579–1588

Nitsch JP (1957) Growth responses of woody plants to photoperiodic stimuli. Proc Am Soc Hortic Sci 70:512–525

Noda Y, Furukawa J, Aohara T, Nihei N, Hirose A, Tanoi K, Nakanishi TM, Satoh S (2016) Short day length-induced decrease of cesium uptake without altering potassium uptake manner in poplar. Sci Rep 6. https://doi.org/10.1038/srep38360

Olsen JE, Junttila O, Nilsen J, Eriksson ME, Martinussen I, Olsen O, Sandberg G, Moritz T (1997) Ectopic expression of oat phytochrome A in hybrid aspen changes critical day length for growth and prevents cold acclimatization. Plant J 12:1339–1350

Qi Z, Hampton CR, Shin R, Barkla BJ, White PJ, Schachtman DP (2008) The high affinity K^+ transporter AtHAK5 plays a physiological role in planta at very low K^+ concentrations and provides a cesium uptake pathway in *Arabidopsis*. J Exp Bot 59:595–607

Rubio F, Santa-Maria GE, Rodriguez-Navarro A (2000) Cloning of *Arabidopsis* and barley cDNAs encoding HAK potassium transporters in root and shoot cells. Physiol Plant 109:34–43

Sterky F, Bhalerao RR, Unneberg P, Segerman B, Nilsson P, Brunner AM, Charbonnel-Campaa L, Lindvall JJ, Tandre K, Strauss SH, Sundberg B, Gustafsson P, Uhlén M, Bhalerao RP, Nilsson O, Sandberg G, Karlsson J, Lundeberg J, Jansson S (2004) A *Populus* EST resource for plant functional genomics. Proc Natl Acad Sci U S A 101:13951–13956

Sylven N (1940) Lang-och kortkagstyper av de svenska skogstraden [Long day and short day types of Swedish forest trees]. Svensk Papperstidn 43:317–324, 332–342, 350–354

Takata D (2013) Distribution of radiocesium from the radioactive fallout in fruit trees. In: Nakanishi TM, Tanoi K (eds) Agricultural implications of the Fukushima nuclear accident. Springer, New York, pp 143–162

International Atomic Energy Agency (2010) Handbook of parameter values for the prediction of radionuclide transfer in terrestrial and freshwater environments, Technical report series No. 472, vol 472. International Atomic Energy Agency, Vienna, pp 99–108

Tuskan GA, DiFazio S, Jansson S, Bohlmann J, Grigoriev I, Hellsten U, Putnam N, Ralph S, Rombauts S, Salamov A, Schein J, Sterck L, Aerts A, Bhalerao RR, Bhalerao RP, Blaudez D, Boerjan W, Brun A, Brunner A, Busov V, Campbell M, Carlson J, Chalot M, Chapman J, Chen GL, Cooper D, Coutinho PM, Couturier J, Covert S, Cronk Q, Cunningham R, Davis J, Degroeve S, Déjardin A, dePamphilis C, Detter J, Dirks B, Dubchak I, Duplessis S, Ehlting J, Ellis B, Gendler K, Goodstein D, Gribskov M, Grimwood J, Groover A, Gunter L, Hamberger B, Heinze B, Helariutta Y, Henrissat B, Holligan D, Holt R, Huang W, Islam-Faridi N, Jones S, Jones-Rhoades M, Jorgensen R, Joshi C, Kangasjärvi J, Karlsson J, Kelleher C, Kirkpatrick R, Kirst M, Kohler A, Kalluri U, Larimer F, Leebens-Mack J, Leplé JC, Locascio P, Lou Y, Lucas S, Martin F, Montanini B, Napoli C, Nelson DR, Nelson C, Nieminen K, Nilsson O, Pereda V, Peter G, Philippe R, Pilate G, Poliakov A, Razumovskaya J, Richardson P, Rinaldi C, Ritland

K, Rouze P, Ryaboy D, Schmutz J, Schrader J, Segerman B, Shin H, Siddiqui A, Sterky F, Terry A, Tsai CJ, Uberbacher E, Unneberg P, Vahala J, Wall K, Wessler S, Yang G, Yin T, Douglas C, Marra M, Sandberg G, Van de Peer Y, Rokhsar D (2006) The genome of black cottonwood, *Populus trichocarpa* (Torr. & Gray). Science 313:1596–1604

Welling A, Kaikuranta P, Rinne P (1997) Photoperiodic induction of dormancy and freezing tolerance in *Betula pubescens*. Involvement of ABA and dehydrins. Physiol Plant 100:119–125

Welling A, Moritz T, Palva ET, Junttila O (2002) Independent activation of cold acclimation by low temperature and short photoperiod in hybrid aspen. Plant Physiol 129:1633–1641

White PJ, Broadley MR (2000) Mechanisms of cesium uptake by plants. New Phytol 147:241–256

Chapter 11
Radiocesium Contamination in Forests and the Current Situation of Growing Oak Trees for Mushroom Logs

Natsuko I. Kobayashi, Ryosuke Ito, and Masaya Masumori

Abstract The radiocesium contamination in mushrooms and in mushroom logs is a matter of concern for the forestry industry. To know the contamination situation precisely and the future for mushroom log production, the ^{137}Cs distribution in six fields cultivating oak trees in Tamura city, Fukushima, was investigated in 2015. The ^{137}Cs concentration in new branches was found to correlate with that in the wood. This result suggests that the ^{137}Cs concentration in tree trunks could be estimated without felling based on the ^{137}Cs concentration in new branches. In addition, trees grown in one of the six fields was found to have very low ^{137}Cs concentration even though the concentration of ^{137}Cs in the soil was high. The impact of the nutritional conditions on the ^{137}Cs absorption in the oak seedlings grown hydroponically was also investigated by a radiotracer experiment.

Keywords *Quercus* · Shiitake mushroom · New branch · Trunk · ^{133}Cs · Tracer

11.1 Introduction

In Japanese cuisine, shiitake mushroom (*Lentinula edodes*) is one of the essential and traditional ingredients that can be grilled, boiled, etc. Dried shiitake is also an important ingredient as it produces a characteristic liquid full of *umami* when it is reconstituted in water.

N. I. Kobayashi (✉)
Isotope Facility for Agricultural Education and Research, Graduate School of Agricultural and Life Sciences, The University of Tokyo, Bunkyo-ku, Tokyo, Japan
e-mail: anikoba@g.ecc.u-tokyo.ac.jp

R. Ito
Laboratory of Radioplant Physiology, Graduate School of Agricultural and Life Sciences, The University of Tokyo, Bunkyo-ku, Tokyo, Japan

M. Masumori
Laboratory of Silviculture, Graduate School of Agricultural and Life Sciences, The University of Tokyo, Bunkyo-ku, Tokyo, Japan

T. M. Nakanishi et al. (eds.), *Agricultural Implications of the Fukushima Nuclear Accident (III)*, https://doi.org/10.1007/978-981-13-3218-0_11

Although shiitake grows naturally in the wild in all corners of Japan, almost all shiitake in the marketplace is grown commercially. Cultivation of shiitake in Japan is considered to have begun in the 1600s. There are two methods for cultivating shiitake: using either logs or sawdust blocks as the cultivation bed. Shiitake produced using logs is reported to have a better flavor, although the productivity is not so high.

The annual production of shiitake in Japan in 2010 was 77,000 t, and the production value was 76 billion yen, which was the highest among all kinds of mushrooms. The production of shiitake has been rather even between regions in Japan. Fukushima prefecture was the seventh largest producer of shiitake; 3500 t was produced in 2010, which accounts for 5% of total Japanese shiitake production. However, as of 2014, the annual production in Fukushima prefecture decreased by approximately 50% due to the accident in the Fukushima Daiichi Nuclear Power Plant, whereas on a national level the annual production decreased only slightly. Shiitake mushroom production using logs has been particularly damaged by the accident as these mushrooms are normally cultivated outdoor.

Besides mushroom production, Fukushima prefecture has been a major producer of mushroom logs. In Japan, several hardwood species, specifically oaks, including *Quercus serrata "Konara"*, *Quercus crispula "Mizunara"*, and *Quercus acutissima "Kunugi"*, are grown for approximately 20 years and used for mushroom logs. Meanwhile, among the 47 prefectures, only nine are self-sufficient in mushroom logs. The prefectures not self-sufficient in logs need to import logs from prefectures that have a surplus, and up until 2011, Fukushima was the prefecture that supplied the most logs to other prefectures. Radiocesium concentration in mushroom logs must be less than 50 Bq/kg in order not to produce shiitake containing more than 100 Bq/kg of radiocesium, which is the Japanese standard limit for general foods. To date, it is actually very difficult to produce mushroom logs containing radiocesium less than 50 Bq/kg in the forest contaminated directly by radioactive fallout. Consequently, oak tree forests in large areas of Fukushima prefecture have not been harvested since the accident, and mushroom growers in many parts of Japan have experienced a shortage of mushroom logs.

11.2 Objective and Research Field

"When can mushroom-log-production resume?" This question is the one that scientists and investigators are frequently asked by forestry workers in Fukushima. Considering that the cultivation period from planting to harvesting spans 20 years, and another 10 years until the shoots from the stump mature, the prospect of restarting log production is essential for forestry workers to know if they have a future in this industry. In this context, the purpose of this research was to assess the prospect of resuming mushroom-log production in contaminated areas of Fukushima. Forest contamination was investigated in three areas (A, B, C) in Tamura city in collaboration with the Fukushima Chuo Forest Cooperative (Miura 2016), located about

Fig. 11.1 Location of the forest investigation. The investigation forest (white triangle) in Tamura city is located to the west of the FDNPP (red triangle). The prefectural boundary of Fukushima is indicated in yellow

20 km from the Fukushima Daiichi nuclear power plant (Fig. 11.1). Air dose rate at a height of 1 m above the ground in the areas A, B, C, was 0.43 ~ 0.5, 0.3 ~ 0.35, and 0.2 ~ 0.25 μSv/h, respectively, in the autumn of 2015. In area A, three sites were provided as the research field (i.e., A-1, A-2, and A-3), and in area C, 2 sites (i.e., C-1 and C-2) were provided. Therefore, six sites in total were established in the three areas (Table 11.1).

11.3 Field Investigation

11.3.1 Sample Collection

Details of the trees sampled in the six sites are shown in Table 11.1. Konara, Mizunara, and Kunugi trees are deciduous trees, and thus they had no leaves in March 2011 when the accident occurred. Planted trees contaminated by direct deposition of radionuclides were investigated in area A. On the other hand, old trees having direct contamination on their surface had been coppiced in area B and C, and

Table 11.1 Identity of the trees sampled in the six sites in Tamura city, Fukushima prefecture

			March, 2011 (at the accident)	Date of coppicing	November, 2015 (at the investigation)	Tree type
Area A						
Site A-1	*Quercus acutissima*	*Kunugi*	4-year-old	–	9-year-old	Planted tree
Site A-2	*Quercus acutissima*	*Kunugi*	11-year-old	–	16-year-old	Planted tree
Site A-3	*Quercus serrata*	*Konara*	Over 9-year-old	–	Over 14-year-old	Planted tree
Area B						
Site B	*Quercus crispula*	*Mizunara*	29-year-old	November, 2011	4-year-old	Coppiced tree
Area C						
Site C-1	*Quercus serrata*	*Konara*	25-year-old	November, 2013	2-year-old	Coppiced tree
Site C-2	*Quercus serrata*	*Konara*	25-year-old	November, 2013	2-year-old	Coppiced tree

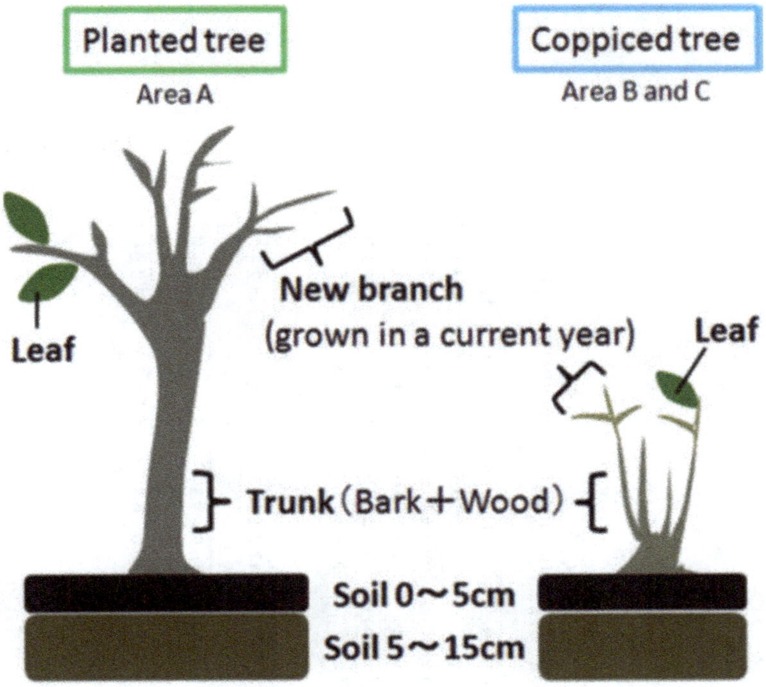

Fig. 11.2 Illustration of the samples analyzed in this study

the coppiced trees developed from the stump were targeted (Table 11.1). From May to September 2015, leaves on branches were sampled monthly in sites A-1, A-2, B, and C-2. In November 2015, planted trees in area A were cut down, and the trunks, the leaves, and the new branches which had grown in the current growing season were sampled (Fig. 11.2). The trunk was further separated into bark and wood (Fig. 11.2). Leaves, new branches, and trunks were also sampled in areas B and C in November 2015 (Fig. 11.2). For soil analysis, samples were collected at a depth of 0–5 cm (surface soil) and 5–15 cm (Fig. 11.2). The number of replicates varied between 14 and 21 depending on the sample type. Radioactivity of the samples was measured using either Ge semiconductor detector (Kobayashi et al. 2016) or NaI scintillation counter (Endo et al. 2015).

11.3.2 Property of the Soil

The bulk density of the soils in site A-1 and A-2 were almost twice as high as the soil in other sites (Table 11.2). This could be because the ground of these two sites was artificially developed, while others have natural ground covered with a thick A_1 layer.

The content of exchangeable ions was determined by the ammonium acetate extraction method. The content of exchangeable potassium, magnesium, and especially calcium in A-1 soil was surprisingly high (Table 11.2). In addition, soil pH in site A-1 was greater than 6 (Table 11.2). The reason for the characteristic soil

Table 11.2 Soil properties in two soil horizons on six sites

	Bulk density (g/cm^3)	Clay content (%)	Cs-137 (kBq/m^2)	Cs-137 (Bq/kg)	exCs-137 (%)	K_2O (mg/100 g)	CaO (mg/100 g)	MgO (mg/100 g)	pH
Soil 0–5 cm									
Site A-1	0.79 ± 0.10	17.5	130 ± 32	3314 ± 778	4.03	27.8 ± 8.0	202.1 ± 41.5	11.3 ± 3.4	6.3 ± 0.3
Site A-2	0.87 ± 0.10	8.5	113 ± 37	2582 ± 798	ND	13.2 ± 2.3	12.1 ± 10.5	3.8 ± 1.5	4.3 ± 0.2
Site A-3	0.39 ± 0.09	23.3	119 ± 31	6376 ± 1973	2.37	19.6 ± 6.7	39.4 ± 16.0	8.9 ± 3.4	4.5 ± 0.2
Site B	0.33 ± 0.05	24.4	86 ± 34	5412 ± 2483	1.20	17.9 ± 4.2	59.7 ± 25.4	9.1 ± 3.4	4.6 ± 0.2
Site C-1	0.42 ± 0.06	26.3	82 ± 25	4024 ± 1496	1.61	17.7 ± 3.6	67.3 ± 30.9	9.5 ± 3.4	4.9 ± 0.2
Site C-2	0.29 ± 0.05	18.8	58 ± 20	4320 ± 2151	1.08	16.7 ± 4.5	57.4 ± 25.8	9.6 ± 3.6	4.5 ± 0.2
Soil 5–15 cm									
Site A-1	1.18 ± 0.13	16.5	33 ± 18	282 ± 160	ND	15.3 ± 7.8	325.5 ± 69.8	6.5 ± 2.6	7.1 ± 0.3
Site A-2	1.56 ± 0.37	7.5	14 ± 11	96 ± 85	ND	9.0 ± 3.1	17.2 ± 18.2	2.2 ± 0.9	5.2 ± 0.2
Site A-3	0.65 ± 0.12	25.7	36 ± 18	564 ± 291	ND	9.8 ± 3.0	9.2 ± 3.4	2.7 ± 0.9	5.1 ± 0.1
Site B	0.58 ± 0.08	27.7	50 ± 73	937 ± 1411	ND	8.0 ± 1.9	16.9 ± 9.3	2.6 ± 1.0	5.2 ± 0.1
Site C-1	0.67 ± 0.12	27.5	22 ± 15	341 ± 246	ND	6.1 ± 2.2	16.6 ± 9.0	2.4 ± 1.0	5.0 ± 0.1
Site C-2	0.53 ± 0.08	22.2	26 ± 24	528 ± 517	ND	5.9 ± 1.9	10.6 ± 6.4	9.6 ± 3.6	4.8 ± 0.1

property in site A-1 should be found in the use history of this field. In fact, site A-1 had been a corn farm before oak trees were planted. For better corn production, this field was limed intensively in the past.

11.3.3 ^{137}Cs Concentrations in Above Ground Parts

Considering the relatively small variation in soil ^{137}Cs concentration among the six sites, the difference in the ^{137}Cs concentration in oak trees was unexpectedly large (Fig. 11.3). Particularly, in site A-1, it was surprising that the ^{137}Cs concentration was mostly lower than 25 Bq/kg in new branches and leaves, and even lower in the wood (Fig. 11.3). On the other hand, the ^{137}Cs concentrations in the new branches and leaves in site A-2 was mostly between 150 Bq/kg and 350 Bq/kg (Fig. 11.3). This means that the radiocesium contamination in trees on site A-1 was one-tenth of those on site A-2, although the soil ^{137}Cs concentration was not different between sites. Contrary, the ^{137}Cs concentration level in the soil in site A-3 was nearly 2.5-fold higher than site A-2, and then the ^{137}Cs concentration level in oak trees in site A-3 was also 2.5-fold higher than that in site A-2 (Fig. 11.3). Characteristically, the ratio between branch ^{137}Cs concentration and soil ^{137}Cs concentration was 0.13 for site A-2 and 0.17 for site A-3, while the ratio was 0.008 for site A-1. A similar relationship between soil and coppiced tree contamination in sites C-1 and C-2 was found. The ^{137}Cs concentration in trees on site C-2 was 2-fold higher than on site C-1, though the soils in these sites were contaminated similarly (Fig. 11.3). In total, current ^{137}Cs concentration level in oak trees cannot be simply determined by the amount of ^{137}Cs deposition in the forest (Table 11.2). Therefore, we aimed to understand the factors that potentially have an affect on ^{137}Cs concentration in oak trees, as described in the following Sect. 11.5.

In area A, ^{137}Cs concentration in each tree part in site A-1 showed low ^{137}Cs concentration, whereas all parts of trees in site A-3 showed high ^{137}Cs concentration (Fig. 11.3). This indicates the possibility that the ^{137}Cs concentration in wood and bark, which are the parts used as the mushroom log, can be estimated nondestructively based on the ^{137}Cs concentration in the new branches or leaves. Figure 11.4 shows the correlation in ^{137}Cs concentration in the different parts of the trees. There was a high correlation ($R^2 = 0.8564$) between the new branches and wood (Fig. 11.4). The ^{137}Cs concentration in wood was approximately one-third of that in the new branches, regardless of the tree age (Fig. 11.4). Similarly, the ^{137}Cs concentration in leaves and wood was positively correlated ($R^2 = 0.8583$) (Fig. 11.4). On the other hand, the correlation between the new branches and bark was lower (Fig. 11.4). Given that the bark provides the pathway for radiocesium contamination into the internal tissues of trees after direct deposition on the surface, the original contamination level in the bark can be assumed to have been correlated with the contamination level in other tree parts. However, over the last 5 years (from 2011 to 2015), surface contamination may have been washed out by rain, or the contaminated bark fell away as the tree grew. Consequently, bark contamination level in each individual tree could have

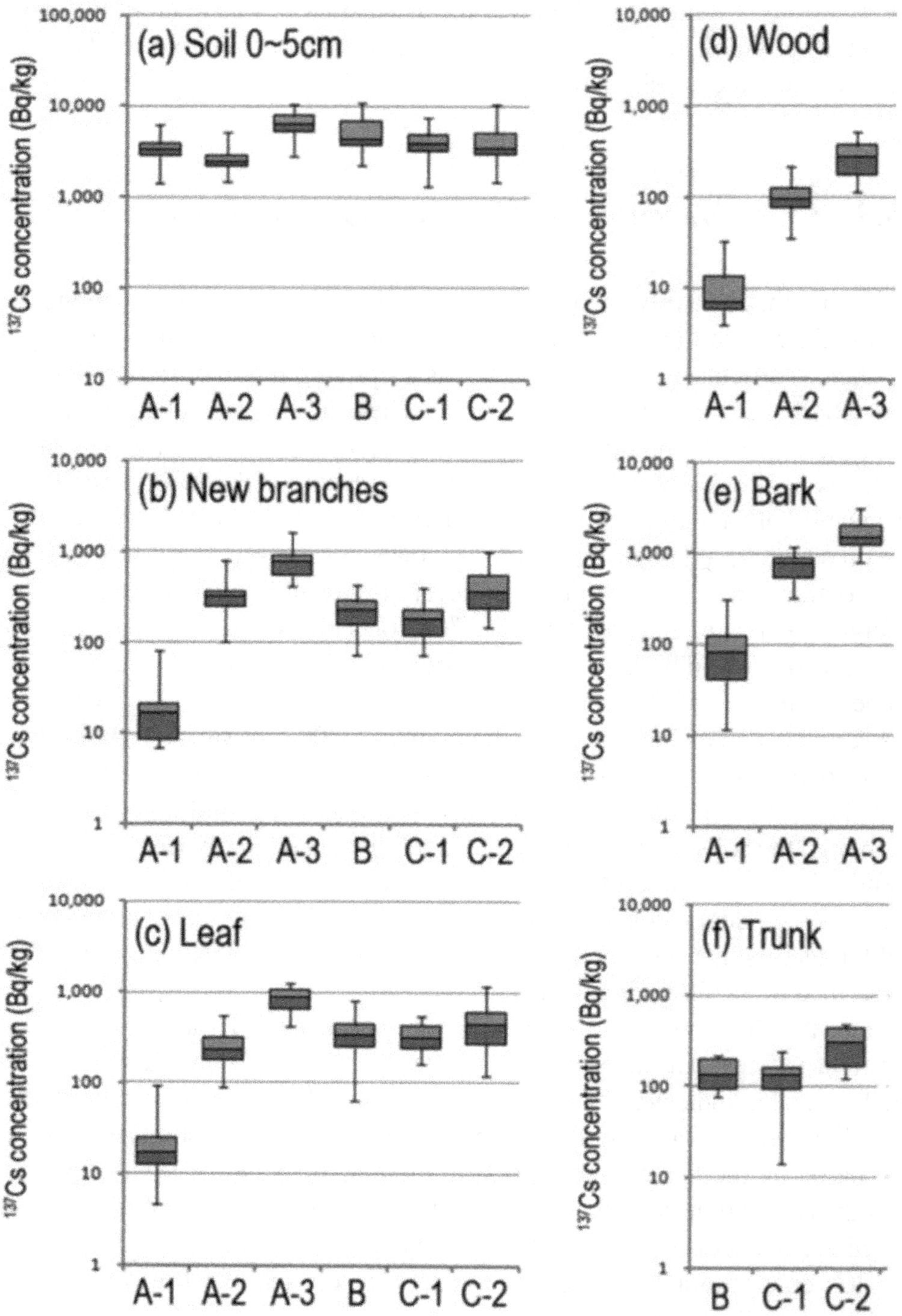

Fig. 11.3 ^{137}Cs concentration in surface soil (**a**), new branches grown in a current growing season (**b**), leaf (**c**), wood (**d**), bark (**e**), and trunk composed of wood and unsuberized young bark (**f**). The hinges refer to the 25th percentile (right gray) and 75th percentile (dark gray), and the whisker lines indicate maximum and minimum values

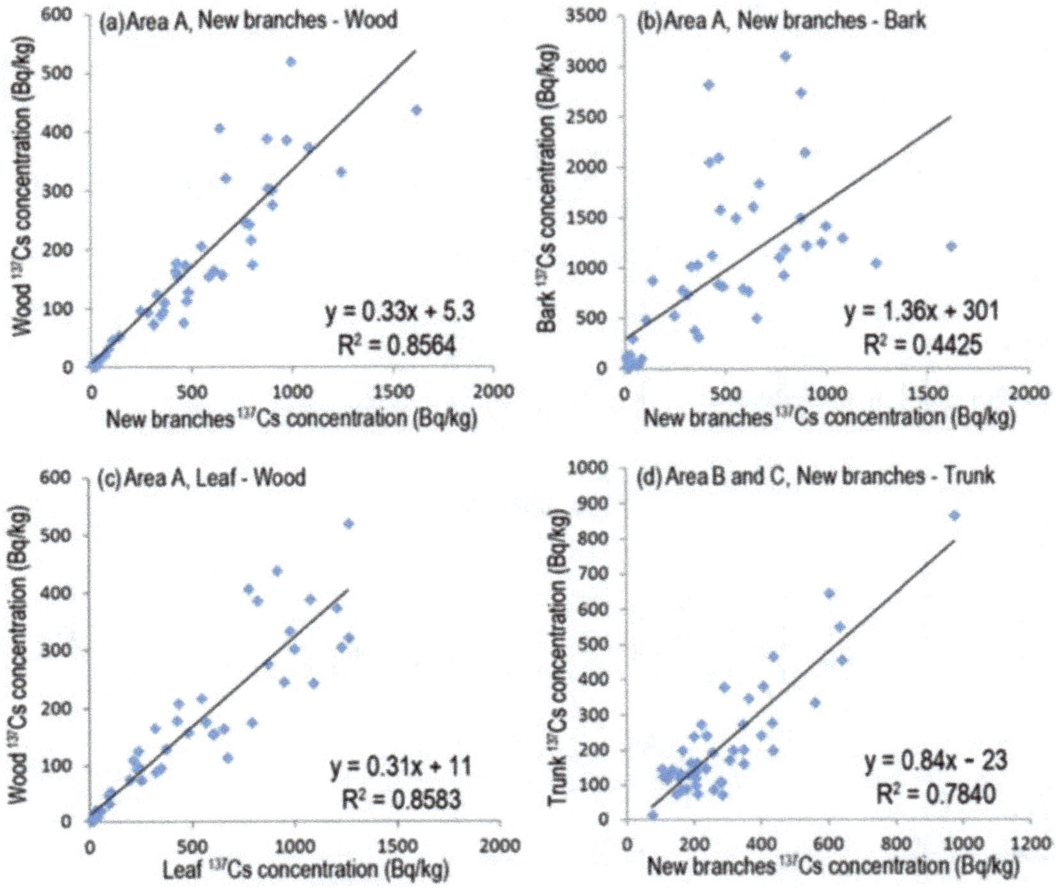

Fig. 11.4 Scatter plots and correlation coefficients between the ^{137}Cs concentrations in new branches and wood (**a**), new branches and bark (**b**), leaves and wood (**c**), and new branches and trunk (**d**)

changed unevenly, and thus reducing the correlation between bark ^{137}Cs concentration and the others tree parts. Therefore, ^{137}Cs concentration in leaves and new branches is a potential indicator to estimate the ^{137}Cs concentration in wood, but leaves and new branches cannot be used to estimate the contamination level in the trunk when radioactive compounds were directly deposited on the bark.

The coppiced trees in areas B and C had young trunks without suberized mature bark, which was not contaminated directly by fallout (Table 11.1). In these trunks, ^{137}Cs concentration was positively and highly correlated with ^{137}Cs concentration in new branches (Fig. 11.4). The slope of the line was 0.84 (Fig. 11.4), suggesting that ^{137}Cs concentration is higher in the younger branch.

11.3.4 *Seasonal Variation in Leaf ^{137}Cs Concentration*

Knowledge about the accumulation and transport of cesium inside a tree body can be essential to think about the long-term changes in radiocesium contamination in trees. In the current study, the seasonal variation of ^{137}Cs concentration in leaves was

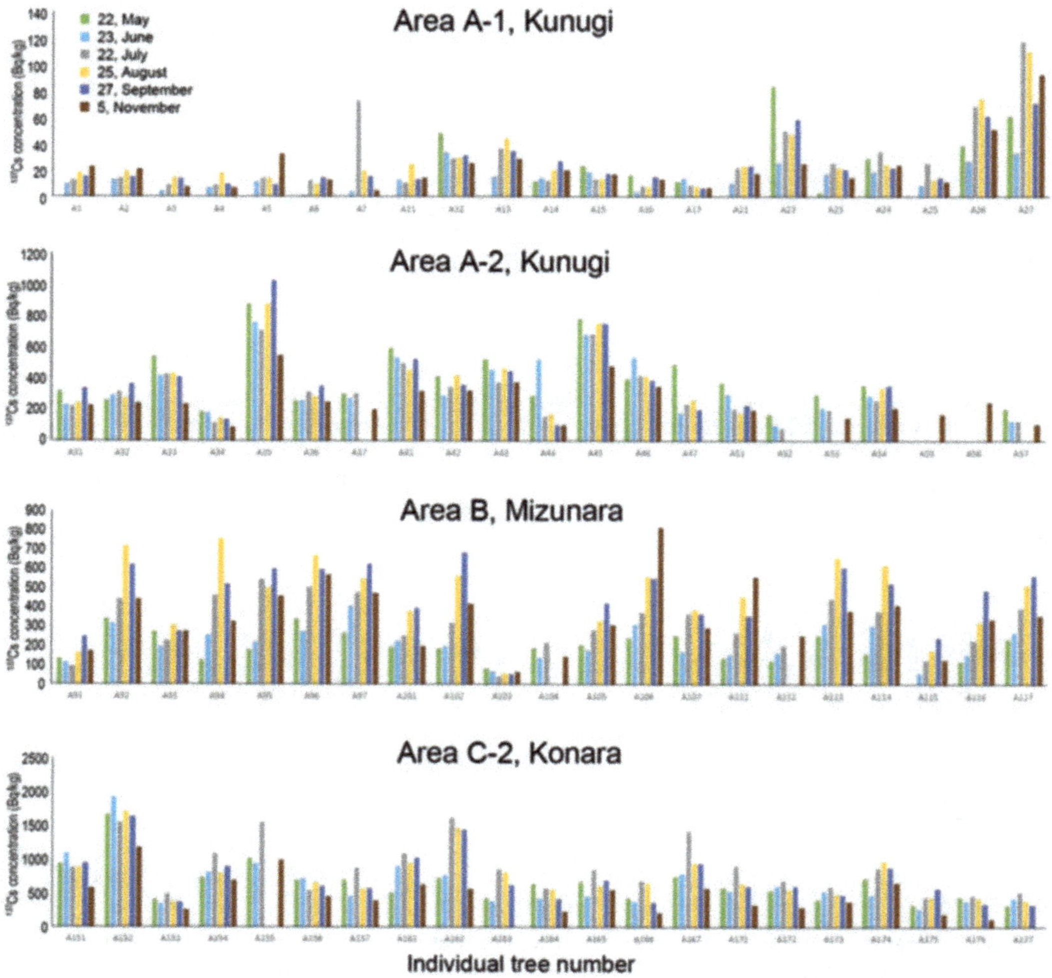

Fig. 11.5 Seasonal variation in leaf ^{137}Cs concentration in 21 individual trees in sites A-1, A-2, B, and C-2. Absence of a color bar indicates that there was no sample available; it does not mean the concentration was too low

investigated. Oak trees in Tamura city come into leaf in May, and the senescent leaves are mostly shed by November (Fig. 11.5). We investigated 42 kunugi trees in area A, 21 mizunara trees in site B, and 21 konara trees in site C-2. No common pattern in the change of ^{137}Cs concentration in leaves through the green period was found. Roughly classified, there were 3 patterns: (1) peaking in summer, (2) continuously decreasing, and (3) no variation. The first pattern predominated in site B, but otherwise, the three patterns were found randomly in each site (Fig. 11.5).

Seasonal variation in mineral concentration in leaves can be determined by root uptake and internal translocation from reserves in various living tissues. In peach trees, reserves have been estimated to supply approximately 40% of the requirement in phosphorous and potassium for new shoot growth during the first 8 weeks after bud-break (Stassen et al. 1983). The quantity of several minerals absorbed via roots was estimated in mature pistachio trees, and potassium uptake was found to occur mostly after spring flush season (Rosecrance et al. 1996). On the other hand, in case of oak trees with ^{137}Cs contamination primarily on their shoot surface and then via root uptake, the ratio between the quantity of ^{137}Cs taken up by roots and the volume of ^{137}Cs reserved within the tissues can be different for each individual tree and also can vary year by year. This situation could cause the variations in the pattern of seasonal change in leaf ^{137}Cs concentration.

11.4 Comparison Between ^{137}Cs Distribution and ^{133}Cs Distribution in Wood, Bark, and New Branches

The clear correlation observed between ^{137}Cs concentration in new branches and wood (Fig. 11.4) indicated that some physiological system determined ^{137}Cs distribution. We propose that it is the same system that regulates ^{133}Cs uptake, accumulation and translocation in oak trees. Thus, the ^{133}Cs distributions in wood, bark, and new branches in trees on site A-2 and A-3 were comparatively analyzed using ICP-MS after the radioactivity in the samples was determined. The result showed that ^{133}Cs concentration in wood positively correlated with that in new branches (R^2, 0.68; slope, 0.22) (Fig. 11.6). Accordingly, the 3: 1 ratio between ^{137}Cs concentration in branches and wood (Fig. 11.4), is assumed to increase over the years to gradually approach a ratio of 5:1 (Fig. 11.6). Nevertheless, we can make an approximate estimation of the ^{137}Cs concentration in wood based on the ^{137}Cs concentrations in new branches. In contrast to wood, the concentration of ^{133}Cs in bark only weakly correlated with ^{133}Cs in new branches (R^2, 0.4836) (Fig. 11.6). Nevertheless, given that the difference in concentration between bark and wood was very small in ^{133}Cs compared to ^{137}Cs (Figs. 11.4 and 11.6), and because wood comprises the largest proportion of the trunk volume, it could be possible that ^{137}Cs concentration in new branches can be used as an indicator of ^{137}Cs concentration in the trunk after the highly contaminated bark tissue falls off and the ^{137}Cs distribution becomes similar to that of ^{133}Cs.

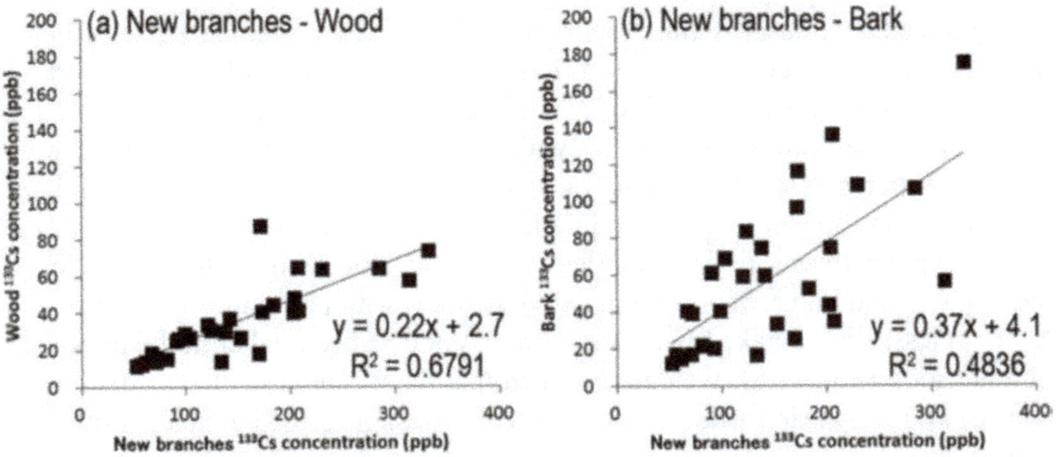

Fig. 11.6 Scatter plots and correlation coefficients of ^{133}Cs concentrations between new branches and wood (**a**), new branches and bark (**b**). The samples were collected in site A-2 and A-3

11.5 Extra Field Investigation to Evaluate the Impact of Field Use History on the Current ^{137}Cs Content in Trees

Because of the unexpectedly low contamination level of kunugi in site A-1 (Fig. 11.3), we tried to identify a factor that could be potentially affecting ^{137}Cs concentration in these trees. The distinctive feature of site A-1 is that it used to be a field that grew corn. Additionally, the trees in site A-1 were the youngest among trees in site A-2 and A-3, indicating the dilution effect due to the higher growth rate needs to be also considered. Therefore, we investigated 9-year-old planted trees in another three sites in Tamura city in December 2015. Site D used to be a farm before the oak trees were planted. Site E and site F are artificial forests covered with bamboo grass (Fig. 11.7). Soil pH was relatively high in site D (Table 11.3), though it was not as high as in site A-1 (Table 11.2). The concentration of ^{137}Cs in surface soil and new branches was measured. In the new branches of both konara and kunugi on site D, the ^{137}Cs concentration was found to be lower than the detection limit (<56 Bq/kg), resulting in the ratio between branch ^{137}Cs concentration and soil ^{137}Cs concentration of less than 0.01 (Table 11.3). On the other hand, the branch-to-soil ratios in sites E and F were higher than 0.04 (Table 11.3). In conclusion, the site condition that is being influenced by the field use history appears to be affecting the ^{137}Cs concentration in trees of the same age. Further studies will be required to elucidate what environmental variable in the site is controlling the ^{137}Cs concentration in trees, and what is the mechanism.

11.6 ^{137}Cs Tracer Experiment Using Hydroponically Grown Young Oak Seedlings

In crop plants, uptake and translocation of cesium are known to be altered depending on the cultivation medium (Nobori et al. 2014). Therefore, the effect of nutritional conditions on cesium transport in oak trees was analyzed by the ^{137}Cs tracer experiment (Kobayashi et al. 2016). Konara trees were hydroponically cultured at 25 °C in the plant growth chamber. Two different types of nutrient solution were used; half-strength Hoagland's solution and Kimura-B. The concentration of minerals in Hoagland's solution is generally higher than in Kimura-B. On the day of the experiment, the trees supplied with rich nutrition by Hoagland's solution developed more leaves and bigger roots compared to the trees with poor nutrition (Fig. 11.8). Nevertheless, ^{137}Cs uptake by roots for 3 h was noticeably smaller in the trees with rich nutrition (Fig. 11.8). We demonstrated that the mechanisms for mediating Cs uptake would appear to be down-regulated in response to the rich nutrition in rhizosphere.

Fig. 11.7 Three sites (D, E and F) where 9-year-old oak trees were sampled in December 2015

The behavior of ^{137}Cs deposited on the surface of a young stem was visually analyzed to consider the radiocesium movement after reaching the vasculature tissues inside the trunk. The ^{137}Cs radiograph showed the rapid translocation of radiocesium toward the leaf and roots within 5 days (Fig. 11.9). Interestingly, not all the root tissue accumulated ^{137}Cs (Fig. 11.9), and furthermore, a little portion of ^{137}Cs was released to the nutrient solution during the next 5 days.

Table 11.3 ^{137}Cs concentration in surface soil, new branches, and their ratio in the 9-year-old oak trees grown in sites D, E, and F

		Cs-137 (Bq/kg)			New branches/ soil ratio	
Name	Soil pH	Soil 0–5 cm	Konara new branches	Kunugi new branches	Konara	Kunugi
Site D	5.5	5348	ND (<56)	ND (<56)	<0.01	< 0.01
Site E	4.9	10,529	453	544	0.043	0.052
Site F	5.0	4981	–	212	–	0.043

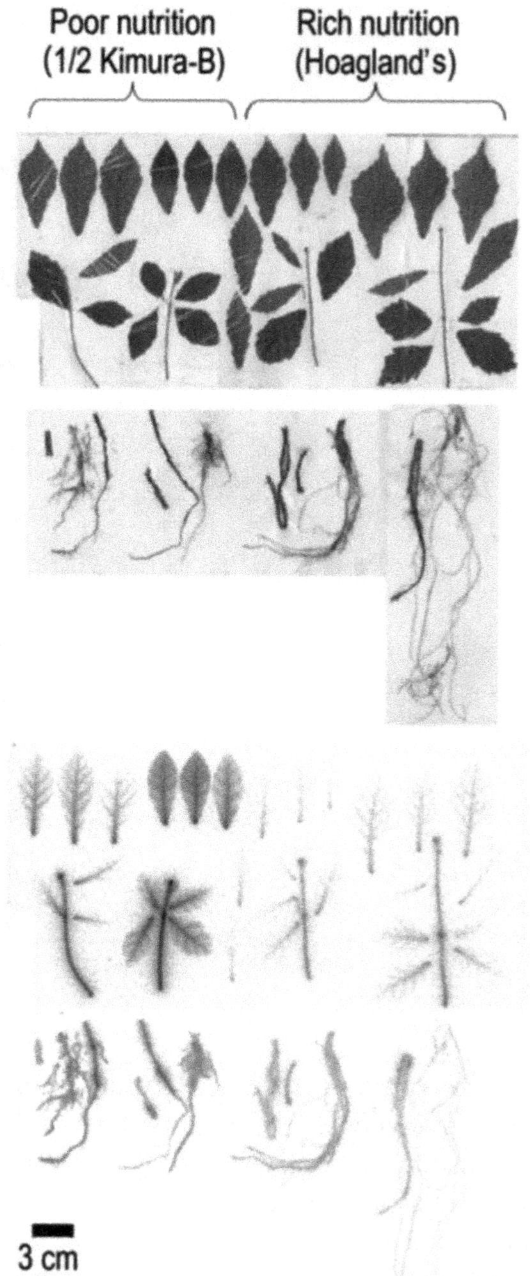

Fig. 11.8 The appearance of young konara trees grown hydroponically and allowed to absorb ^{137}Cs for 3 h (upper panel) and their autoradiographs (lower)

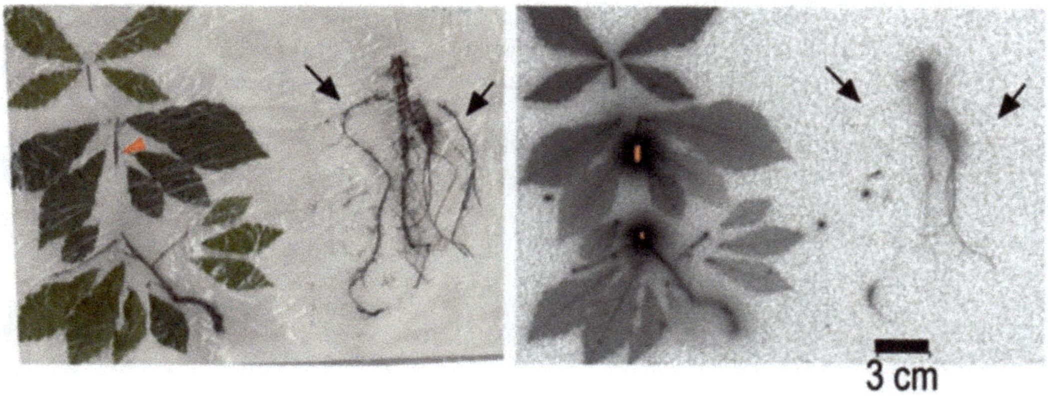

Fig. 11.9 A young konara tree on which a droplet containing ^{137}Cs was added (red arrowhead, left panel). The autoradiograph was taken after 5 days (right panel). Note, there are roots without ^{137}Cs accumulation (black arrows)

11.7 Conclusion

The contamination level in oak trees is practically in proportion to the contamination level in the soil in which the tree is growing, which can be a mirror of the amount of radiocesium deposition after the nuclear power plant accident. However, there can be additional environmental factors modulating the radiocesium concentration in trees 4 years after the accident. One factor could be a chemical property of the soil, including pH and nutrient content. Further investigation to identify the factors can allow the selection of low contaminated trees even today, and take measures to gradually reduce the radiocesium concentration in the trunk of the hardwood trees grown in the contaminated forest.

Acknowledgement This project was carried out with cooperation from the following members; Dr. Satoru Miura (Forest Research and Management Organization), Dr. Nobuhito Sekiya (Mie University), Dr. Daisuke Takata (Fukushima University), Dr. Stephan Bengtsson (Formerly at Fukushima University), Dr. Kazuhisa Yamasaki (The University of Tokyo), Dr. Naoto Nihei (The University of Tokyo), Dr. Keitaro Tanoi (The University of Tokyo), and Mr. Ryosuke Ito (Former master-course student). This project was supported in part by The Specific Research Grant 2015 for East Japan Great Earthquake Revival by The New Technology Development Foundation.

References

Endo I, Ohte N, Iseda K, Tanoi K, Hirose A, Kobayashi NI, Murakami M, Tokuchi N, Ohashi M (2015) Estimation of radioactive 137-cesium transportation by litterfall, stemflow and throughfall in the forests of Fukushima. J Environ Radioact 149:176–185

Kobayashi NI, Sugita R, Nobori T, Tanoi K, Nakanishi TM (2016) Tracer experiment using $^{42}K^+$ and $^{137}Cs^+$ revealed the different transport rates of potassium and caesium within rice roots. Funct Plant Biol 43:151–160

Miura S (2016) The effects of radioactive contamination on the forestry industry and commercial mushroom-log production in Fukushima, Japan. In: Agricultural implications of the Fukushima nuclear accident. The first three years. Springer, Tokyo. https://doi.org/10.1007/978-4-431-55828-6_12

Nobori T, Kobayashi NI, Tanoi K, Nakanishi TM (2014) Effects of potassium in reducing the radiocesium translocation to grain in rice. Soil Sci Plant Nutr 60:772–781

Rosecrance RC, Weinbaum SA, Brown PH (1996) Assessment of nitrogen, phosphorus, and potassium uptake capacity and root growth in mature alternate-bearing pistachio (*Pistacia vera*) trees. Tree Physiol 16:949–956

Stassen PJC, Du Preez M, Stadler JD (1983) Reserves in full-bearing peach trees. Macro-element reserves and their role in peach trees. DFGA 33:200–206

Chapter 12
Radiocesium Dynamics in Wild Mushrooms During the First Five Years After the Fukushima Accident

Toshihiro Yamada

Abstract Dynamics of radiocesium in wild mushrooms, especially in mycorrhizal fungi, in forest ecosystems were investigated for 5 years after the Fukushima nuclear accident, in relation to substrates such as litter, soil and wood debris. Some mushroom species contained a high level of radiocesium in the first or second year, and then the radiocesium content decreased. Changes in radiocesium activities were ambiguous for many other mushrooms. Radiocesium accumulation with time was not common contrary to expectations. Reduction of radiocesium activities in litter and increase in mushrooms and soils, i.e. transfer of radiocesium from litter to mushrooms and soils, was recognized in the first and second year, but it was not obvious in subsequent years. Radiocesium accumulated in several mushroom species, especially in mycorrhizal fungi, while radiocesium in the other mushrooms did not exceed those in the neighboring forest litter. Similar differences in radiocesium level among mushroom species were observed in relation to ^{40}K levels, though ^{137}Cs/^{40}K ratio in mushrooms was lower than in O horizon, but at the same level of the A horizon in general. These facts suggested differences in the mechanisms of cesium accumulation. Residual ^{137}Cs due to nuclear weapons tests or the Chernobyl accident still remained in mushrooms and soils. From the ratio of the past residual ^{137}Cs, it was suggested that the residual ^{137}Cs was tightly retained in the material cycles of forest mushroom ecosystem, whereas ^{137}Cs emitted from the Fukushima accident was still fluid.

Keywords Chernobyl nuclear accident · Nuclear weapons tests · Radioactive fallout · Radiocesium · The University of Tokyo Forests · Transfer factor · Wild mushrooms

---

T. Yamada (✉)
The University of Tokyo Chichibu Forest, Graduate School of Agricultural and Life Sciences, The University of Tokyo, Chichibu, Saitama, Japan
e-mail: yamari@uf.a.u-tokyo.ac.jp

T. M. Nakanishi et al. (eds.), *Agricultural Implications of the Fukushima Nuclear Accident (III)*, https://doi.org/10.1007/978-981-13-3218-0_12

Abbreviations

Cs	cesium
DW	dry weight
FW	fresh weight
K	potassium
NPP	nuclear power plant
NWT	nuclear weapons test
TF	transfer factor
UTF	the University of Tokyo Forest

12.1 Introduction

Radioactive material released from the Fukushima Daiichi nuclear power plant (F1-NPP) accident spread over a wide area of East Japan. Wild mushrooms often contain a high level of radiocesium in even lower contaminated areas. The University of Tokyo has seven research forests located in East Japan, 250–660 km from F1-NPP, where radiocesium contamination is low (0.02–0.12 μSv/hr. of air dose rate 1 year after the accident). These forests are used for many activities including research, education, forest management and recreation.

Radiocesium contaminated wild and cultivated mushrooms is a major concern for consumers of these forest products. Fungi, including mushrooms, are also one of the major and important components of the forest ecosystem. Radioactive contamination of mushrooms should be considered not only from the viewpoint of food but also from the viewpoint of its effects on plants and animals through its circulation in the forest ecosystem. Therefore, we have surveyed radiocesium contamination in relation to mushrooms in the University of Tokyo Forests.

Mushrooms have been reported to accumulate radiocesium (Byrne 1988; Kammerer et al. 1994; Mascanzoni 1987; Muramatsu et al. 1991; Sugiyama et al. 1990, 1994). The transfer factors (TF) for radiocesium in mushrooms were reported to be 2.6–21 in culture tests (Ban-nai et al. 1994). However, the radiocesium concentration ratio in mushrooms relative to the soil was rather low and the ratio was often <1 in a field study (Heinrich 1992). Symbiotic mycorrhizal mushrooms tend to have higher TF of ^{137}Cs than the saprobic fungi in general, though different mushroom species also have widely varying degrees of radiocesium activity (Heinrich 1992; Sugiyama et al. 1993).

Another feature of fungi is the considerable proportion of ^{137}Cs in forest soil is retained by the fungal mycelia, and fungi are considered to prevent the elimination of radiocesium from ecosystems (Brückmann and Wolters 1994; Guillitte et al. 1994; Vinichuk and Johanson 2003; Vinichuk et al. 2005). Thus, fungal activity is likely to contribute substantially to the long-term retention of radiocesium in the organic layers of forest soil by recycling and retaining radiocesium between fungal mycelia and soil (Muramatsu and Yoshida 1997; Steiner et al. 2002; Yoshida and Muramatsu 1994, 1996). In fact, some examples of long-term ^{137}Cs radioactivity persistence in mushrooms in forests and transfer to animals have been reported, whereas that in plants had short ecological half-lives (Fielitz et al. 2009; Kiefer et al. 1996; Zibold et al. 2001).

In previous reports (Yamada 2013; Yamada et al. 2013), radiocesium contamination of wild mushrooms in the University of Tokyo Forests half a year after the Fukushima accident has been summarized. We found rapid uptake of radiocesium in one species of mushroom after the Fukushima accident and residual contamination from atmospheric nuclear weapons tests (NWT) or the Chernobyl accident. In the current study, the dynamics of radiocesium were surveyed over a 5 year period in wild mushrooms and their substrates (litter, soil or wood debris) in relatively low-contaminated forest areas, and features of the dynamics of mushroom contamination were elucidated, paying attention to accumulation and retention of radiocesium in mushroom related forest ecosystems. The raw data of our surveys were presented in Yamada et al. (2018).

12.2 Research Sites and Sampling

Mushrooms appeared in the autumn of 2011–2015 (Table 12.1) and their presumptive substrates, i.e., the O horizon (organic litter layer, called A_0 horizon in Japan), the A horizon (mineral layer and accumulated organic matter), and the C/O horizon

Table 12.1 Lists of radiocesium-measured mushrooms

Research forest	Lifestyle	Species	Japanese name
UTHF (Hokkaido)	M	*Lyophyllum connatum*	Oshiroishimeji
	M	*Suillus grevillea*	Hanaiguchi
UTCF (Chichibu)	M	*Russula emetica*	Dokubenitake
	M	*Tricholoma saponaceum*	Mineshimeji
	S	*Bondarzewia berkeleyi*	Oomiyamatonbimai
	S	*Hericium erinaceum*	Yamabushitake
	S	*Sarcomyxa edulis* (synonym *Panellus serotinus*)	Mukitake
	S	*Trametes versicolor*	Kawaratake
FIWSC (Fuji)	M	*Amanita caesareoides*	Tamagotake
	M	*Chroogomphus rutilus*	Kugitake
	M	*Lactarius hatsudake*	Hatsutake
	M	*Lactarius laeticolor*	Akamomitake
	M	*Lyophyllum shimeji*	Honshimeji
	M	*Suillus grevillea*	Hanaiguchi
	M	*Suillus luteus*	Numeriiguchi
	M	*Suillus viscidus*	Shironumeriiguchi
	S	*Armillaria mellea*	Naratake
	S	*Pholiota lubrica*	Chanametsumutake
	S	*Hypholoma sublateritium*	Kuritake
	S	*Lentinula edodes*	Shiitake
	S	*Pholiota microspora*	Nameko
	S	*Pleurotus ostreatus*	Hiratake
UTCBF (Chiba)	M	*Catathelasma imperial*	Oomomitake

M mycorrhizal fungi, *S* saprobic fungi

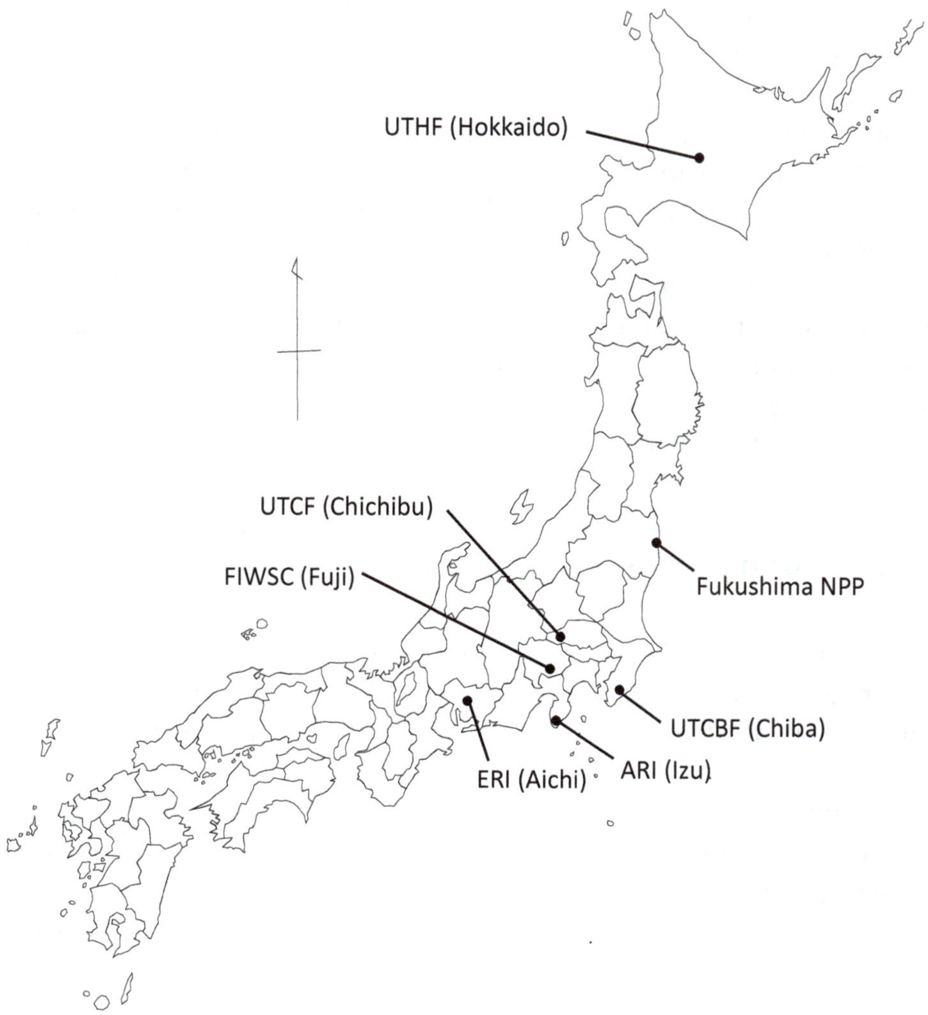

Fig. 12.1 Locations of the University of Tokyo Forests from where samples were obtained
Fukushima NPP, Fukushima Daiichi nuclear power plant
UTHF The University of Tokyo Hokkaido Forest (Hokkaido), 660 km from F1-NPP, *UTCF* The University of Tokyo Chichibu Forest (Chichibu), 250 km from F1-NPP, *FIWSC* Fuji Iyashinomori Woodland Study Center (Fuji) (formerly Forest Therapy Research Institute, FTRI), 300 km from F1-NPP, *UTCBF* The University of Tokyo Chiba Forest (Chiba), 260 km from F1-NPP, *ARI* Arboricultural Research Institute (Izu), 360 km from F1-NPP, *ERI* Ecohydrology Research Institute (Aichi), 420 km from F1-NPP

(mineral layer with a small quantity of organic matter, which is little affected by pedogenic processes (Soil Survey Staff 2014)) of the soil, or mushroom logs were collected from six (in 2011), 5 (in 2012) and 4 (between 2013 and 2015) research forests shown in Fig. 12.1. Figures 12.2 and 12.3 show examples of samples, the appearance of the environment where samples were collected and sample preparation for radioactivity measurement. The concentrations of ^{134}Cs, ^{137}Cs and ^{40}K were determined using a germanium semiconductor detector (GEM-type, ORTEC, SEIKO EG&G, Tokyo, Japan). Distribution of radiocesium deposition and γ-ray air dose rate in 2011 was presented in a previous report (Yamada 2013).

Fig. 12.2 Sampling of mushrooms and soils
(**a**) Deciduous mixed forest where *Suillus grevillea* mushrooms were collected in Hokkaido (UTHF); (**b**) *S. grevillea* mushrooms in Hokkaido; (**c**) litter layer (O horizon) under *S. grevillea* mushrooms; (**d**) surface soil (A horizon) under *S. grevillea* mushrooms; (**e**) Slices of *Amanita caesareoides* mushrooms collected in Fuji (FIWSC); (**f**) Mushroom and soil samples were placed into U-8 containers. (Photo by K. Iguchi (**a**, **b**, **c**, **d**) and H. Saito (**e**))

12.3 Gamma Ray Air Dose Rate at the Mushroom Collection Sites (Fig. 12.4)

Gamma ray air dose rate (μSv/h) 1 m above ground level was measured with a dose rate meter (TC100S, Techno AP Co. Ltd., Japan) using a CsI (Tl) scintillation detector. Although considerable variation in dose rate was observed among UTFs due to environmental variation such as geological features, trends of changes and levels in dose rate were similar within each UTF. Similar levels of pre-Fukushima contamination from nuclear weapons tests and the Chernobyl accident were estimated from $^{137}Cs/^{134}Cs$ ratio in soils in Chiba (UTCBF) and Fuji (FIWSC). Initial air dose rate in Fuji, however, was somewhat lower than that in Chiba, probably due to geological features. Dose rate slightly decreased in Chiba with time, whereas the decrease was not clear in Fuji. However, in 2015, dose rate in Fuji was similar to the dose rate recorded in Chiba. The original dose rate before the Fukushima accident was thought to be low in Fuji and Chiba. Although contamination due to the Fukushima accident did not reach Hokkaido, the dose rate was higher in Hokkaido (UTHF) than that in Fuji and Chiba. One year after Fukushima accident, the dose rate in Chichibu (UTCF) was higher than that in other UTFs, and was over 0.1 μSv/h, especially in high mountain areas, then gradually reduced by about half by 2015. Dose rate may decrease further in Chichibu, as the dose rate due to the Fukushima accident was estimated at approximately 50 nGy/h (0.05 μSv/h equivalent dose rate of radiocesium) in Chichibu (Minato 2011), whereas the dose rate in other UTFs appeared to become almost stable by 2015.

Fig. 12.3 Examples of collected mushrooms and forests where mushrooms grew
(**a**) Deciduous mixed forest where *Russula emetica* mushrooms were collected in Chichibu (UTCF); (**b**) *Tricholoma saponaceum* mushrooms on flat land in Chichibu; (**c**) Soil profile in Fuji (FIWSC). O horizon and C/O horizon were observed; (**d**) Saprobic mushrooms, *Pholiota microspora* (left) and *Pleurotus ostreatus* (right), cultivated on the wood logs in Fuji. Wild *Armillaria mellea* mushrooms were also collected; (**e**) Mixed forest of *Pinus densiflora* and *Malus toringo* on Yamanaka-ko lakeside in Fuji; *Suillus luteus* and *Lactarius hatsudake* mushrooms were collected; (**f**) *Abies homolepis* forest in Fuji showing *Pholiota lubrica* mushrooms on a felled tree (**g**) and on soil (**h**), and mycorrhizal *Lactarius laeticolor* mushrooms on soil (**i**) were collected. (Photo by K. Takatoku (**b**) and H. Saito (**c**))

12.4 Dynamics of Radiocesium in Each of the University of Tokyo Forests (Fig. 12.5)

12.4.1 Litter and Soil Layer

Hokkaido (UTHF): We believe no Fukushima-derived contamination reached Hokkaido because ^{134}Cs was not detected. ^{137}Cs was often below the detection limit, and its concentration in the A horizon was similar to the concentration in the O horizon, indicating the contamination in Hokkaido was old from the viewpoint of transfer to the soil. Similarly, ^{137}Cs was regularly detected in mushrooms even at low levels, but ^{134}Cs was not detected. It indicated that radiocesium in mushrooms was from the pre-Fukushima fallout.

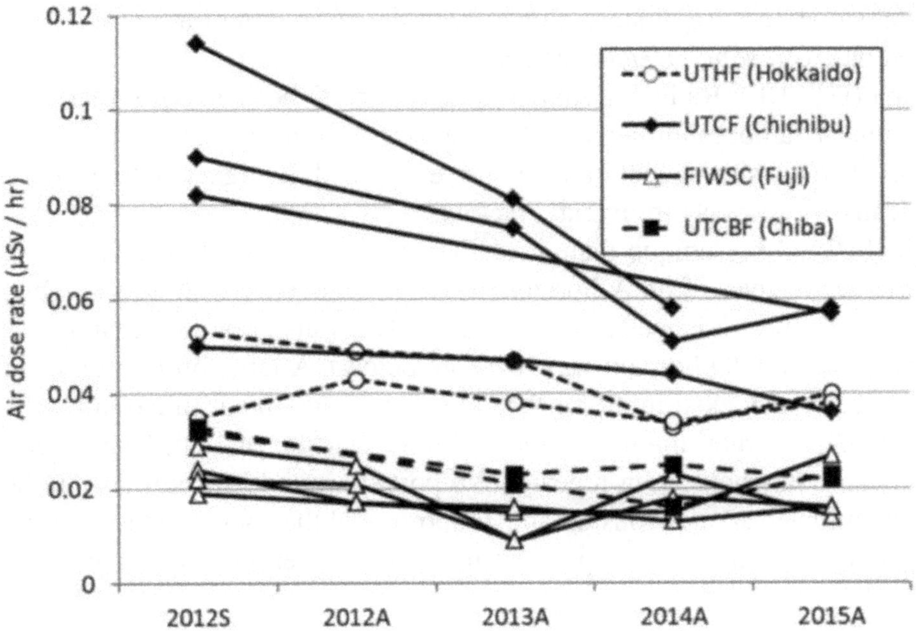

Fig. 12.4 Change in air dose rate 1 m above ground at each University of Tokyo Forest
S spring, *A* autumn

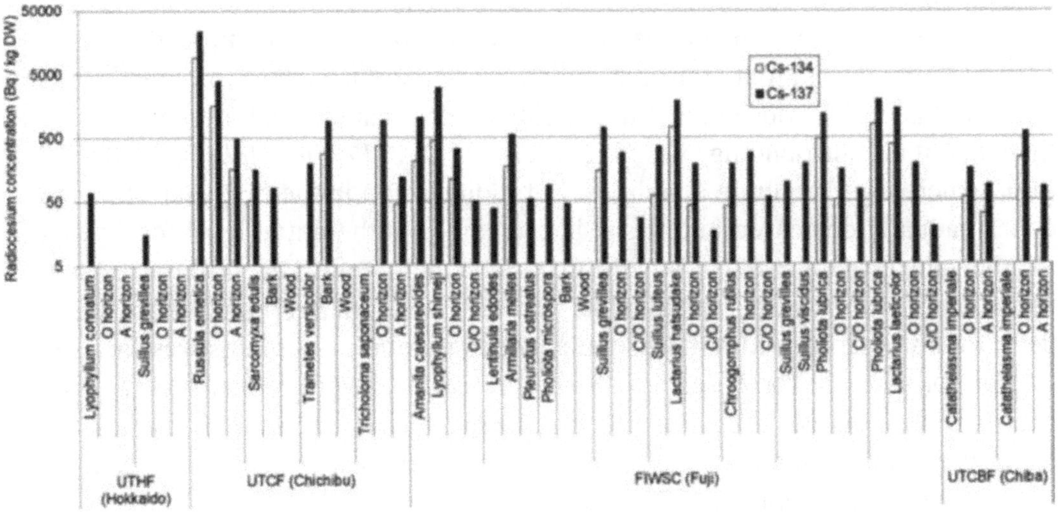

Fig. 12.5 Radiocesium concentration in mushrooms and soils in each Research Forest in 2013
No data: radiocesium concentration was below the detection limit

Chichibu (UTCF): Radiocesium levels in the O horizon were high (200–4400 Bq/kg DW) half a year after the accident, and then decreased relatively rapidly. At the same time, the A horizon also contained ^{134}Cs (20–120 Bq/kg DW), indicating a rapid transfer to the A horizon because ^{134}Cs derived from past emissions had already decayed. Subsequent transfer to the A horizon was recognized for example in *Tricholoma saponaceum*-collected site, however, transfer was generally small. It was possible that radiocesium was mobilizing to lower regions such as valleys or colluvial slope because of steep slopes, with the exception of a few sites (e.g., the flat land where *T. saponaceum* was collected).

Chiba (UTCBF): Radiocesium decreased in the O horizon with time. A certain proportion of radiocesium appeared to transfer into the A horizon even by 2012, however, no clear subsequent transfer was recognized; It might reach a stable condition because of local environmental factors.

Fuji (FIWSC): A unique feature of FIWSC is that a C/O horizon of volcanic Scoria exists instead of an A horizon. FIWSC is covered with Scoria which is a volcanic immature soil ejected from Mt. Fuji. The transfer of radiocesium from the O horizon to the C/O horizon was low (see below). A large proportion of mycorrhizal mycelia may exist in the surface litter layer, and this resulted in mushrooms accumulating a larger amount of radiocesium. Outside of Fuji, heavily contaminated mushrooms have been repeatedly reported around Mt. Fuji despite being a low-contaminated area. A considerable proportion of the contamination was thought to be derived from nuclear weapons testing and the Chernobyl accident.

12.4.2 Mushrooms

Russula emetica in Chichibu had a high level of radiocesium. Soil analyzed from the *R. emetica*-collection site was also highly contaminated compared with other sites in Chichibu; fallout from the radioactive plume appeared to have deposited here by chance. The dose rate of this highly contaminated site, however, was lower than that of the surrounding sites. The level of ^{137}Cs in *Pholiota lubrica*, collected in Fuji, which absorbed quite a high level of radiocesium in the first year of the accident, gradually decreased in one site but remained at the initial level for 4 years in another site. Dynamics of ^{137}Cs in the O horizon might reflect the difference because mycelia of *P. lubrica* was spread widely in the O horizon. Six months after the accident, Fukushima-derived radiocesium concentration in mushrooms was lower than that of soils except for *P. lubrica*. Some mycorrhizal mushrooms such as *Suillus grevillea*, *S. viscidus*, *Amanita caesareoides*, *Lyophyllum shimeji* and *Lactarius laeticolor* in Fuji contained less ^{134}Cs compared with ^{137}Cs. It was concluded that the past contamination remained (See Sect. 12.8).

Trametes versicolor in Chichibu had a low radiocesium concentration in 2011; the majority of the radiocesium seemed to be derived from the Fukushima accident judging from the proportion of ^{134}Cs. The radiocesium content in *T. versicolor* was high between 2012–2014, indicating the accumulation in mycelia, but decreased in 2015. In other saprobic mushrooms, a high concentration of radiocesium was detected in *Sarcomyxa edulis* (synonym *Panellus serotinus*) in the first year of the accident, then the content decreased in 2013 and 2014 to the same level found in *T. versicolor*. Litter and soils of the sites, where both mushrooms were collected, were contaminated with radiocesium. Several saprobic mushrooms were collected and surveyed in Fuji; In *Lentinula edodes*, *Pleurotus ostreatus*, *Armillaria mellea* and *Pholiota microspora*, radiocesium level was much higher compared with bark and wood as substrates, except for bark in 2012. Radiocesium concentration was low in *L. edodes* and *P. ostreatus* but accumulated in *A. mellea*. Saprobes are thought to absorb radiocesium in proportion to the contamination level of the substrate. However, absorption seemed low compared with some mycorrhizal fungi. High radiocesium content in *A. mellea* might be due to the wide distribution of its mycelia in litter and soil, like *P. lubrica* and several mycorrhizal fungi.

12.5 Dynamics of Radiocesium in the Same Sampling Sites (Figs. 12.6 and 12.7)

Over a four year period (2011–2015), the decrease in radiocesium concentration by physical decay was 0.912 and 0.262 for ^{137}Cs and ^{134}Cs, respectively (calculated from the half-life of both isotopes). ^{137}Cs content of the O horizon gradually decreased with time more than the rate of physical decay in general, whereas the changes of ^{137}Cs level were ambiguous in several sites of Chiba and Fuji. ^{137}Cs was shown to migrate very slowly into the A horizon in Belarus soils after the Chernobyl accident (Kammerer et al. 1994; Pietrzak-Flis et al. 1996; Rühm et al. 1998). In all sites visited in the current study, obvious transfer of radiocesium from the O horizon as well as reduction of radiocesium was not observed in the A or C/O horizon. In the case of mushrooms, *Pholiota lubrica* in Fuji (1 site) and *Catathelasma imperiale* in Chiba (1 site) showed a constant reduction in radiocesium level. The radiocesium concentration in European mushrooms increased for a few years after the Chernobyl accident (Borio et al. 1991; Smith and Beresford 2005); one case of *P. lubrica* and *Suillus grevillea* in Fuji showed a similar pattern, with an increase in radiocesium once during 2011–2012 and a decrease after 2012. In other sites or other mushroom

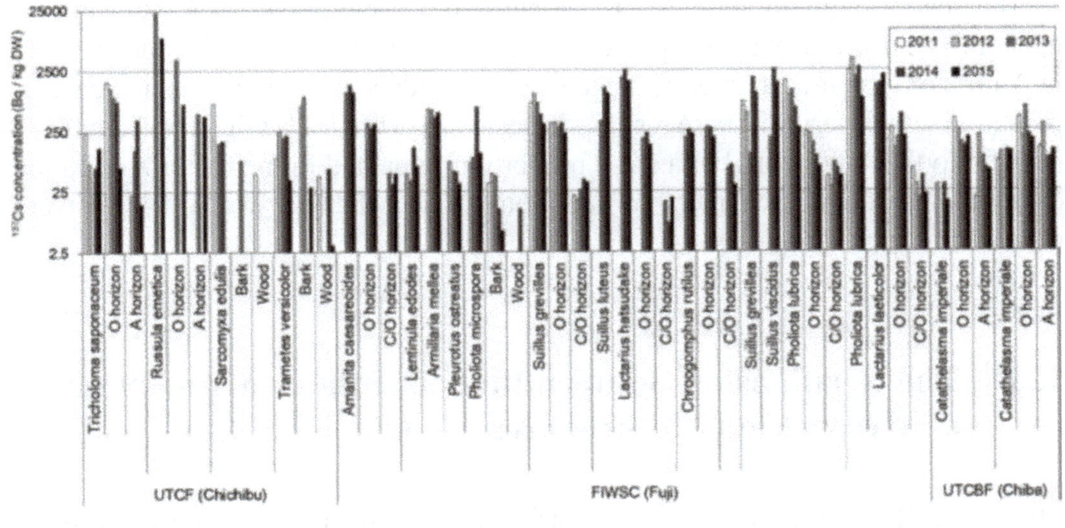

Fig. 12.6 Changes in ^{137}Cs concentration in mushrooms and soils from same sampling sites
No data: either no mushrooms were collected on the site or the radiocesium concentration was below the detection limit

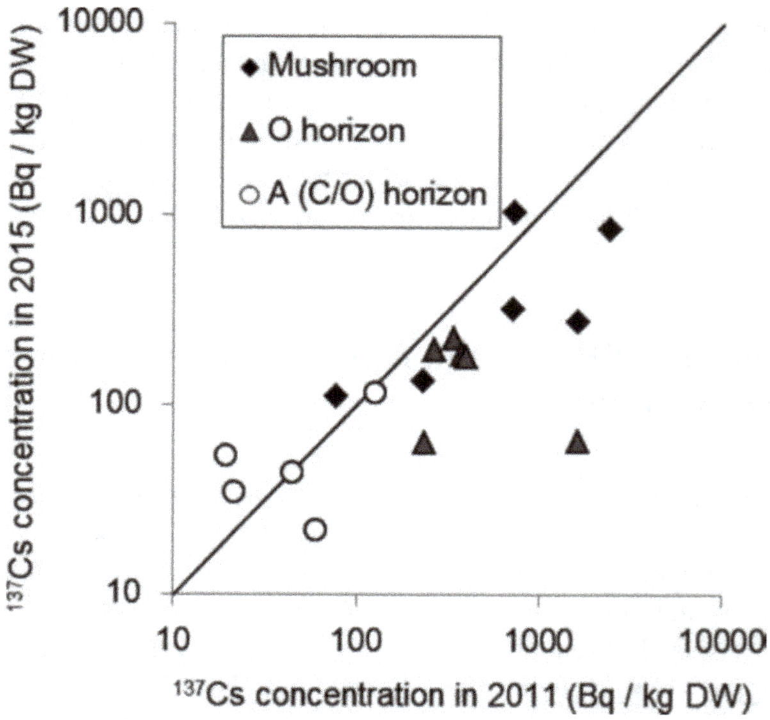

Fig. 12.7 Scatter diagram indicating ^{137}Cs changes from 2011 to 2015 in mushrooms and soils from the same sampling sites
The oblique solid line (Y = X) indicates the same ^{137}Cs concentration between 2011 and 2015

species, a reduction of radiocesium was not obvious. Because radiocesium activity at each soil depth changes with time, radiocesium activity in different fungal species at different mycelial depths are also expected to vary with time (Rühm et al. 1998; Yoshida and Muramatsu 1994). The variation observed among sites of our field study may be due to geographic and pedological conditions.

The scatter diagram (Fig. 12.7) also showed a decrease of ^{137}Cs with time in general. The decrease was conspicuous especially in O horizon. ^{137}Cs concentration in A horizon was low at an early stage of post-accident and no obvious increase or reduction was observed. These results suggested a part of ^{137}Cs migrated from the O horizon to the A horizon, but a large proportion remained in the O horizon. In the case of mushrooms, considerable variations of the changes in ^{137}Cs level were observed between species.

12.6 The Relationship Between Radiocesium Contamination of Mycorrhizal Mushrooms and Soils (Fig. 12.8)

Mushroom/O or A (C/O) horizon ratio of ^{137}Cs was compared in mycorrhizal fungi, and found to be high in Fuji and low in Chiba. Additional data is necessary on the same mushroom species, for example between Chichibu and Fuji, to reveal what

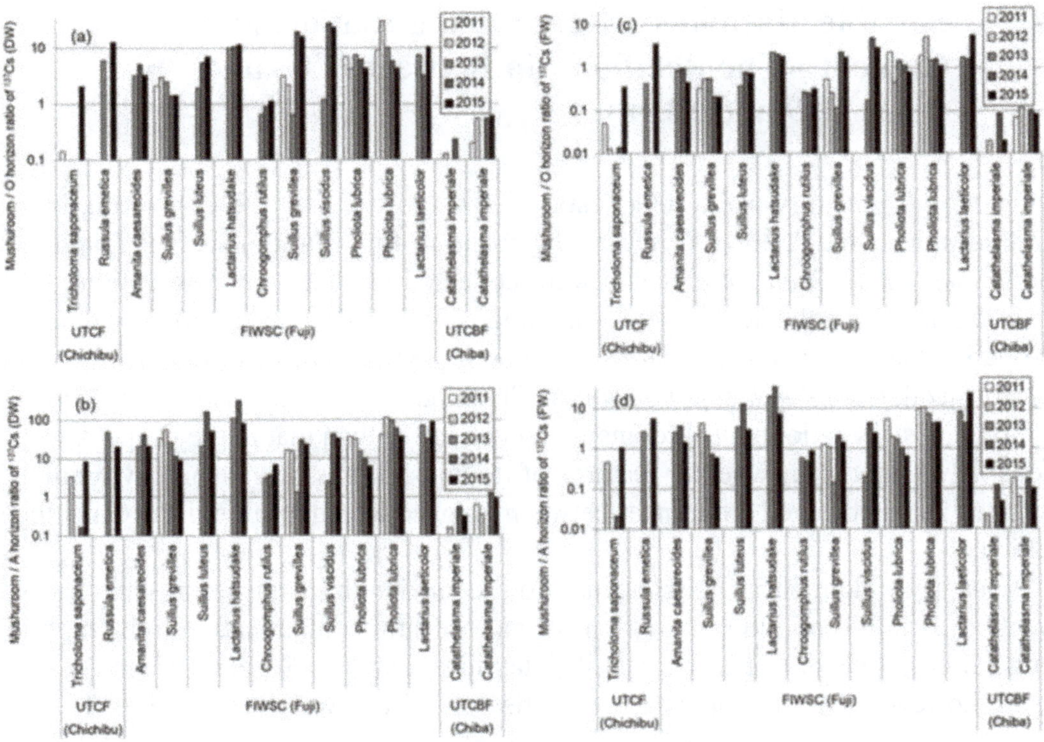

Fig. 12.8 Mushroom/soils ratio of ^{137}Cs concentration
Mushroom/O horizon ratio (**a**, **c**) or mushroom/A (C/O) horizon ratio (**b**, **d**) of ^{137}Cs concentration on dry weight basis (**a**, **b**) or on fresh weight basis (**c**, **d**)
No data: either no mushrooms were collected on the site or the radiocesium concentration was below the detection limit

environmental factors caused such differences. The mushroom/O or A (C/O) horizon ratio of >1 was common in Chichibu and Fuji on a dry weight basis. The ratio on a fresh weight basis was approximately equal to one with a wide range. These findings corresponded with the results of a field study in Europe (Heinrich 1992). Heavily contaminated *R. emetica* detected in Chichibu did not have a high ratio compared with other mycorrhizal fungi. High contamination could have been due to heavy soil contamination rather than a biological feature of this species. *P. lubrica* in Fuji showed a clear reduction over time for both mushroom/O horizon and mushroom/C/O horizon ratios. In other sites or other mushroom species, reduction in the mushroom/O or A (C/O) horizon ratios were not obvious. A wide ratio range was observed even within the same species, and no obvious radiocesium accumulation was observed in mushrooms over time.

12.7 Possible Mechanism Determining Radiocesium Content – The Relationship Between ^{137}Cs and ^{40}K (Figs. 12.9 and 12.10)

Mushrooms generally had a lower ratio of ^{137}Cs to ^{40}K than the O horizon, but a similar ratio to the A horizon; several mycorrhizal fungi in Fuji such as *Lactarius hatsudake* and *L. laeticolor*, which were collected in 2013, were exceptions. For *Pholiota lubrica* collected in 2011, this fungus absorbed radiocesium very quickly probably due to an abundance of its mycelia in the O horizon. On the contrary, the mycorrhizal *Tricholoma saponaceum* in Chichibu and *Catathelasma imperiale* in Chiba, ^{137}Cs/^{40}K ratio in mushrooms was much lower than that in the O and A horizons. It was not clear whether features of the mushrooms or the soil environment resulted in the observed differences between fungi collected from Fuji and Chichibu/Chiba.

On a dry weight basis, ^{40}K concentration seemed high in mushrooms and low in the O or A (C/O) horizons. A reason for radiocesium contamination to be high in mushrooms appears to be because of potassium richness (Seeger 1978). The ^{40}K level, however, was similar for the triparties on a fresh weight basis (Fig. 12.10), suggesting no special mechanism of K absorption. Becauese most K exists as ions in the cytoplasm, the difference was due to a high water content in mushrooms (a water content of 90–95% is common). The high ^{137}Cs/^{40}K ratio observed in *Russula emetica* was probably induced by heavy soil contamination. On the other hand,

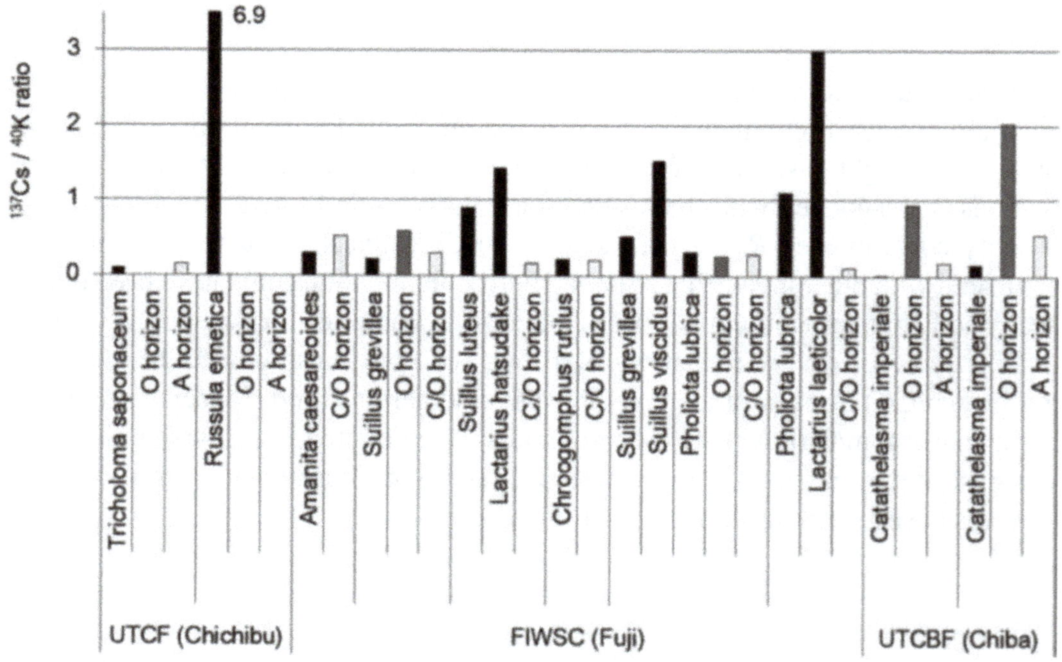

Fig. 12.9 ^{137}Cs/^{40}K ratio in mushrooms and soils in 2015
■, Mushroom; ▩, O horizon; □, A (C/O) horizon. ^{137}Cs/^{40}K ratio in *Russula emetica* was 6.9
No data: radiocesium concentration was below the detection limit

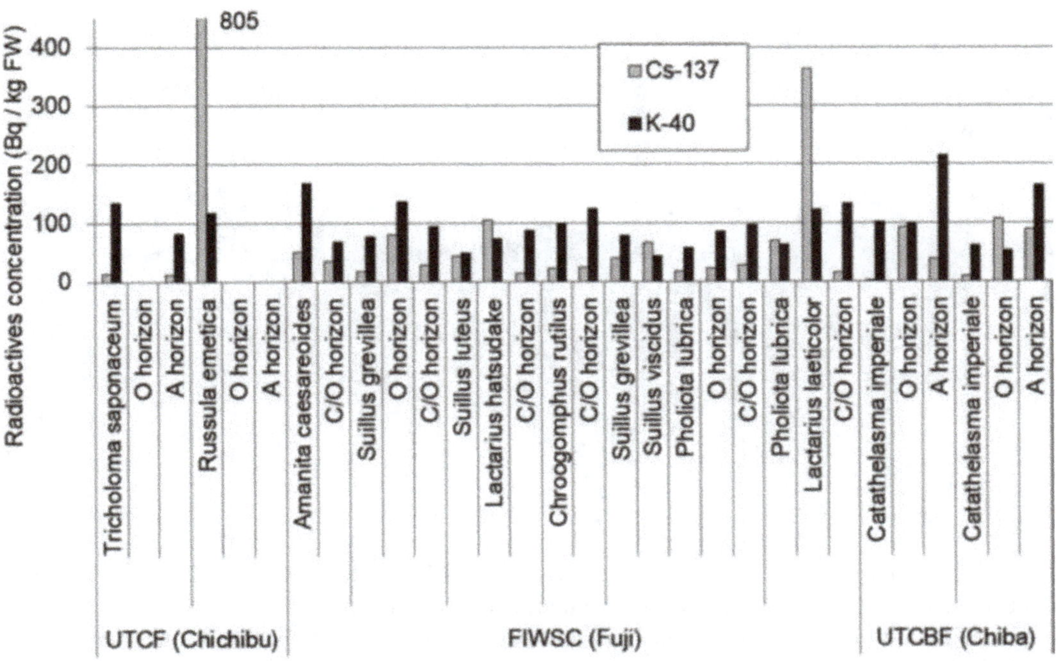

Fig. 12.10 Comparison of ^{137}Cs and ^{40}K concentration on a fresh weight basis in 2015
^{137}Cs concentration in *Russula emetica* was 805 Bq/kg FW
No data: radiocesium concentration was below the detection limit

mushroom/O or A (C/O) horizon ratios of ^{137}Cs for *R. emetica* (See Sect. 12.6) was not higher compared with Fuji mushrooms, but much higher compared with mycorrhizal fungi in Chichibu and Chiba. Some physiological or ecological mechanisms for Cs accumulation might work also in the case of *R. emetica*.

12.8 Features of Radioactive Contamination with Different Date of Fallout (Fig. 12.11)

A high uptake of ^{137}Cs by mushrooms, derived from nuclear weapons tests (NWT), was observed in Japan from the 1950s to 1960s (Muramatsu and Yoshida 1997; Sugiyama et al. 1994; Yoshida and Muramatsu 1996). This contamination originated from the global fallout by NWT, which peaked in 1963 (Komamura et al. 2006), and by the Chernobyl accident in 1986. NWT affected the wild mushrooms in Japan more than the Chernobyl accident. The contribution of the Chernobyl accident was estimated to be in the range of 7–60% and 10–30% on average in each study (Igarashi and Tomiyama 1990; Muramatsu et al. 1991; Shimizu et al. 1997; Yoshida and Muramatsu 1994; Yoshida et al. 1994). In the current study, ecological features of radioactive contamination in mushrooms and in soils were discussed by comparing the contamination from the Fukushima accident with those from NWT and the Chernobyl accident.

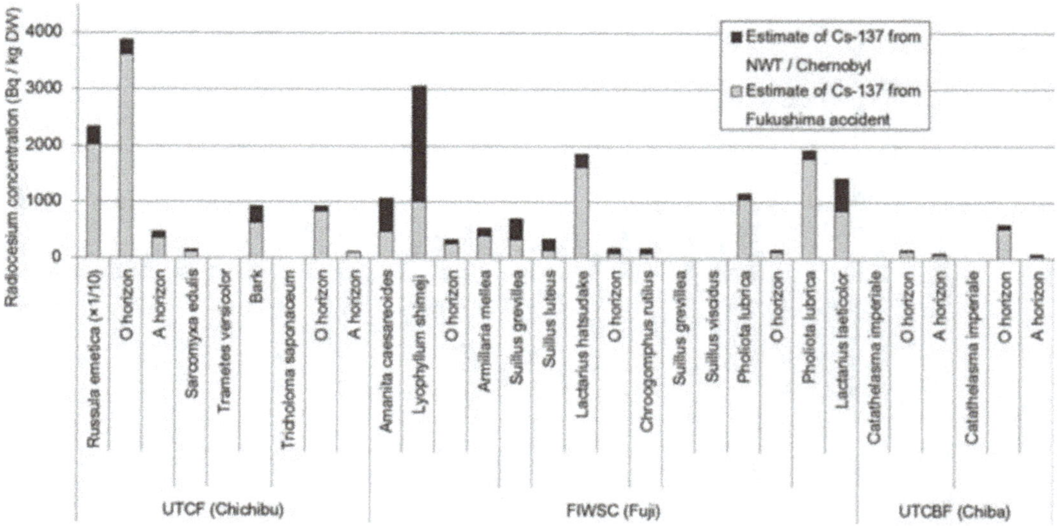

Fig. 12.11 Contribution of nuclear weapons tests (NWT), the Chernobyl accident and the Fukushima accident to total radiocesium contamination in 2013
Radiocesium concentration in *Russula emetica* was shown as one tenth of the actual value (i.e., total 23,400 Bq/kg DW)
No data: radiocesium concentration was below the detection limit

Shortly after the Fukushima accident, a large proportion of total ^{137}Cs in the O horizon of soils from Chichibu, Fuji and Chiba were derived from the Fukushima accident, and the ^{134}Cs/^{137}Cs ratio was constant among these research Forests. It showed a similar percentage contribution of contamination before the Fukushima accident. The mean contribution of the Fukushima accident to total contamination was roughly 88% in autumn 2011. This value decreased with time; 86% in 2012, 77% in 2013 and 65% in 2014. These results suggest that radiocesium released from the Fukushima accident moved relatively quickly out of the O horizon, whereas most the past residual ^{137}Cs remained in the material cycle system on the soil surface. For example, ^{137}Cs might have been sequestered inside mycorrhizal mycelia. However, it is far from a quantitative evaluation, because of the unstable occurrence of mushrooms between years and locations. In the A or C/O horizon, the mean ratio of Fukushima ^{137}Cs to total ^{137}Cs increased from 59% in 2011 to 73% in 2012, then decreased and stabilized at the equivalent level to the O horizon of about 65% in 2013 and 2014. The changes in the ratio appeared to be due to the transfer of ^{137}Cs from the O horizon, and somewhat to the transfer out of the A horizon.

The proportion of pre-Fukushima ^{137}Cs is high in mycorrhizal fungi, such as *Suillus grevillea*, *S. luteus*, *S. viscidus*, *Amanita caesareoides*, *Lyophyllum shimeji* and *Lactarius laeticolor* sampled in Fuji. Sugiyama et al. (2000) reported high ^{137}Cs activities in *P. lubrica* and *S. grevillei* collected around Mt. Fuji in 1996. These fungal species can be characterized by their ability to retain radiocesium. Specifically, more than half the ^{137}Cs was derived from pre-Fukushima fallout in *S. grevillea*, *A. caesareoides* and *L. shimeji*. Further, *Catathelasma imperiale* in Chiba had a low concentration of ^{137}Cs, but the ratio of the pre-Fukushima ^{137}Cs was also high. Thus, the range of radiocesium concentrations found in mycorrhizal fungi is large. It is unusual that the contribution of pre-Fukushima ^{137}Cs fallout remained high under the influence of fallout from the Fukushima accident. The mechanisms remain unclear how fungi with a high turnover rate of cells and tissues can retain pre-Fukushima ^{137}Cs, in which the ratio is much higher than in the soil substrate. ^{137}Cs deposited over a few decades may continue to be circulated in a closed system of fungal mycelium, which prevents its loss to the lower soil horizons.

12.9 Conclusion

In this chapter, some of the dynamics of radiocesium contamination in the forest ecosystem in relation to mushrooms was revealed. Radiocesium accumulation in several mycorrhizal mushrooms was similar to that reported after the Chernobyl accident, but not all mushrooms were contaminated equally. Biology and ecology of mushrooms, geographical, geological and pedological features may affect radiocesium dynamics in forests. Monitoring data of radiocesium concentration could evaluate the transfer of radiocesium from the litter to the soil layer or mushrooms and will provide useful information on the mechanisms of radiocesium accumulation in relation to potassium, and the selective retention of absorbed radiocesium in mushrooms. The number of samples and period of monitoring, however, was insufficient. Long-term monitoring of ^{137}Cs is necessary to clarify more precisely the dynamics of the contamination, though monitoring of ^{134}Cs is now becoming difficult because of its short half-life.

Acknowledgments I sincerely thank the staff of the University of Tokyo Forests for collecting and preparing the samples, and Drs. N. I. Kobayashi, K. Tanoi and T. M. Nakanishi for measuring the radioactivity in samples and for their valuable comments.

References

Ban-nai T, Yoshida S, Muramatsu Y (1994) Cultivation experiments on uptake of radionuclides by mushrooms. Radioisotopes 43:77–82 (in Japanese with English Summary)

Borio R, Chiocchini S, Cicioni R, Esposti PD, Rongoni A, Sabatini P, Scampoli P, Antonini A, Salvadori P (1991) Uptake of radiocesium by mushrooms. Sci Total Environ 106:183–190

Brückmann A, Wolters V (1994) Microbial immobilization and recycling of ^{137}Cs in the organic layers of forest ecosystems: relationship to environmental conditions, humification and invertebrate. Sci Total Environ 157:249–256

Byrne AR (1988) Radioactivity in fungi in Slovenia, Yugoslavia, following the Chernobyl accident. J Environ Radioact 6:177–183

Fielitz U, Klemt E, Strebl F, Tataruch F, Zibold G (2009) Seasonality of ^{137}Cs in roe deer from Austria and Germany. J Environ Radioact 100:241–249

Guillitte O, Melin J, Wallberg L (1994) Biological pathways of radionuclides originating from the Chernobyl fallout in a boreal forest ecosystem. Sci Total Environ 157:207–215

Heinrich G (1992) Uptake and transfer factors of ^{137}Cs by mushrooms. Radiat Environ Biophys 31:39–49

Igarashi S, Tomiyama T (1991) Radionuclide concentrations in mushrooms. Annu Rep Fukui Pref Inst Public Health 29:70–73 (In Japanese with English Summary)

Kammerer L, Hiersche L, Wirth E (1994) Uptake of radiocaesium by different species of mushrooms. J Environ Radioact 23:135–150

Kiefer P, Pröhl G, Müller G, Lindner G, Drissner J, Zibold G (1996) Factors affecting the transfer of radiocaesium from soil to roe deer in forest ecosystems of southern Germany. Sci Total Environ 192:49–61

Komamura M, Tsumura A, Yamaguchi N, Fujiwara H, Kihou N, Kodaira K (2006) Long-term monitoring and analysis of ^{90}Sr and ^{137}Cs concentrations in rice, wheat and soils in Japan from 1959 to 2000. Bull Natl Inst Agro Environ Sci No. 24-1-21 (In Japanese with English Summary)

Mascanzoni D (1987) Chernobyl's challenge to the environment: a report from Sweden. Sci Total Environ 67:133–148

Minato S (2011) Distribution of dose rates due to fallout from the Fukushima Daiichi reactor accident. Radioisotopes 60:523–526

Muramatsu Y, Yoshida S (1997) Mushroom and radiocesium. Radioisotopes 46:450–463 (In Japanese)

Muramatsu Y, Yoshida S, Sumiya M (1991) Concentrations of radiocesium and potassium in basidiomycetes collected in Japan. Sci Total Environ 105:29–39

Pietrzak-Flis Z, Radwan I, Rosiak L, Wirth E (1996) Migration of ^{137}Cs in soils and its transfer to mushrooms and vascular plants in mixed forest. Sci Total Environ 186:243–250

Rühm W, Steiner M, Kammerer L, Hiersche L, Wirth E (1998) Estimating future radiocaesium contamination of fungi on the basis of behaviour patterns derived from past instances of contamination. J Environ Radioact 39:129–147

Seeger R (1978) Kaliumgehalt höherer Pilze. Z Lebensm Unters Forsch 167:23–31

Shimizu M, Anzai I, Fukushi M, Nyuui Y (1997) A study on the prefectural distribution of radioactive cesium concentrations in dried *Lentinula edodes*. Radioisotopes 46:272–280 (In Japanese with English Summary)

Smith JT, Beresford NA (2005) Radioactive fallout and environmental transfers. In: Smith JT, Beresford NA (eds) Chernobyl – catastrophe and consequences. Springer, Berlin

Soil Survey Staff (2014) Keys to soil taxonomy, 12 edn. United States Department of Agriculture, Natural Resources Conservation Service

Steiner M, Linkov I, Yoshida S (2002) The role of fungi in the transfer and cycling of radionuclides in forest ecosystems. J Environ Radioact 58:217–241

Sugiyama H, Iwashima K, Shibata H (1990) Concentration and behavior of radiocesium in higher basidiomycetes in some Kanto and the Koshin districts, Japan. Radioisotopes 39:499–502 (In Japanese with English Summary)

Sugiyama H, Shibata H, Isomura K, Iwashima K (1994) Concentration of radiocesium in mushrooms and substrates in the sub-alpine forest of Mt. Fuji Japan. J Food Hyg Soc Jpn 35:13–22

Sugiyama H, Terada H, Isomura K, Tsukada H, Shibata H (1993) Radiocesium uptake mechanisms in wild and culture mushrooms. Radioisotopes 42:683–690 In Japanese with English Summary

Sugiyama H, Terada H, Shibata H, Morita Y, Kato F (2000) Radiocesium concentrations in wild mushrooms and characteristics of cesium accumulation by the edible mushroom (*Pleurotus ostreatus*). J Health Sci 46:370–375

Vinichuk MM, Johanson KJ (2003) Accumulation of ^{137}Cs by fungal mycelium in forest ecosystems of Ukraine. J Environ Radioact 64:27–43

Vinichuk MM, Johanson KJ, Rosén K, Nilsson I (2005) Role of the fungal mycelium in the retention of radiocaesium in forest soils. J Environ Radioact 78:77–92

Yamada T (2013) Mushrooms: radioactive contamination of widespread mushrooms in Japan. In: Nakanishi TM, Tanoi K (eds) Agricultural implications of the Fukushima nuclear accident. Springer, Tokyo

Yamada T, Murakawa I, Saito T, Omura K, Takatoku K, Iguchi K, Inoue M, Saiki M, Saito H, Tsuji K, Tanoi K, Nakanishi TM (2013) Radiocesium accumulation in wild mushrooms from low-level contaminated area due to the Fukushima-Daiichi nuclear power plant accident – a case study in the University of Tokyo Forests. Radioisotopes 62:141–147 (In Japanese with English Summary)

Yamada T, Omura K, Saito T, Igarashi Y, Takatoku K, Saiki M, Murakawa I, Iguchi K, Inoue M, Saito H, Tsuji K, Kobayashi NI, Tanoi K, Nakanishi TM (2018) Radiocesium contamination of wild mushrooms collected from the University of Tokyo Forests over a six-year period (2011–2016) after the Fukushima nuclear accident. Misc Inf Univ Tokyo For 60:31–47

Yoshida S, Muramatsu Y (1994) Accumulation of radiocesium in basidiomycetes collected from Japanese forests. Sci Total Environ 157:197–205

Yoshida S, Muramatsu Y (1996) Environmental radiation pol1ution of fungi. Jpn J Mycol 37:25–30 (In Japanes with English Summary)

Yoshida S, Muramatsu Y, Ogawa M (1994) Radiocesium concentrations in mushrooms collected in Japan. J Environ Radioact 22:141–154

Zibold G, Drissner J, Kaminski S, Klemt E, Miller R (2001) Time-dependence of the radiocaesium contamination of roe deer: measurement and modeling. J Environ Radioact 55:5–27

Chapter 13
The Spatial Distribution of Radiocesium Over a Four-Year Period in a Forest Ecosystem in North Fukushima After the Nuclear Power Station Accident

Masashi Murakami, Takahiro Miyata, Natsuko Kobayashi, Keitaro Tanoi, Nobuyoshi Ishii, and Nobuhito Ohte

Abstract We have investigated the redistribution dynamics of radiocesium deposited after the nuclear power station accident in March 2011 in a forested catchment located in North Fukushima over a four-year period (2012–2015). At the catchment scale, ^{137} Cs accumulation decreased drastically by 50% of the estimated initial accumulation during the first 2 years. Cs-137 accumulation in the forest floor occurred in the litter layers and the surface part of mineral soils and have accounted for about 90% of the total catchment scale accumulation. The internal ^{137}Cs cycle among the soil-plant system was also identified as a retention mechanism and was biologically dynamic. Monitoring the decreasing and retaining mechanisms of radioactivity at the ecosystem scale will be required for effective forest and water resource management.

Keywords ^{37}Cs · Inventory · Environmental half-life · Forest ecosystem · Internal cycle · Fukushima

---

M. Murakami · T. Miyata
Graduate School of Science, Chiba University, Chiba, Japan

N. Kobayashi · K. Tanoi
Graduate School of Agricultural and Life Sciences, The University of Tokyo, Tokyo, Japan

N. Ishii
National Institute of Radiological Sciences, Chiba, Japan

N. Ohte (✉)
Department of Social Informatics, Graduate School of Informatics, Kyoto University, Kyoto, Japan
e-mail: nobu@i.kyoto-u.ac.jp

T. M. Nakanishi et al. (eds.), *Agricultural Implications of the Fukushima Nuclear Accident (III)*, https://doi.org/10.1007/978-981-13-3218-0_13

13.1 Introduction

The explosions at the Fukushima Daiichi Nuclear Power Station in March 2011 released a large amount of radioactive materials, especially iodine (^{131}I and ^{133}I) and cesium (^{134}Cs and ^{137}Cs), into the environment (Chino et al. 2011; Steinhauser et al. 2014). Half-lives of ^{131}I and ^{133}I are only 8.03 days and 20.8 h, respectively, but the half-lives of ^{134}Cs (2.07 years) and ^{137}Cs (30.1 years) are longer. Because of the long half-life of ^{137}Cs, it is expected that ^{137}Cs will pollute the surrounding natural environment for many years into the future.

About 70% of the Fukushima prefectural territory is covered by forests. The local communities have traditionally utilized the forests extensively not only for timber production but also for firewood and charcoal production before the 1960s. The collection of edible wild plants and mushrooms have also been a part of the traditional lifestyle of the residents for a very long time. It has been wildly recognized in Japan that local people have generally respected the environment and products of the forest ecosystem of this region (Fukushima Prefecture 2017). Therefore, it is critical for local people to precisely understand the quantity and distribution of radiocaesium in the forests and surrounding environment.

The first phase of the governmental surveys revealed that a major portion of the deposited radiocesium was trapped in the canopy of coniferous forests and in the litter layer on the forest floor of deciduous forests (Ministry of Education, Culture, Sports, Science and Technology; Ministry of Agriculture, Forestry and Fisheries 2012; Hashimoto et al. 2012). Within the forest ecosystem, radioactive materials deposited on the tree canopies subsequently moved to the forest floor by precipitation (Kinnersley et al. 1997; Kato et al. 2012) and litter fall (Bunzl et al. 1989; Schimmack et al. 1993; Hisadome et al. 2013). The movement of radiocesium from the canopy to the forest floor has gradually decreased (Hashimoto et al. 2013). Radiocesium has been shown to be easily adsorbed by clay minerals and soil organic matter (Kruyts and Delvaux 2002), which can be transported by eroded soil, particulates, and dissolved organic matter through hydrological channels, streams, and rivers (e.g., Fukuyama et al. 2005; Wakiyama et al. 2010).

Dissolved radiocesium, which is relatively free from soil adsorption, can also be taken up by microbes, algae, and plants in soils and aquatic ecosystems. By propagating through the food web in the forest ecosystem, it was expected that radiocesium would eventually be introduced into insects, worms, fish, birds and mammals. Many previous reports on the distribution and transfer of radionuclides have focused on their bioaccumulation and the transition between trophic levels (Kitchings et al. 1976; Rowan and Rasmussen 1994; Wang et al. 2000).

Our four-year monitoring study since early 2012 at a forested headwater catchment in the northern part of Fukushima have demonstrated that radiocesium movement has been most drastic during the early years after the accident (Ohte et al. 2013, 2016). Observed results have shown that radiocesium has continuously moved from the forest canopies to the forest floor with throughfall, stemflow and litter fall

(Endo et al. 2015). It was found that fallen leaves and throughfall waters in the deciduous tree stands contain a certain amount of radiocesium since 2012. Since the attachment of deposited radiocesium did not occur directly onto leaves at the time of the accident in March 2011, because deciduous trees did not have new leaves, radiocesium inclusion in leaves in 2012 indicated that the translocation of radiocesium occurred in leaves from within the tree body (Endo et al. 2015). One of the possible sources of radiocesium transported to leaves was the deposited radiocesium in the litter layer on the forest floor. ^{137}Cs in the organic layers (i.e, L, F and H horizons) of the deciduous stands ranged from 15 to 50 KBq m^{-2}; these organic layers were the largest compartment where ^{137}Cs accumulated in the forest ecosystem (Murakami et al. 2014). For radiocesium to be transferred via the food web in the same catchment ecosystem, the primary pathway was found to be the food chains originating from leaf and wood detritus, which was highly contaminated with radiocesium. Additionally, it was found that the decrease in ^{137}Cs concentration was through trophic interactions, which was suggestive of biological dilution and not the accumulation of ^{137} Cs (Murakami et al. 2014).

During the past 4 years, while the total radiocesium accumulation has continuously decreased, its redistribution has proceeded by active transfers among compartments, such as transferring from canopy to litter layers on the forest floor. To develop a long-term future perspective in forest management, it is important to prepare a spatiotemporal prediction of radiocesium accumulation in the forest ecosystem. The aim of this paper is to summarize the time sequential changes in ^{137}Cs accumulation in major compartments of the forest ecosystem and to discuss the mechanism behind the estimation of the catchment-scale environmental half-life of ^{137}Cs accumulation.

13.2 Material and Method

13.2.1 Study Site

The study site was located at the headwater part of the Kami-Oguni River catchment in Date City, Fukushima prefecture, about 50 km northwest of the stricken nuclear power station (Fig. 13.1). Mean annual temperature in this region was 13 °C in 2012–2013 (Japan Meteorological Agency, 2012–2013), and the annual precipitation was 912 mm in 2012–2013 in this study site. Snow accumulation was observed every year, but annual maximum snow depth has been less than 60 cm during the period 1961–2017.

The monitoring catchment (18.9 ha) has been set at the headwater part of the Kami-oguni river. The catchment is mainly covered by second-growth forests consisting mainly of deciduous tree species (e.g. *Quercus serrata*, *Acer pictum*, *Zelkova serrata*, etc.). Some Japanese red pines (*Pinus densiflora*) were found at the ridge

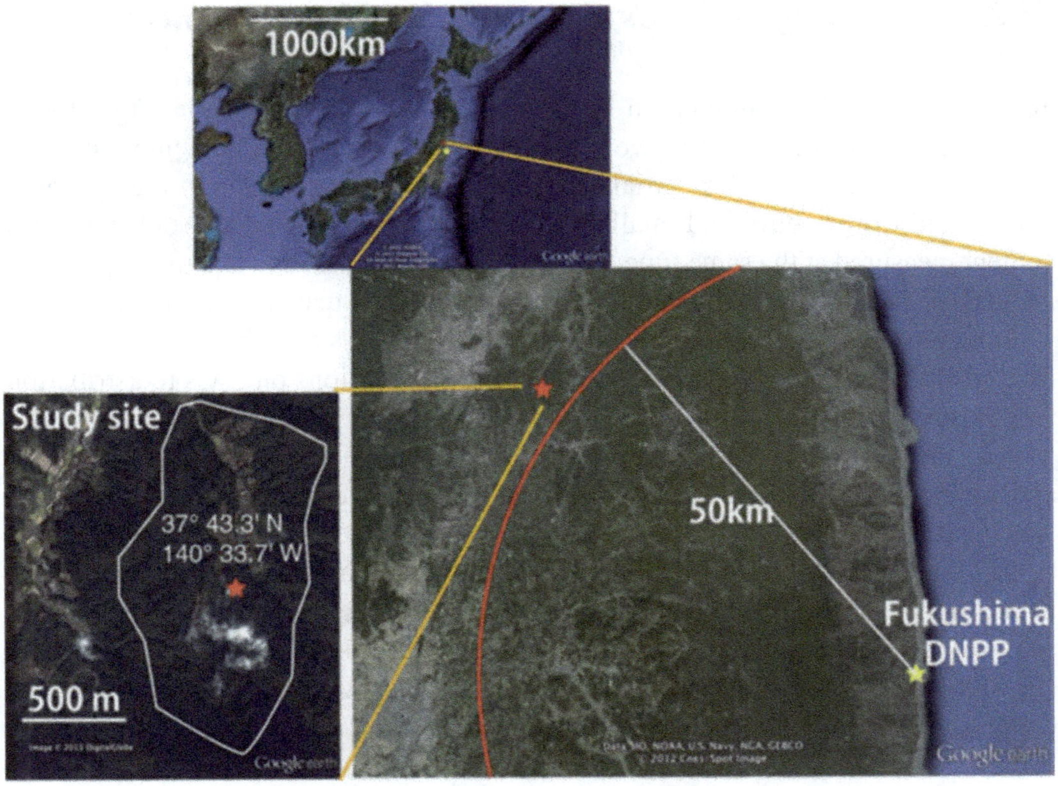

Fig. 13.1 Maps indicating the location of the study site at the Kami-Oguni River catchment. The study area was delineated with the geographic coordinates. These maps were attributed to Zenrin, Kingway Ltd., US Dept. of State Geographer, Mapabc.com, DATA SIO, NOAA, U.S. Navy, NGA, GEBCO, Cnes/Spot Image, and DigitalGlobe. (After Murakami et al. 2014)

part of the catchment. Deciduous trees in the second-growth forest have traditionally been utilized for charcoal and firewood production. Tree density of the second growth forests varied from 800 to 1300 ind. ha^{-1}. Additionally, Japanese cedar (*Cryptmeria japonica*) plantations for timber production are situated at the lower part of the hill slopes. The age of the cedar plantation was about 50 years old; the stand density was 2100 ind. ha^{-1}.

Two rectangular plots (20 × 20 m) in the second growth forest dominated by *Q. serrata* and one plot (10 × 40 m) in the Japanese cedar plantation was prepared for the vegetation survey. The height and diameter of all trees (at breast height) were measured to estimate the above-ground vegetation biomass.

According to a radioactivity survey report using an airborne survey device, the air dose rate and estimated ^{137}Cs deposition were 1.9–3.8 μSv h^{-1} and 100–300 kBq m^{-2} in August 2011, respectively, which was 5 months after the accident. The dose rate has decreased to 0.5–1.0 μSv h^{-1} at (Ministry of Education, Culture, Sports, Science and Technology Japan 2017).

13.2.2 Sampling and ^{137}Cs Concentration Measurements

To determine ^{137}Cs accumulation in the major compartments of the target forest ecosystem, litter, soil and trees were sampled in May, July and September of each year from 2012 to 2015. Terrestrial and aquatic organisms including fungi, grasses, insects, soil worms, birds and small mammals were also sampled and had their ^{137}Cs content measured at the same time. Because these terrestrial and aquatic organisms' area-based biomasses were very small, we considered that their accumulation of ^{137}Cs would be negligible for the catchment scale discussion. Thus, we defined the major compartments of the forest catchment for discussions on ^{137}Cs redistribution within this ecosystem (Fig. 13.2).

All samples were dried for more than 48 h at 60 °C and powdered using a mortar and pestle. At least 200 mg of each sample was collected and ^{137}Cs concentration was measured.

Litter and soil samples were collected from three points in the *Q. serrata* dominated stands and from two points in the *C. japonica* stands. Litter was separately sampled from the L, F and H horizons of the organic layer using a square frame (20 × 20 cm, 15 × 15 cm or 10 × 10 cm). Soil samples were taken from horizons at depths of 0–5 cm, 5–10 cm and 10–15 cm.

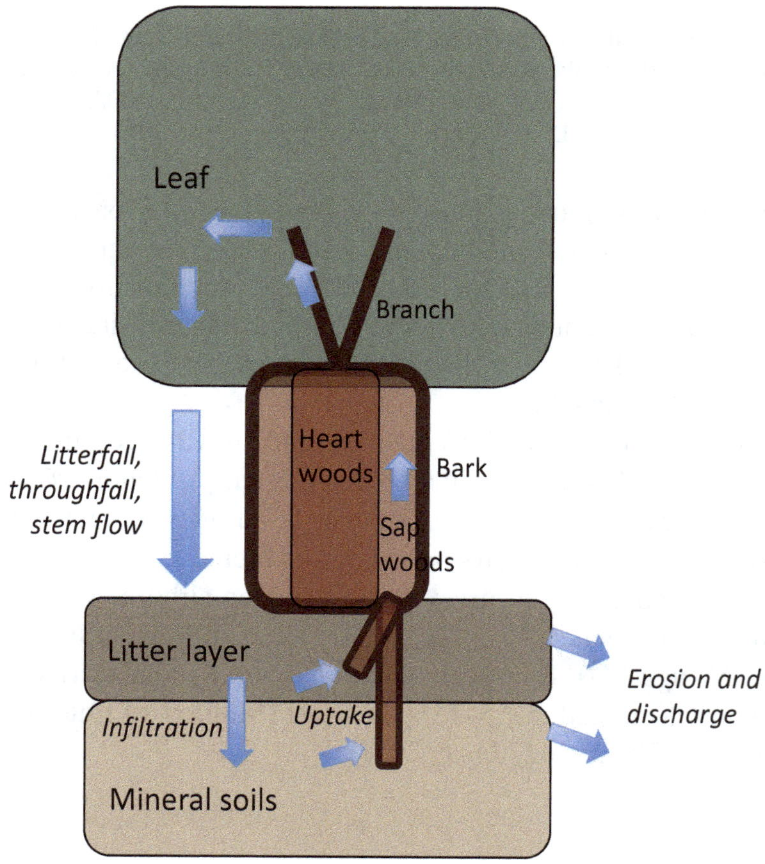

Fig. 13.2 Sampled substances for determining the ^{137}Cs distribution and redistribution mechanisms (in Italics)

As tall trees in the stand were dominated by *Q. serrata*, two individuals of *Q. serrata* were cut down each year from 2012 to 2014 (one individual for 2015). One individual of *C. japonica* was also cut down from the *C. japonica* stands every year from 2012 to 2015. Leaves, branches, bark and wood (heartwood and sapwood) were sampled at every 2 m from the root side of sampled trees.

Gamma-ray spectrometry was conducted using germanium semiconductor detectors (Seiko EG&G) for ^{137}Cs concentration measurements of all samples. An efficiency calibration of the detectors was made using volume radioactivity standard gamma sources (MX0333U8, Japan Radioisotope Association). Measuring accuracy was confirmed with the standard reference material JSAC-0471 (the Japan Society for Analytical Chemistry). The measured values were corrected for sampling day.

13.2.3 Estimation of ^{137}Cs Accumulation and Its Environmental Half-Life

For litter layers and soil on the forest floor, the ^{137}Cs accumulation value of each compartment (Bq m^{-2}) was estimated by multiplying the ^{137}Cs concentration value by the dry weight of each sample and then dividing by the sampling area (the square frame for sampling).

Dry weight biomass of each tree part (i.e., leaves, branches, bark, heartwood and sapwood) was estimated from the vegetation survey using allometry equations developed for the forests in Fukushima and proposed by Kajimoto et al. (2014) and Hosoda et al. (2010). Then, the area based ^{137}Cs accumulation as tree biomass (Bq m^{-2}) was estimated by multiplying the average ^{137}Cs concentration value with the dry weight of each component.

For estimation of total accumulation in the study catchment, we calculated the areal ratios of the *Q. serrata* dominated stands, the *C. japonica* stands and stream channels from the aerial photo image of the catchment. Then, the area-based accumulation of each compartment was estimated.

13.3 Results

13.3.1 Annual Changes of ^{137}Cs Accumulation in Litter Layers, Soils and Trees

The total accumulation of ^{137}Cs in forest floors decreased during the period 2012–2015 at both the *C. japonica* plantation and at the *Q. serrata* dominated stands (Fig. 13.3), while the initial distribution was significantly different between the two stands. This was because the deciduous trees had not leafed out when the accident

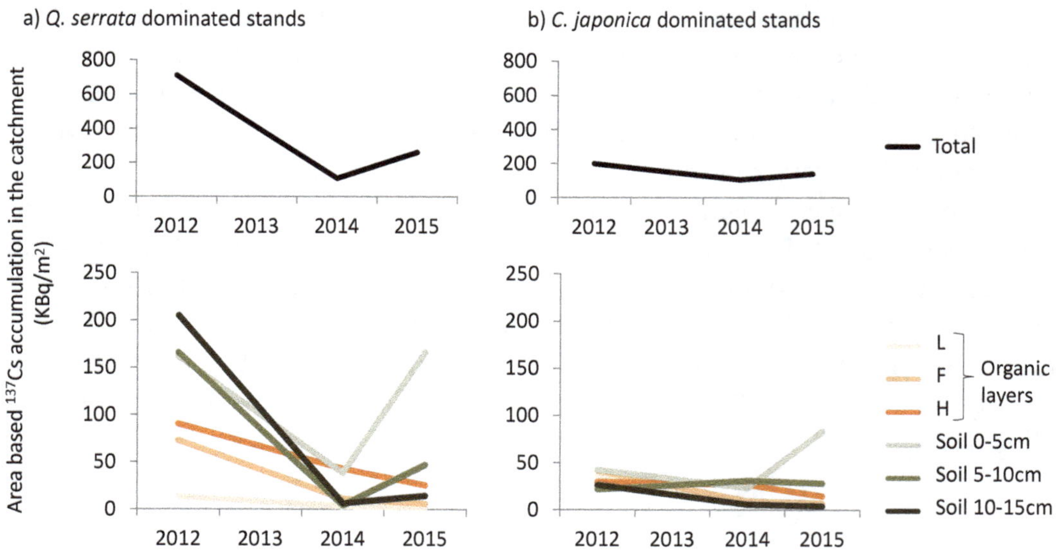

Fig. 13.3 Area based ^{137}Cs accumulation of litter layers and soil profile in (**a**) the *Q. serrata* dominated stands, and (**b**) the *C. japonica* dominated stands. (After Miyata 2017)

occurred in March 2011. ^{137}Cs accumulation in the litter layer of the cedar plantation site in 2012 was approximately 60% of that in the *Q. serrata* dominated stands (Fig. 13.3). This indicated that the canopies of the cedar trees captured ^{137}Cs deposition effectively. While significant decreases were found in most layers from 2012 to 2014 at both stands, ^{137}Cs accumulation in soils at a depth of 0–5 cm increased during 2014–2015.

While the ^{137}Cs accumulation was greater in the fresh leaves of the canopy than the other parts of the *C. japonica* tree, it was greater in the bark for the *Q. serrata* dominated stands (Fig. 13.4). The ^{137}Cs accumulation in the cedar bark was significantly lower than that of *Q. serrata*. This indicated that the dense canopy of the cedar trapped the ^{137}Cs deposition effectively and reduced the amount of deposition detected beneath the canopy.

It was notable that ^{137}Cs accumulation in the heartwood and sapwood increased gradually during the period of 2012–2015, although the total accumulation of ^{137}Cs in the bodies of trees decreased (Fig. 13.4).

13.3.2 Changes in ^{137}Cs Accumulation in Each Compartment of the Catchment

Since 2012, ^{137}Cs accumulation in most compartments decreased significantly in this catchment, except in surface soils and in heartwood and sapwood (Fig. 13.5). The largest accumulation has continuously existed in the litter layer and the near-surface mineral soils (0–15 cm). About 90% of total ^{137}Cs accumulation of the catchment was held in the forest floor, and about 10% of that was retained in the

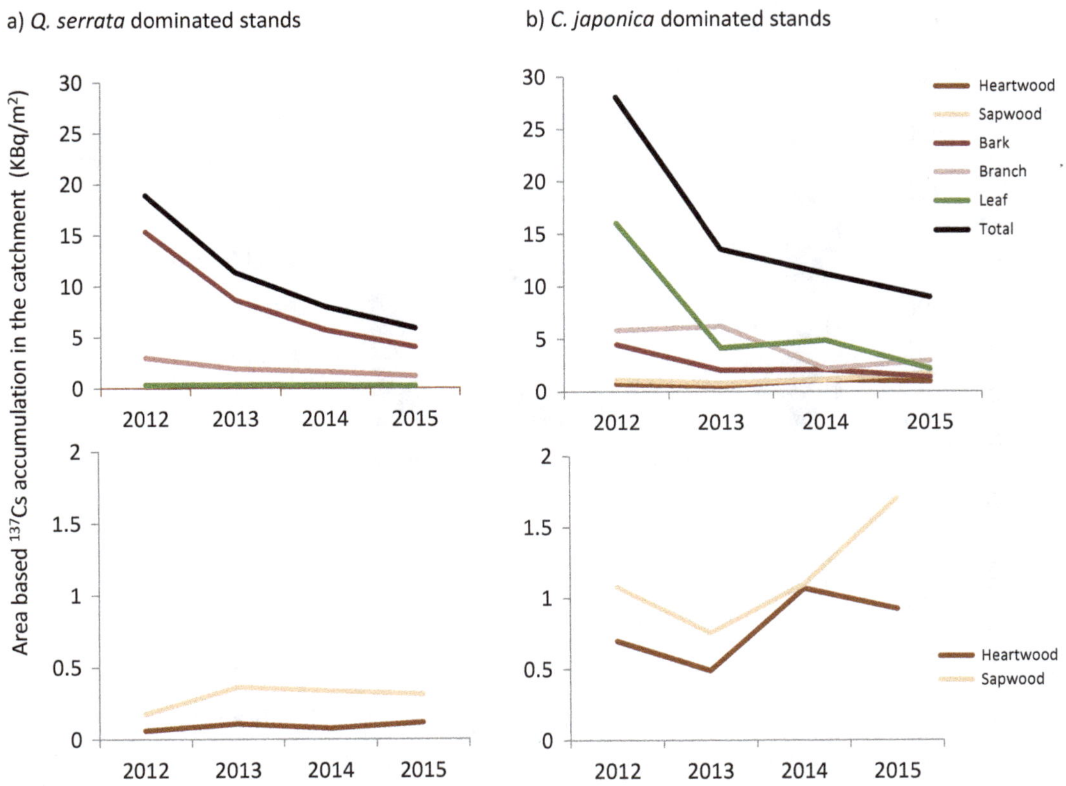

Fig. 13.4 Area based ^{137}Cs accumulation within the tree body in (**a**) the *Q. serrata* dominated stands, and (**b**) the *C. japonica* dominated stands. Close-up on the changes of ^{137}Cs accumulation in the heartwood and sapwood of both stands were also shown. (After Miyata 2017)

above-ground living tree biomass. While the total accumulation has decreased one-sixth between 2012 and 2014, the proportion of the accumulation in the shallow mineral soils (0–5 cm) has increased during the same period. The accumulation itself has also increased in the 5–10 cm soils during 2014–2015 (Fig. 13.5).

13.4 Discussion

13.4.1 Redistribution of the ^{137}Cs Accumulation

The increase in ^{137}Cs accumulation in wood parts for both *Q. serrata* and *C. japonica* stands (Fig. 13.4) indicated that circulation occurred between trees and soils. ^{137}Cs in the tree body was transported to leaves, and then it moved to the forest floor through litterfall. Decomposition of litterfalls releases ^{137}Cs into the root zone within the organic and shallow mineral soil layers. This ^{137}Cs could be absorbed by trees through their root system. This cycle has acted as a retention system of ^{137}Cs in the forest.

At the interface between the litter layer and mineral soils, ^{137}Cs has moved down from the H horizon to the surface part of the mineral soils in recent years (Fig. 13.3).

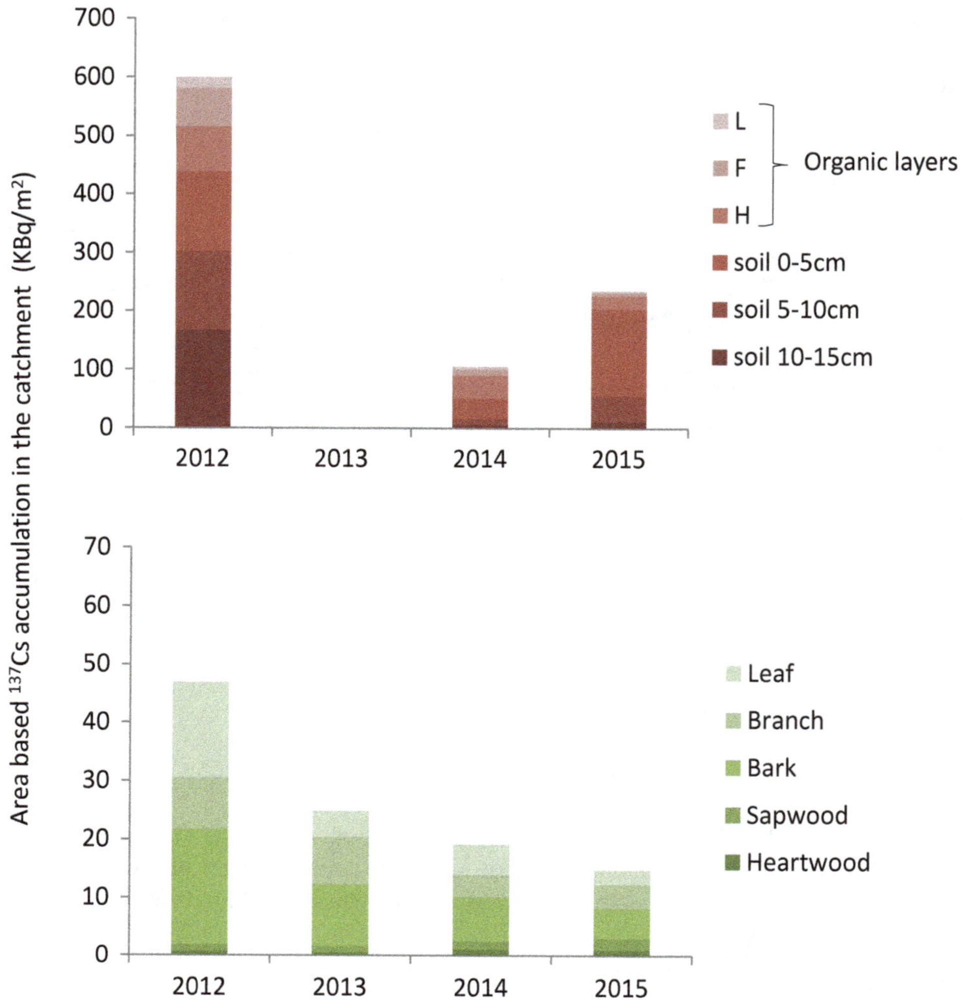

Fig. 13.5 Area based ^{137}Cs accumulation in organic layers, soil horizons (upper panel) and aboveground biomass (lower panel) in the study catchment including both the *Q. serrata* dominated stands and the *C. japonica* dominated stands. (After Miyata 2017)

Significant increases in ^{137}Cs accumulation in soil depths of 0–5 cm during 2014–2015 could partly be explained by this mechanism. However, other accumulation mechanisms are required to be able to fully understand this drastic increase. One of the possible mechanisms was lateral input from the upslope parts onto the actual sampling points. Drifting of litter, for example, from upslope can supply extra ^{137}Cs in addition to the vertical movement.

Several surveys have previously been conducted on the slope movement of deposited ^{137}Cs and it was found that the movement with drifting litter and surface soil occurred occasionally by surface water flows and drifting snow accumulation (Kashihara 2014; Takada et al. 2017). As the soil layer for most of the catchment was established on the slopes, the ^{137}Cs movement with drifting litter along hill slopes still needs to be researched quantitatively to estimate the catchment scale ^{137}Cs accumulation precisely.

13.4.2 *Catchment-Scale Environmental Half-Life of the ^{137}Cs Accumulation*

The environmental half-life of ^{137}Cs accumulation in the entire catchment ecosystem was estimated from the fitted curve applied to yearly change based on the data in Fig. 13.4; the half-life was estimated to be about 2 years since July 2012 when the survey was initiated. This estimation was significantly shorter than the physical half-life of ^{137}Cs, because of the washout effect on the deposited ^{137}Cs in the first few years immediately after the accident. After 2014, the annual discharge of ^{137}Cs through stream flow from this catchment was estimated to be 2–3 orders of magnitude lower than the initial deposited amount of ^{137}Cs in this area (Iseda 2015). It was clear that the rate of decrease in ^{137}Cs accumulation in watershed has further declined since then.

13.5 Perspective

The internal ^{137}Cs cycle among the soil-plant system was clearly identified as a retention mechanism and was biologically dynamic. The increasing trend of ^{137}Cs in sapwood and heartwood of *Q. serrata* has slowed. This might indicate that the ^{137}Cs cycle between trees and soils was approaching a steady state condition.

Currently, the fixation with clay mineral in soils and the biological retention by the internal cycle among soil and plants are major factors retaining ^{137}Cs in this forest catchment. Mechanisms that decrease ^{137}Cs accumulation are the occasional discharge through streams via soil and litter particles during storm events, and physical decay. It is essential to continue to monitor the decreasing and retaining mechanisms carefully at the ecosystem level, because this information will be needed for careful consideration on radioactivity controls in forest and water resources management.

Acknowledgement All data were collected during a research projects supported by a grant (24248027, 16H04934) for scientific research from the Ministry of Education, Culture, Sports, Science and Technology and with a grant for environmental researches from The Sumitomo Foundation. The authors would like to thank Mr. Chonosuke Watanabe for his kind help during the field work.

References

Bunzl K, Schimmack W, Kreutzer K, Schierl R (1989) Interception and retention of Chernobyl-derived ^{134}Cs, ^{137}Cs and ^{106}Ru in a spruce stand. Sci Total Environ 78:77–87. https://doi.org/10.1016/0048-9697(89)90023-5

Chino M, Nakayama H, Nagai H, Terada H, Katata G, Yamazawa H (2011) Preliminary estimation of release amounts of ^{131}I and ^{137}Cs accidentally discharged from the Fukushima Daiichi nuclear power plant into the atmosphere. J Nucl Sci Technol 48:1129–1134. https://doi.org/10.1080/18811248.2011.9711799

Endo I, Ohte N, Iseda K, Tanoi K, Hirose A, Kobayashi NI, Murakami M, Tokuchi N, Ohashi M (2015) Estimation of radioactive 137-cesium transportation by litterfall, stemflow and throughfall in the forests of Fukushima. J Environ Radioact 149:176–185. https://doi.org/10.1016/j.jenvrad.2015.07.027

Fukushima Prefecture (2017) Forests of Fukushima. in Division of Forest Planning, editor. Fukushima Prefecture, Fukushima. http://www.pref.fukushima.lg.jp/sec/36055a/shinrinkeikaku.html (referred on 2017/07/28)

Fukuyama T, Takenaka C, Onda Y (2005) ^{137}Cs loss via soil erosion from a mountainous headwater catchment in Central Japan. Sci Total Environ 350:238–247. https://doi.org/10.1016/j.scitotenv.2005.01.046

Hashimoto S, Ugawa S, Nanko K, Shichi K (2012) The total amounts of radioactively contaminated materials in forests in Fukushima, Japan. Sci Rep 2. https://doi.org/10.1038/srep00416

Hashimoto S, Matsuura T, Nanko K, Linkov I, Shaw G, Kaneko S (2013) Predicted spatio-temporal dynamics of radiocesium deposited onto forests following the Fukushima nuclear accident. Sci Rep 3. https://doi.org/10.1038/srep02564

Hisadome K, Onda Y, Kawamori A, Kato H (2013) Migration of radiocaesium with litterfall in hardwood-Japanese red pine mixed forest and sugi plantation. J Jpn Soc 95:267–274 (In Japanese with English abstract)

Hosoda K, Mitsuda Y, Iehara T (2010) Differences between the present stem volume tables and the values of the volume equations, and their correction. Jpn J For Plan 44:23–39 (in Japanese)

Iseda K (2015) Study on changes in discharge forms of ^{137}Cs from a forest catchment in northern Fukushima using information on mechanisms of suspended solid discharge. University of Tokyo, Tokyo (in Japanese)

Japan Meteorological Agency Japan (2017) Tables of Monthly Climate Statistics. http://www.data.jma.go.jp/obd/stats/data/en/smp/index.html, Japan Meteoro- logical Agency (referred on 2017/0728)

Kajimoto T, Takano T, Saito T, Kuroda K, Fujiwara K, Komatsu M, Kawasaki T, Ohashi S, Seino Y (2014) Methods for assessing the spatial distribution and dynamics of radiocesium in tree components in forest ecosystems. Bull FFPRI 13:113–136 (in Japanese)

Kashihara M (2014) Changes in spatial distribution of radiocesium deposited from the nuclear power plant accident with surface soil movement in a forest of Fukushima. Graduation thesis of University of Tsukuba, Tsukuba, Japan (in Japanese)

Kato H, Onda Y, Gomi T (2012) Interception of the Fukushima reactor accident-derived ^{137}Cs, ^{134}Cs and ^{131}I by coniferous forest canopies. Geophys Res Lett 39:L20403. https://doi.org/10.1029/2012GL052928

Kinnersley RP, Goddard AJH, Minski MJ, Shaw G (1997) Interception of caesium-contaminated rain by vegetation. Atmos Environ 31:1137–1145. https://doi.org/10.1016/S1352-2310(96)00312-3

Kitchings T, Digregorio D, P Van Voris (1976) A review of ecological parameters in vertebrate food chains. In: Proceedings of the fourth national symposium on radioecology. Ecological Society of America, Oregon State University, Corvallis, Oregon, pp 304–313

Kruyts N, Delvaux B (2002) Soil organic horizons as a major source for radiocesium biorecycling in forest ecosystems. J Environ Radioact 58:175–190. https://doi.org/10.1016/s0265-931x(01)00065-0

Ministry of Education, C., Sport, Science and Technology and Ministry of Ministry of Agriculture, Forestry and Fisheries (2012) Study report on distribution of radioactive substances emitted by the accident of the Fukushima Daiichi Nuclear Power Plant. Tokyo

Ministry of Education, C., Sports, Science and Technology Japan (2017) Extension site of the distribution map for radiation dose. http://ramap.jmc.or.jp/map/. Ministry of Education, Culture, Sports, Science and Technology Japan (referred on 2017/07/28)

Miyata T (2017) Inventory estimation of 137Cs in a forested ecosystem originates from the nuclear power plant accident. Master thesis of graduate School of Science, Chiba University, Chiba, Japan (in Japanese)

Murakami M, Ohte N, Suzuki T, Ishii N, Igarashi Y, Tanoi K (2014) Biological proliferation of cesium-137 through the detrital food chain in a forest ecosystem in Japan. Sci Rep 4. https://doi.org/10.1038/srep03599

Ohte N, Murakami M, Iseda K, Tanoi K, Ishii N (2013) Diffusion and transportation dynamics of ^{137}Cs deposited on the forested area in Fukushima after the nuclear power plant accident in March 2011. In: Nakanishi T, Tanoi K (eds) Agricultural implications of the Fukushima nuclear accident. Springer, New York, pp 177–186

Ohte N, Murakami M, Endo I, Ohashi M, Iseda K, Suzuki T, Oda T, Hotta N, Tanoi K, Kobayashi NI, Ishii N (2016) Ecosystem monitoring of radiocesium redistribution dynamics in a forested catchment in Fukushima after the nuclear power plant accident in March 2011. In: Nakanishi TM, Tanoi K (eds) Agricultural implications of the Fukushima nuclear accident: the first three years. Springer, Tokyo, pp 175–188

Rowan DJ, Rasmussen JB (1994) Bioaccumulation of radiocesium by fish: the influence of physicochemical factors and trophic structure. Can J Fish Aquat Sci 51:2388–2410. https://doi.org/10.1139/f94-240

Schimmack W, Förster H, Bunzl K, Kreutzer K (1993) Deposition of radiocesium to the soil by stemflow, throughfall and leaf-fall from beech trees. Radiat Environ Biophys 32:137–150. https://doi.org/10.1007/BF01212800

Steinhauser G, Brandl A, Johnson TE (2014) Comparison of the Chernobyl and Fukushima nuclear accidents: a review of the environmental impacts. Sci Total Environ 470:800–817. https://doi.org/10.1016/j.scitotenv.2013.10.029

Takada M, Yamada T, Takahara T, Endo S, Tanaka K, Kajimoto T, Okuda T (2017) Temporal changes in vertical distribution of ^{137}Cs in litter and soils in mixed deciduous forests in Fukushima, Japan. J Nucl Sci Technol 54:452–458. https://doi.org/10.1080/00223131.2017.1287602

Wakiyama Y, Onda Y, Mizugaki S, Asai H, Hiramatsu S (2010) Soil erosion rates on forested mountain hillslopes estimated using ^{137}Cs and $^{210}Pb_{ex}$. Geoderma 159:39–52. https://doi.org/10.1016/j.geoderma.2010.06.012

Wang WX, Ke C, Yu KN, Lam PKS (2000) Modeling radiocesium bioaccumulation in a marine food chain. Mar Ecol Prog Ser 208:41–50

Chapter 14
Parallel Measurement of Ambient and Individual External Radiation in Iitate Village, Fukushima

Yoichi Tao, Muneo Kanno, Soji Obara, Shunichiro Kuriyama, Takaaki Sano, and Katsuhiko Ninomiya

Abstract This report is based on a survey we conducted in March–June 2017 to measure radiation levels in Iitate Village after the 2011 Fukushima Daiichi nuclear power plant accident. Six NPO members including two evacuees who returned after the evacuation order was lifted in April 2017 took part in the study. Each participant worked in the area each day, carrying an ambient radiation measuring device, two personal radiation measuring devices, and a GPS receiver to record their movement and compare ambient and individual external doses. No matter where they were, results showed that for those taking the same daily route for the same amount of time, the individual dose was about 11~30% lower than that of the ambient dose; the ratio of ambient to individual external dose was 100 to 70–89. Further measurement and additional data are needed to protect the health of villagers, and to identify areas that require further decontamination.

Keywords Fukushima Iitate Village · Fukushima Daiichi Nuclear Accident · Ambient dose · Individual dose · Lifestyle pattern

---

One Sentence Summary: When measuring individual dose, it is essential to understand where an individual stays in a contaminated area and for how long.

Y. Tao (✉) · M. Kanno
Approved Specified Non-Profit Corporation for Resurrection of Fukushima, Fukushima, Japan
e-mail: taoyoichi@bridge.ocn.ne.jp

S. Obara · S. Kuriyama · T. Sano · K. Ninomiya
Approved Specified Non-Profit Corporation for Resurrection of Fukushima, Tokyo, Japan

T. M. Nakanishi et al. (eds.), *Agricultural Implications of the Fukushima Nuclear Accident (III)*, https://doi.org/10.1007/978-981-13-3218-0_14

14.1 Introduction

Iitate Village is located 30–50 km northwest of Fukushima Daiichi nuclear power plant. The entire village was designated as deliberate evacuation area in April 2011. On March 31st, 2017, the evacuation order was lifted with the exception of the Nagadoro district. Figure 14.1 shows where the evacuation orders (dotted pattern) have been lifted in March and April 2017 (gray areas are still under evacuation orders).

Prior to the survey in March–June 2017, we have been conducting monthly regional ambient radiation dose measurements in Iitate Village for 6 years. We used two cars equipped with a five/three-inch NaI scintillation survey meters, and *Geiger-*Mueller counters connected to GPS, and have displayed the results on maps (e.g.,

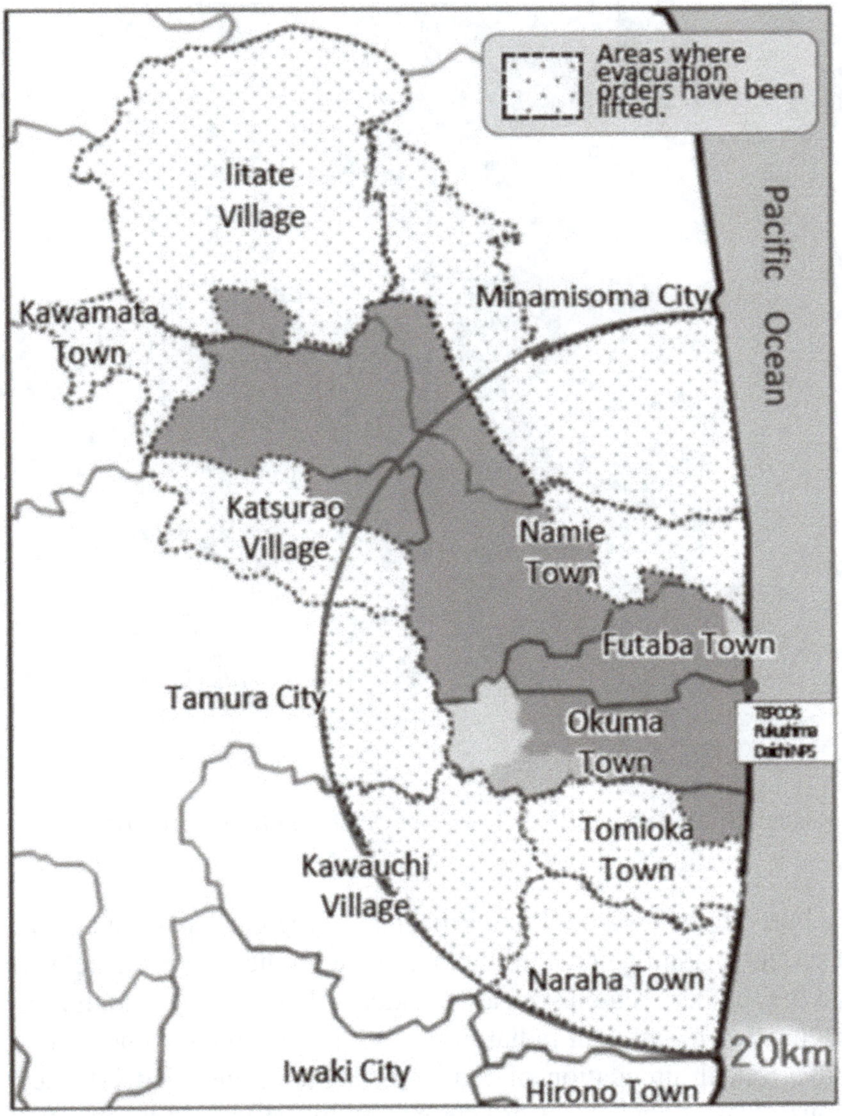

Fig. 14.1 The location and status of Iitate Village. (Source: "Fukushima Revitalization Station" http://www.pref.fukushima.lg.jp/site/portal-english/)

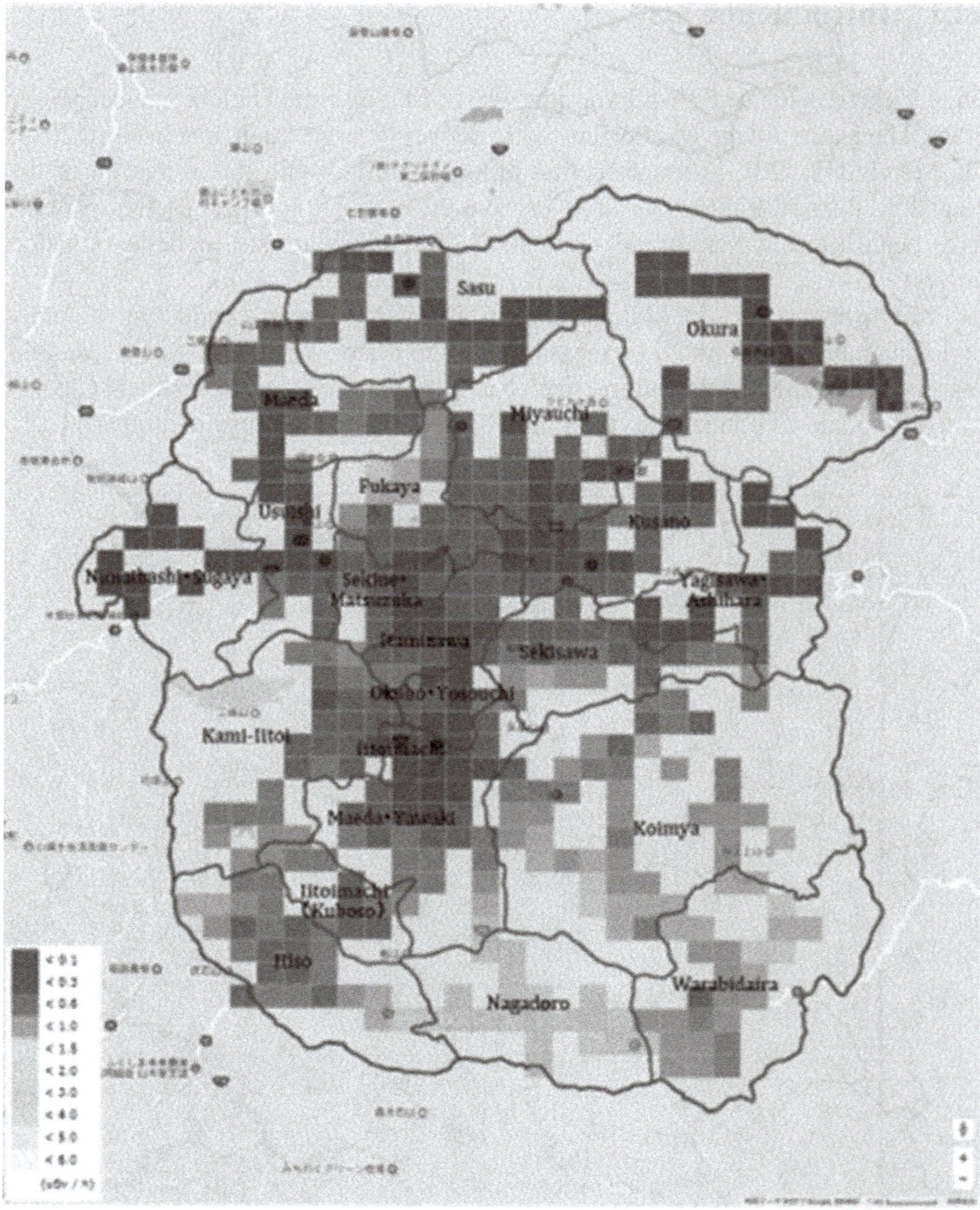

Fig. 14.2 Radiation dose rate in Iitate village from October to December 2016

Fig. 14.2). Forty evacuees from Iitate Village have regularly participated in the radiation measurement. Ambient radiation dose in the northern region is relatively lower than that in the southern region such as Nagadoro, Hiso and Warabidaira. Figure 14.3 depicts attenuation of ambient radiation due to the passage of time, which is faster than the theoretical attenuation of radiation. We also installed 20 fixed units in mountainous areas as well as in private homes in the village to measure ambient dose and conducted a 24-h continuous measurement. We have made precise measurements of participants out walking in the forests and fields. From these various

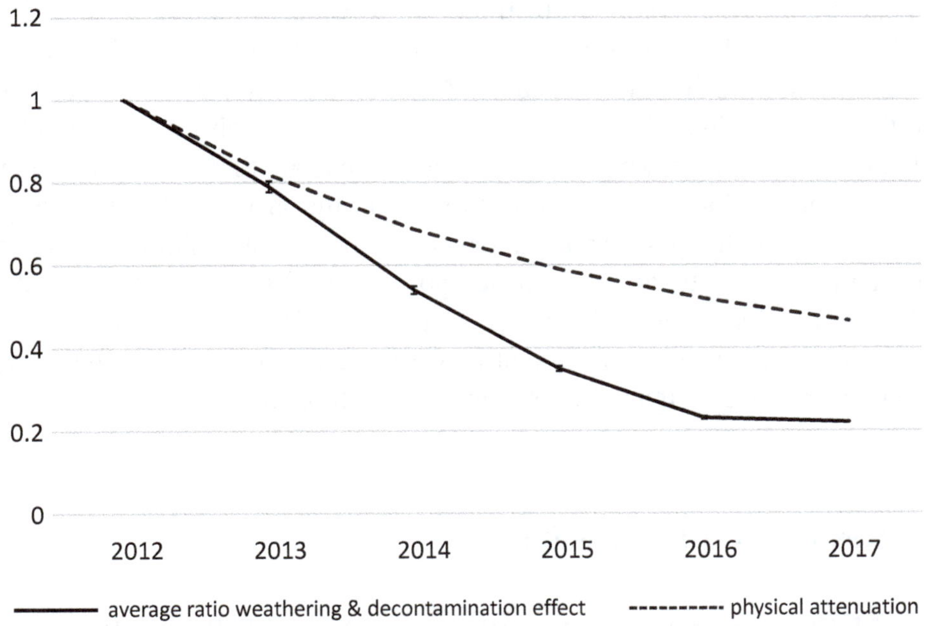

Fig. 14.3 Relative air radiation dose rate from 2012 to 2017 in Iitake village

measurements, we have accumulated ambient dose data for Iitate Village (Resurrection of Fukushima, NPO, Resurrection of Life and Industry in Fukushima, Empathy and Collaboration 2015; Resurrection of Fukushima, NPO, Monitoring Jouhou (Information) 2012–2017). Our purpose is to provide residents with accurate radioactivity data so that they will have the information necessary to decide if they should move back to the village permanently or move elsewhere to rebuild their lives.

14.2 Methods

Parallel Measurement of Ambient and Individual External Radiation is based on a survey that we conducted March–June 2017. Two evacuees from Iitate Village and four members of our NPO took part. All participants worked in the area each day. Each participant carried an ambient radiation measuring device, two types of personal radiation measuring devices, and a GPS (Global Positioning System) receiver, allowing us to record their movement and to compare ambient and individual external doses. To ensure the accuracy of ambient dose, participants spent some time beside fixed radiation monitoring equipment that the central government, Fukushima Prefectural Office and Iitate Village authorities installed at various locations throughout the village to assure cross-calibration.

We have utilized the following instruments: NaI scintillation survey meters, ALOKA TCS-172; electronic pocket dosimeter PDM-501 (Hitachi Healthcare) and electronic personal D-Shuttle dosimeter (Chiyoda Technol Corporation, Joint development with the National Institute of Advanced Industrial Science and Technology (AIST). By connecting the recording devices to a TCS-172, we recorded the dose every second and then calculated the average hourly dose. All participants carried the TCS-172 in their hand, at a height of about 1 m above the ground while moving on foot. While working in the field or on the farm, the measurement devices were placed on a table or tripod, not directly on the ground. While driving, the devices were put on the passenger seat or on the luggage rack. For a comparison of dosimeter measurements, all participants wore both a PDM-501 and a D-Shuttle. The PDM-501 was set to record the dose every 5 min. The average hourly dose was then calculated. The dose on the D-shuttle was calculated every hour. In this report, we explain three cases. Figure 14.4 shows each participant's monitoring route.

14.3 Results and Discussion

Figure 14.5 (Participant A), Fig. 14.6 (Participant B) and Fig. 14.7 (Participant A) show the overall results of the daily measurements of Participants A and B. The participants stayed in Iitate Village during the day and slept not far away in Ryozen, Ishida District, Date City, where the dose was lower than that in Iitate. The dose in the southern part of Iitate, such as in Nagadoro, Hiso, and Komiya, was higher than that in the northern areas like Sasu and Matsuzuka

Figure 14.5 depicts the results of measurements taken from 10:00 a.m. on April 15 to 12:00 p.m. on April 16, 2017. During this period, the individual started outdoor activities at 10:00 a.m. in Hiso, then went to Sasu around 12:00 p.m. and worked in the field there until 6:00 p.m. He left the measurement devices in the Sasu office until 5:00 a.m. the following morning; the next day, the 16th, he was in Ishida, Date City, working inside and out from 5 a.m. to 8 a.m. He then went back to Sasu and worked in the field there until 12:00 p.m. The solid line shows the ambient dose measured on the TCS-172. The dotted line in the figure shows 70% of the ambient dose. It approximately matches the gray bars, representing the measurement of the pocket dosimeter PDM-501. Measurements from the D-Shuttle dosimeter are shown as a dark blue bar in the chart. The D-Shuttle measurements tended to be somewhat higher than those of the PDM-501 and of the dispersion of the readings.

Figure 14.6 shows the results of measurements from 1:00 p.m. on June 17 to 3:00 p.m. on June 18th, 2017. This individual left Sasu by car around 1:00 p.m., drove to Hiso and Nagadoro in the southern part of Iitate Village, then went on to Itamisawa and Matsuzuka in the central part of the village. He spent approximately 30 min outside the car in each place. He stayed in a lodging house in Ishida, Date City (not far from Iitate Village) from 5:00 p.m. to 8:00 a.m. the following morning. He worked in the field in Sasu from 8:00 a.m. to 12:00 p.m., then had lunch in the Sasu office. After lunch, he went to Komiya and returned to Sasu. We found that, as on the previous day, the D-shuttle calibration was higher than that of the DPM501, which was approximately 70% of the ambient dose.

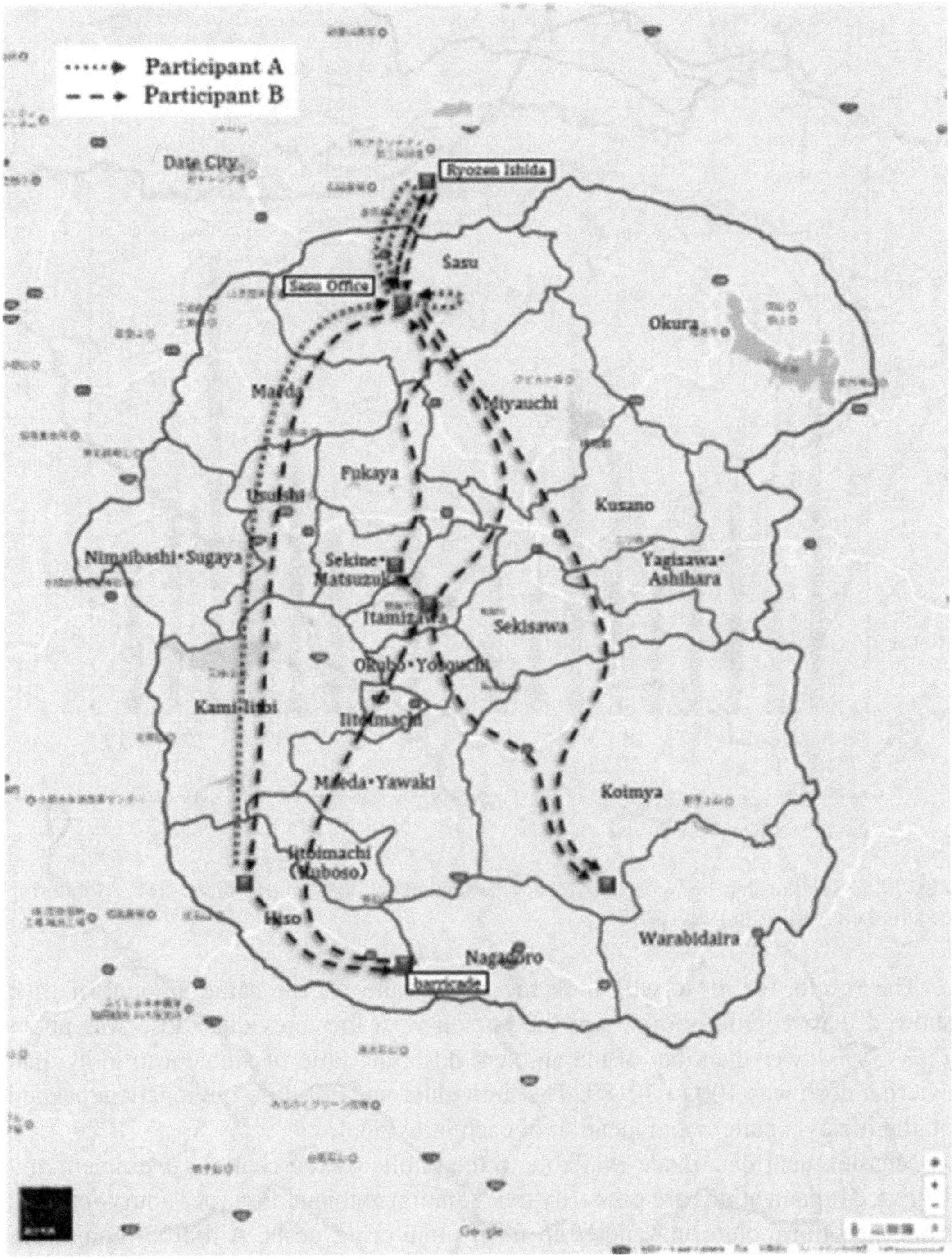

Fig. 14.4 Participant A's monitoring route on April 15th–16th, 2017 and participant B's route on June 17th–18th, 2017

Figure 14.7 indicates the results of measurements taken from 9:00 a.m. to 1:00 p.m. on September 3rd, 2017. This individual started outdoor activities at 9:00 a.m. in Nagadoro and stayed in a community house around 11:00 p.m. to noon, then went to Sasu.

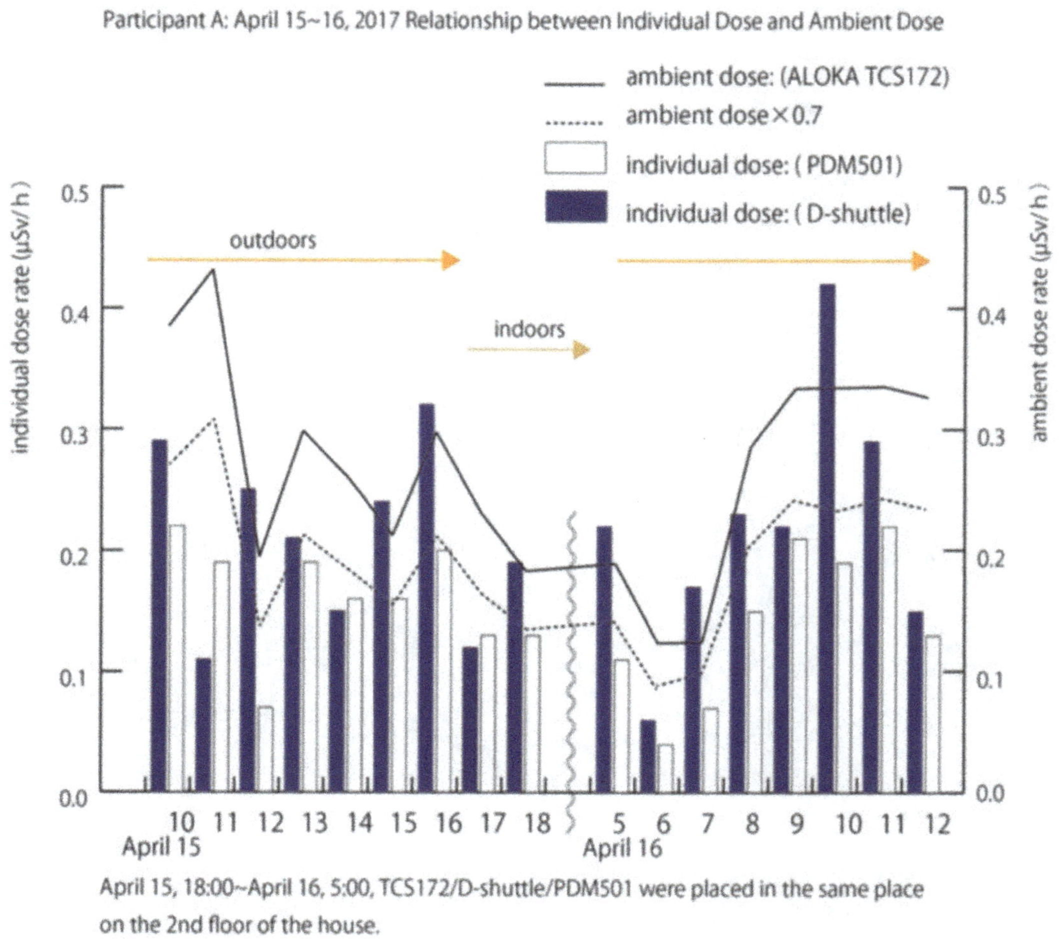

Fig. 14.5 Relationship between individual (Participant A) dose and ambient dose. Monitoring occurred on April 15th–16th, 2017

The results for those who took the same route for the same amount of time showed that regardless of where the person was, the individual dose was about 11%~30% lower than that of the ambient dose; the ratio of ambient to individual external dose was 100 to 70–89. The individual integral dose obviously depended on the lifestyle pattern and location of each individual.

Measurement data made available to the public by the central government and affiliated organizations are primarily the radiation ambient dose (per hour) obtained from airborne monitoring and from fixed monitoring posts. Airborne monitoring data is valuable to understand the average ambient dose within a radius of 300 m (Nuclear Regulation Authority 2017). This data was useful for policymaking as well as measuring radiation levels in highly contaminated areas right after the Fukushima Daiichi nuclear power plant accident (Yuuki et al. 2014).

We found a high correlation between the individual dose and ambient dose simultaneously measured in the same location. Needless to say, the ambient dose was very different at each point, whether in a house, forest, field or on the street, even in places located within a 300 m radius. When we study individual dose in a

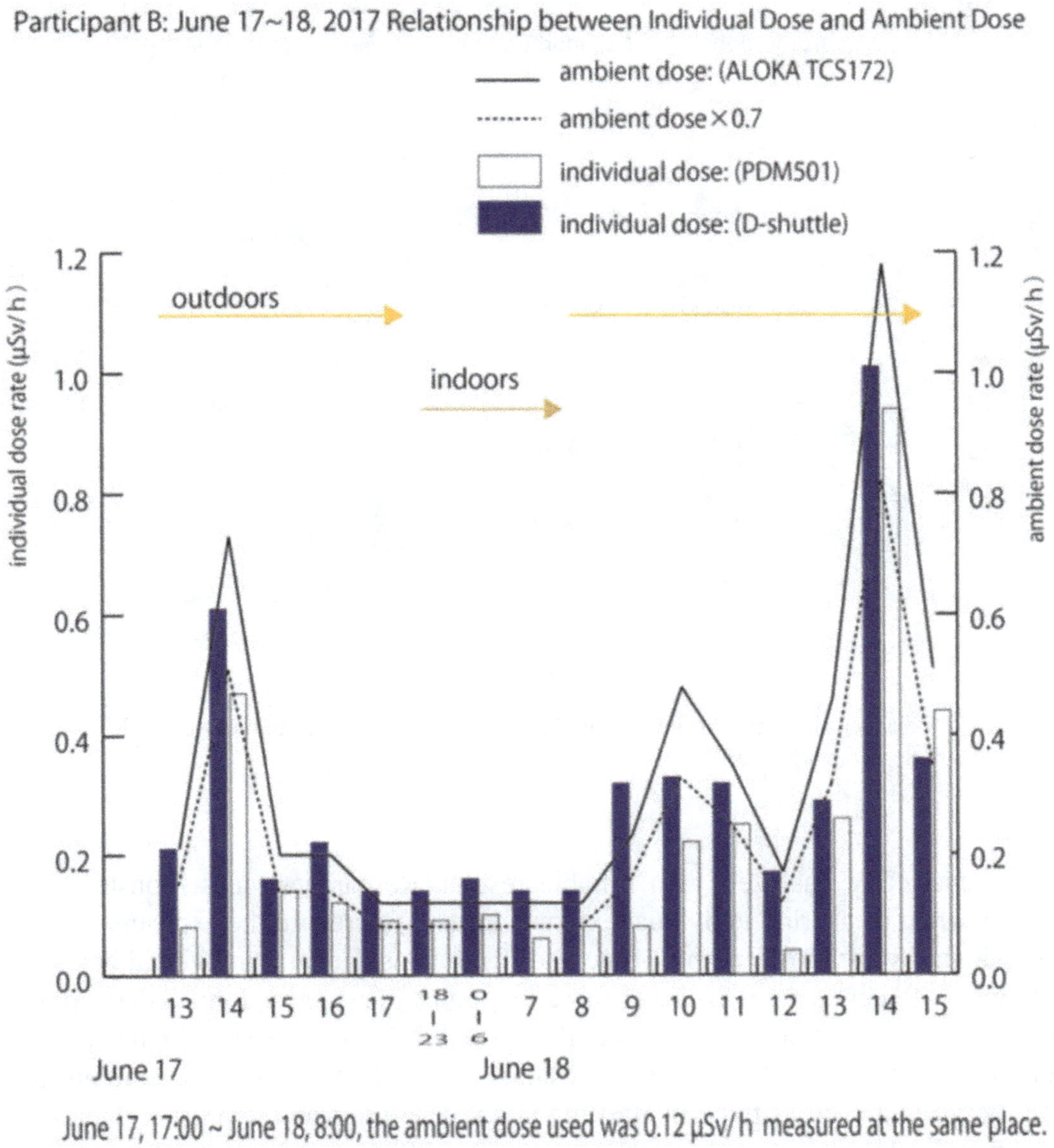

Fig. 14.6 Relationship between individual (Participant B) dose and Ambient Dose. Monitoring occurred on June 17th–18th, 2017

contaminated area, it is essential to be aware of the individual's location, how long he or she stays there and the ambient dose of radiation at the site. A comparison between ambient dose and individual dose without consideration of those elements may cause misunderstanding.

Some studies have reported the relationship between ambient dose obtained from airborne monitoring and individual dose (Ishikawa et al. 2016; Naito et al. 2017; Miyazaki and Hayano 2017a, b). We doubt the relevancy of using airborne monitoring data in relation to individual dose, as was done, for example, for the Date City's glass badge monitoring (Miyazaki and Hayano 2017a, b). This combination cannot

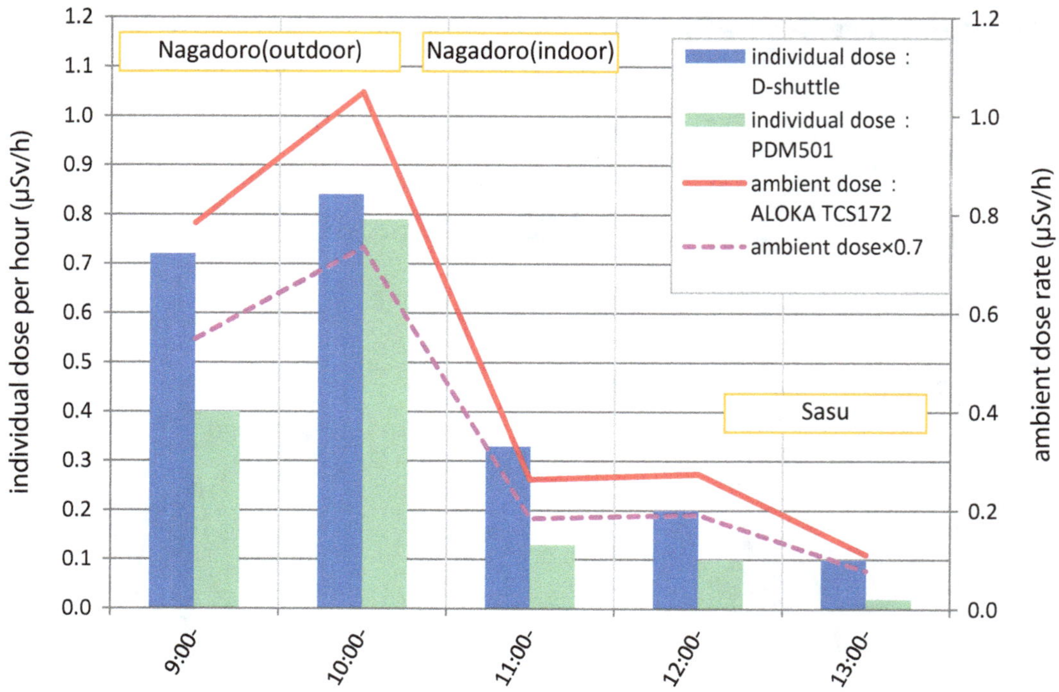

Fig. 14.7 Relationship between individual (Participant A) dose and ambient dose. Monitoring occurred on September 3rd, 2017

be scientifically analyzed with reliable results. Researchers must consider two points when measuring individual external dose: (1) to record the exposure to each individual by location/length of time and (2) to consider only individual data, not the mean. Recent studies ignoring these two points have caused misunderstanding of the current situation in Fukushima (Kornei 2017). Besides, we would like to point out that airborne monitoring data are usually higher than ambient dose monitoring on the ground such as fixed monitoring posts. The airborne monitoring data is an estimated value from 1 m above the ground. The difference between airborne monitoring and ground monitoring from 2013 to 2016 will be larger in Iitate Village by our estimate. Many Date city residents who received glass badges did not wear them outdoors and simply sent them back to the local government office every 3 months after receiving replacements (The minutes of the proceedings of Date council 2016; Date City council News 2017). In other words, it is not known how many of those glass badges were never worn outside during those 3 months. A comparison between individual dose and ambient dose in a rather wide area without accurate information on location and length of time as well as a system to maintain individual dose records is not useful for understanding the current situation in Fukushima. Even worse, it can mislead residents in Fukushima into thinking that the exposure dose is lower than it actually is. We must offer proper scientific facts to prevent misunderstanding.

14.4 Conclusion

When the evacuation order was lifted in April 2017, some evacuees from Iitate Village returned to work there. Some resumed farming, others reopened their businesses, even though there are still quite a few areas where relatively high radioactivity is observed. Not all returnees are staying overnight in Iitate. Individual dose is highly dependent on each person's actual activities, and those activities vary. Although the radiation level in decontaminated areas has gone down in Iitate Village, the level in the forests and mountains is higher than in the residential areas because they have not been decontaminated. More thorough radiation measurement is needed for the safety of residents and for further analysis.

Acknowledgments We would like to thank the High Energy Accelerator Research Organization (KEK) for providing the high spec measuring equipment and the calibration. We also are grateful to the Iitate Village Office and the evacuated residents for giving us the opportunity to jointly make monthly measurements for more than 5 years. Mr. Tadashi Ogawa, president of Knowledge Design, has visualized our monthly measurements on maps from the very first stage of our project. Professor Masaru Mizoguti, Graduate School of Agricultural and Life Science / the University of Tokyo, cooperated in taking measurements in fields and forests. Ms. Itsuko Yano supported in arranging and recording. The Mitsui & Co. Environment Fund has given activity grants. We greatly appreciate their contribution. We hope that more related organizations and individuals will become interested in and work on accurate data collection and accumulation.

References

Date City council News, (Date Shigikai Dayori) vol 45(10) (2017). http://www.city.fukushima-date.lg.jp/uploaded/attachment/28009.pdf. Accessed 20 July 2017

Ishikawa T, Yasumura S, Ohtsuru A, Sakai A, Akahane K, Yonai S, Sakata R, Ozasa K, Hayashi M, Ohira T, Kamiya K, Abe M (2016) An influential factor for external radiation dose estimation for residents after the Fukushima Daiichi Nuclear Power Plant accident—time spent outdoors for residents in Iitate Village. J Radiol Prot 36:255. Accessed 20 July 2017

Kornei K (2017) Fukushima residents exposed to far less radiation than thought. Science. https://doi.org/10.1126/science.aal0641. Accessed 20 July 2017

Miyazaki M, Hayano R (2017a) Individual external dose monitoring of all citizens of Date City by passive dosimeter 5–51 months after the Fukushima NPP Accident (series): I. Comparison of individual dose with ambient dose rate monitored by aircraft surveys. J Radiol Prot 37(1). IOPscience (2016). Accessed 20 July 2017

Miyazaki M, Hayano R (2017b) Individual external dose monitoring of all citizens of Date City by passive dosimeter 5 to 51 months after the Fukushima NPP accident (series): II. Prediction of lifetime additional effective dose and evaluating the effect of decontamination on individual dose. J Radiol Prot 37:623. IOPscience (2017). Accessed 14 Aug 2017

Naito W, Uesaka M, Kurosawa T, Kuroda Y (2017) (. Accessed 20 July 2017) Measuring and assessing individual external doses during the rehabilitation phase in Iitate village after the Fukushima Daiichi nuclear power plant accident. J Radiol Prot 37:606

Nuclear Regulation Authority, Results of the Eleventh Airborne Monitoring and Airborne Monitoring out of the 80km Zone of Fukushima Dai-ichi NPP February 13, 2017. http://radioactivity.nsr.go.jp/en/contents/12000/11830/24/11th%20Airborne_eng.pdf. Accessed 20 July 2017

Resurrection of Fukushima, NPO, Monitoring Jouhou (Information) (2012–2017) Monitoring information. http://rad.fukushima-saisei.jp/mesh.html. Accessed 20 Aug 2017

Resurrection of Fukushima, NPO, Resurrection of Life and Industry in Fukushima, Empathy and Collaboration (2015) Resurrection of Fukushima activity report http://www.fukushima-saisei.jp/app-def/S-102/madei/wp-content/uploads/2015/06/Presentation-package-as-of-June-10-2015.pdf. Accessed 20 Aug 2017

The minutes of the proceedings of Date council, December 8, (2016) 196–198. http://www.kaigi-roku.net/kensaku/datecity/datecity.html. Accessed 20 July 2017

Yuuki Y, Maeshima M, Hirata R, Matsui M (2014) Airborne radiation monitoring in the Fukushima Daiichi nuclear power plant accident, Oyo technical report, pp 106–115. Accessed 20 July 2017

Chapter 15
Mobility of Fallout Radiocesium Depending on the Land Use in Kasumigaura Basin

Shuichiro Yoshida, Sho Shiozawa, Naoto Nihei, and Kazuhiro Nishida

Abstract Although Lake Kasumigaura stores twice as much radiocesium compared to the direct fallout onto the lake surface, the additional source of radiocesium has not been determined. The present study examined the major source of radiocesium deposited in the lake based on surveys of the dry beds of the rivers flowing into the lake. The basin of four rivers, two of which flow through an urbanized region and the other two through a rural region to Lake Kasumigaura, were selected. The radioactivity per unit area of the dry river bed and the top of the river bank was measured. On the dry river beds of the rivers flowing from the urbanized area, the deposition of radiocesium per unit area was found to be much higher than the direct fallout per unit area, revealing a considerable amount of radiocesium had been discharged from the urbanized upstream of the rivers by flooding events. On the other hand, rivers flowing from the rural area stored almost the same amount of radiocesium as the direct fallout. These observations revealed that the urbanized areas located upstream to Kasumigaura Lake were a major additional source of radiocesium contamination in the lake.

Keywords Radiocesium contamination · Watershed · Water source · Land use · GIS

15.1 Introduction

The accident at the Fukushima-Daiichi Nuclear Power Station caused expansive radioactive contamination over farmlands, forests and cities not only in Fukushima Pref. but also in broad areas of eastern Japan. Because cesium is known to be strongly adsorbed by soil minerals, the mobility of the fallout onto the soil surface should be low. However, the fallout onto paved surfaces is less adsorbed and can be carried by the surface water flow. Kakamu et al. (2012) reported that the reduction in the dose rate from April to November 2011 depended highly on ground surface

S. Yoshida (✉) · S. Shiozawa · N. Nihei · K. Nishida
Graduate School of Agricultural and Life Sciences, The University of Tokyo, Tokyo, Japan
e-mail: agyoshi@mail.ecc.u-tokyo.ac.jp

T. M. Nakanishi et al. (eds.), *Agricultural Implications of the Fukushima Nuclear Accident (III)*, https://doi.org/10.1007/978-981-13-3218-0_15

type. The ratios of environmental radiation dose rate of soil and asphalt on 11 November compared with 11 April were 47.9% and 70.7%, respectively. The fallout onto the soil surface could move down to deeper layers with the percolating water, causing a reduction in dose rate. However, the fallout onto the asphalt could not move into the deeper structure of the asphalt but stayed near the surface, causing less change in dose rate during that period. This suggested that most of the fallout onto asphalt was scoured by rainfall soon after the accident, becoming the main source of the contamination in the water system. However, the movement of radiocesium fallout in the water system had not been directly monitored by any studies until the monitoring framework was built. Therefore, indirect evidence was required to prove the difference in the radiocesium mobility depended on the land use. Among the studies, Shiozawa (2016) measured the radiocesium storage density (kBq m^{-2}) in a contrasting pair of irrigation reservoirs, one of which is downstream of a forest and the other located below premises of a factory. The radiocesium storage per unit area (kBq m^{-2}) in the reservoirs depended highly on the land use of the upper basin. The reservoir collecting the water from the factory stored five times as much radiocesium as the direct fallout onto the reservoir surface. However, the reservoir collecting the water from the forest had the same amount of radiocesium as the direct fallout. This study clearly suggested that the paved ground surface adsorbed less radiocesium than the soil, causing an outflow of radiocesium downstream via the rivers or canals.

Lake Kasumigaura is the most important water source for the southern part of Ibaraki Pref., and the major land use of its basins is diverse. The monitoring by the Ministry of the Environment showed that the radiocesium concentration of the mud in the lake bottom gradually increased in the first year after the accident. This suggested not only direct fallout but also the inflow from the rivers caused contamination of the lake. However, the major source of the contamination is not clearly established by the official monitoring. Substantial migration of suspended particles along the river occurs due to heavy rain events when the water level in the rivers rises up to the dry river beds. During these rare events, sediment in the dry river beds can be replaced by the soils or sands flowing down from the upper basin. Because most of the fallout radiocesium attaches to suspended particles, if the amount of radiocesium per unit area of dry river bed is higher than the fallout, radiocesium has probably flowed down from the upper basin. Thus, the ratio of the amount of radiocesium per unit area of dry river bed to the fallout per unit area should be a measure to assess the influence of the upper basin as the source of radiocesium.

From these considerations, the basin of four rivers, two of which flow through an urbanized region and the other two flow through a rural region to Lake Kasumigaura, were selected for this study. The radioactivity per unit area of the dry river bed and the top of the river bank were surveyed to reveal the difference in the migration of radiocesium fallout from the upper stream, and if the migration of radiocesium depended on land use.

15.2 Methods

15.2.1 Characteristics of the Study Area

Four watersheds of Hanamuro (H1~H3), Sakura (S1, S2), Koise (K1, K2) and Bizen (B) rivers were subject to our survey (Fig. 15.1). Most of the watersheds of the rivers were determined with the aid of an elevation map. However, because the upper stream of Hanamuro River functions as the main river for drainage of rainwater from the urbanized area of Tsukuba City, the watershed H3 was determined from the sewage network map provided by the Sewage Management Division of Tsukuba City. The information was introduced to Arc Map10.4 (ESRI Inc.) to facilitate further analyses. Figure 15.2 provides the results of the airborne monitoring in the distribution survey of radioactive cesium performed by The Ministry of Education, Culture, Sports, Science and Technology (MEXT) in July 2011. The inventory values on the map were corrected to those on 21 Oct. 2016.

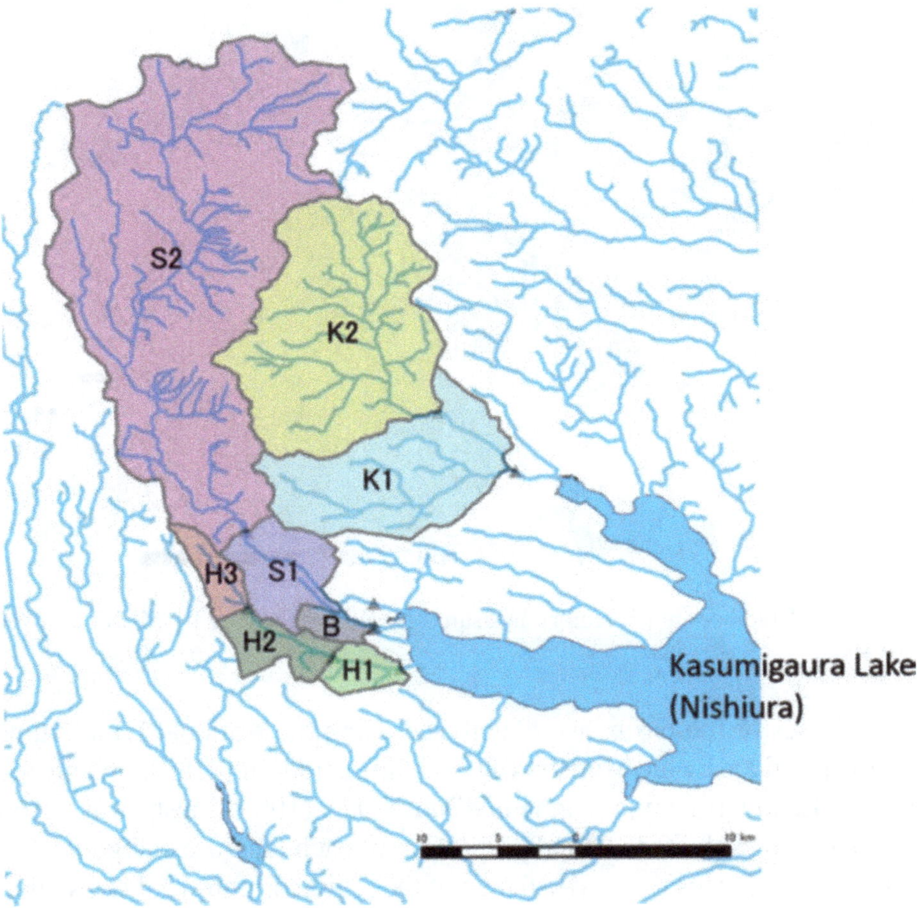

Fig. 15.1 Basins of the rivers for the survey

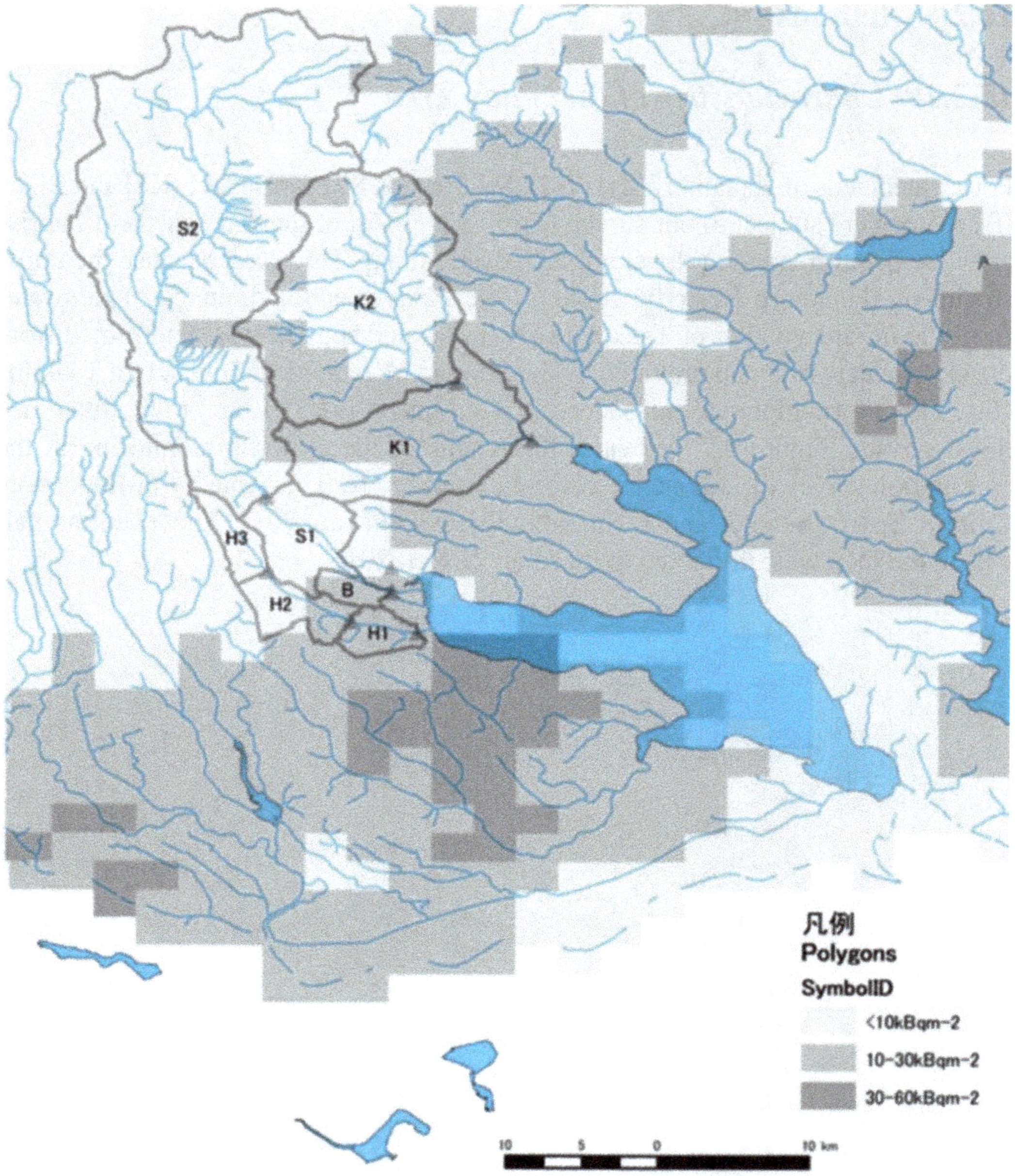

Fig. 15.2 Map of radiocesium inventory measured by airborne survey by MEXT (provided by JAEA 2014)

The land uses of the basins were analyzed using the National Land Numerical Information provided by the Ministry of Land, Infrastructure, Transport and Tourism (MILT). The Land Utilization Segmented Mesh Data (100 m mesh) was introduced to ArcMap10.4 to evaluate the areal proportion of each land use category.

Figure 15.3 illustrates the land use map of the region. Figure 15.4 shows the areal percentage of each land use in the watersheds. Hanamuro River flows through the urbanized area of Tsukuba City and Tsuchiura City. More than 40% of the land is used for buildings. As previously mentioned, the rain falling on roofs and paved ground surfaces are collected by drains that connect to sewers which discharges the rainwater directly to the river. Bizen Riv. also flows through the center of Tsuchiura

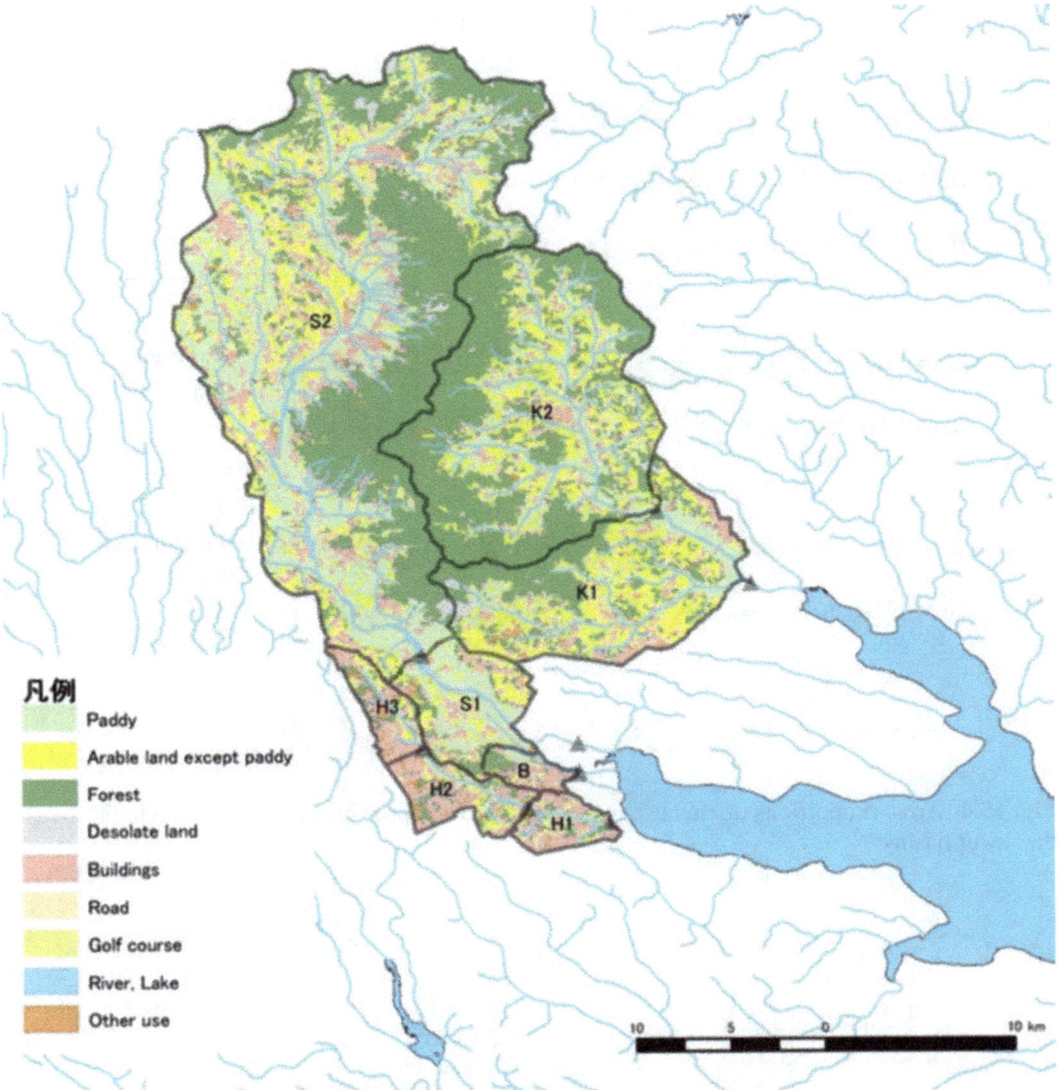

Fig. 15.3 Land use map of the basins

City, where more than 40% of the watershed is used for buildings. In contrast, Sakura Riv. and Koise Riv. flow from the rural areas, and most of the land is used for farming or forestry.

15.2.2 Measurement Apparatus of Deposited Radiocesium per Unit Area (kBq M^{-2})

The radioactivity per unit area (inventory kBq m^{-2}) was measured by the apparatus shown in Fig. 15.5. A commercial NaI scintillation probe and survey meter (Model 5530/5000, National Physical Instrument, USA) was used as the detector. To improve the directionality, the probe was insulated by lead plate except in the

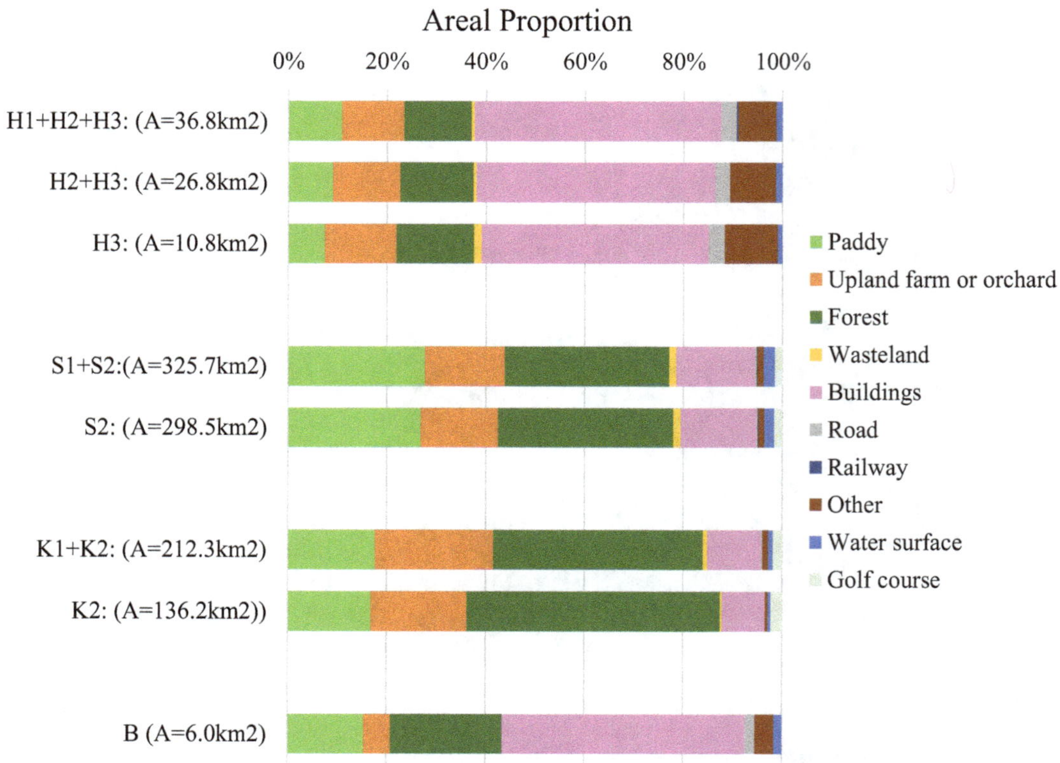

Fig. 15.4 Areal proportions of land use in the basins. (The upper basins are included in the area of the lower basins)

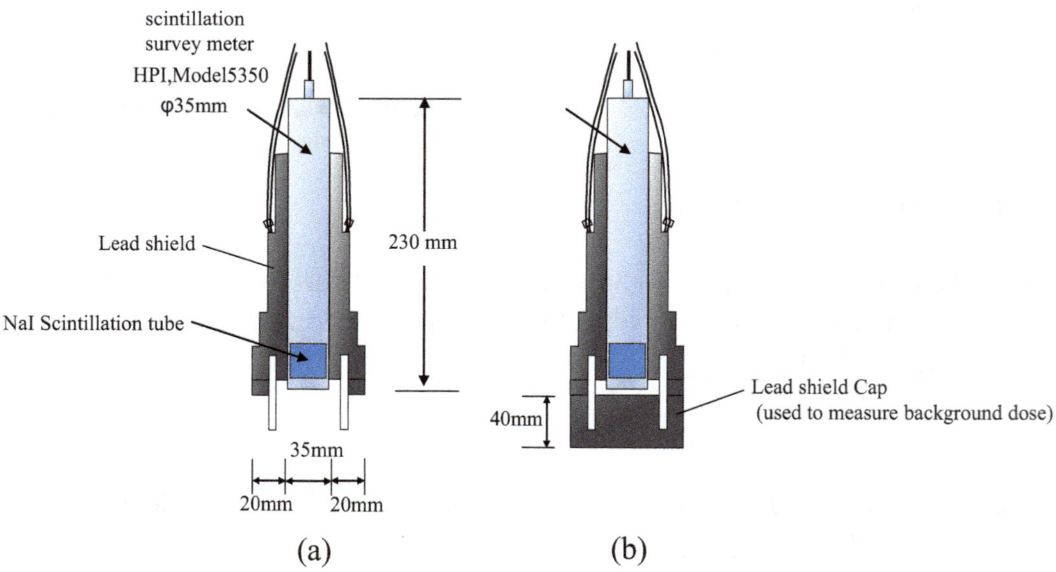

Fig. 15.5 Apparatus for measuring radioactivity per unit area (inventory) in soil. (**a**) To measure the gamma beam radiation from the radiocesium deposited to the ground surface. (**b**) To measure the background radiation alone

direction the measurements were taken (Fig. 15.5a). Radioactivity (counts of gamma beam) was measured three times from 1 m above the ground at each location. Measurements were also conducted when the opening was closed with a lead cap to eliminate background radiation (Fig. 15.5b).

The radioactivity per unit area of the submerged bottom of the river (kBq m^{-2}) was measured by the apparatus shown in Fig. 15.6. The same commercial NaI scintillation survey probe (Model 5530, National Physical Instrument, USA) was covered by a rubber cap to waterproof it. Because the attenuation of a gamma beam in water is a thousand times higher than in air, round foamed polystyrene boards were attached to the tip of the rod to avoid attenuation of the gamma beam by the water between the probe and the bottom mud. Meanwhile, the insulation against the background radiation was not necessary due to the attenuation by water. The measurements were conducted by hanging the apparatus from the middle of pedestrian and vehicle bridges and lowering it to the bottom of the river. Radioactivity was measured three times at each location.

Undisturbed soil samples (diameter = 50 mm, length = 100 mm) were collected at the river beds of Hanamuro Riv. (H1, H2, H3) to compare the radioactivity with the aforementioned method. The vertical profiles of radiocesium concentration were determined by measuring the samples with a NaI scintillation analyzer on Jan 8 and 10, 2017.

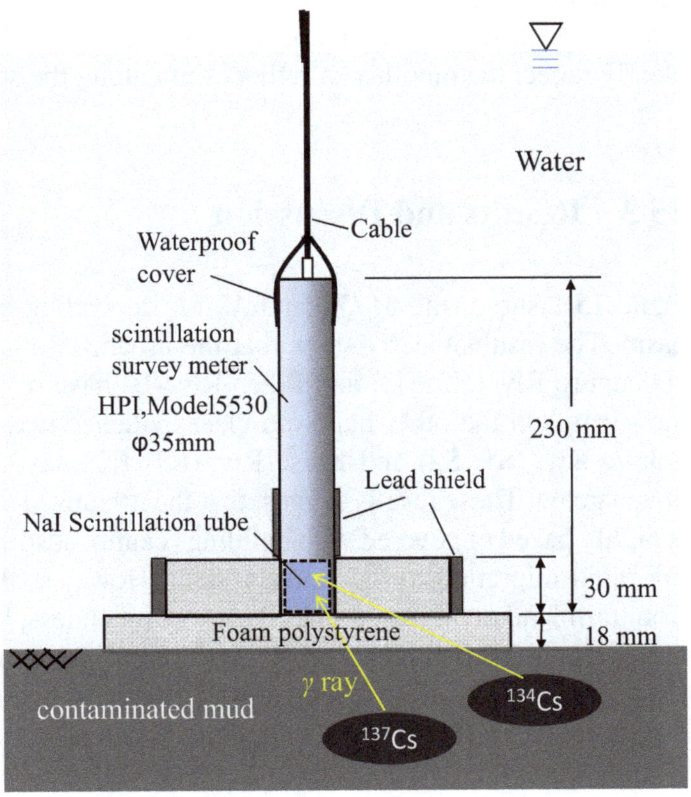

Fig. 15.6 Apparatus for measuring radioactivity per unit area (inventory) in the submerged river bottom

15.2.3 Measurement and Analysis of Radioactivity

The amount of deposited radiocesium per unit area was measured along the cross sections of the rivers on 27 Sept. and 21 Oct. 2016. All the measured cross sections were located near bridges, above which the corresponding watersheds spread. Radiocesium in the river water did not reach the unpaved top of the river bank, and only direct fallout radiocesium was captured by the soil. The dry river bed is the domain where the river water flows exclusively during flooding events, in which suspended particles can deposit. The river bottom is always under water which carries in contaminated particles but also carries contaminated particles out.

When the measured amounts of radiocesium per unit area at the top of the bank, dry river bed and submerged river bottom are denoted by M_a, M_b and M_c, respectively, the ratios M_b/M_a and M_c/M_b probably reflect the migration of radiocesium from the upper basin along the rivers. If M_b/M_a is larger than one, radiocesium must have migrated from the upper stream of the river when the discharge of the river increased. Because the water level does not frequently reach the upper river bed, sediment on the dry river bed is thought to remain for more than a few years. Thus, high radioactivity of the dry river bed is expected to remain for many years after the migration of radiocesium. If M_c/M_a is larger than one, radiocesium must have migrated from the upper stream of the river as well. However, even if it is close to one, it does not necessarily mean the absence of migration from the upper stream, because the contaminated sediment at the river bottom does not stay for long but moves down stream continuously. Therefore, M_b/M_a rather than M_c/M_a should clearly reflect the mobility of radiocesium along the streams.

15.3 Results and Discussion

Table 15.1 shows the M_b/M_a and M_c/M_a depending on the land use of the upper basin. The results clearly show that the urbanized middle and upper watershed of Hanamuro Riv. (H2, H3) and Bizen Riv. (B) have released much radiocesium into the rivers. On the other hand, no clear evidence was found that the watershed of Sakura Riv. (S1, S2) and Koise Riv. (K1, K2) have supplied radiocesium to the downstream. These results suggest that the urbanized area where the ground surface is highly paved or covered with buildings cannot adsorb nor hold radiocesium, causing rapid migration to the water system. However, the migration of radiocesium from farmland and forest to the rivers was much less, because the soil strongly held radiocesium. Yoshioka et al. (2013) started to monitor the outflow of radiocesium from a paddy field in Fukushima on 28 April 2011, showing 3.3% of the inventory was released in the first year after the accident. Eguchi (2017) summarized the reports on radiocesium balance in paddy fields from 2011 to 2015. The balance relative to the radiocesium inventory ranged between +0.24 and – 2.96% and the mean was −0.52%. On the other hand, the radiocesium outflow from urbanized regions

Table 15.1 Measured radioactivity at the end of basin of the rivers

Basin	Areal proportion of building site and road/Total area of the basin	M_a: Top of the bank kBq m^{-2}	M_b: Dry river bed kBq m^{-2}	M_c: Submerged river bottom kBq m^{-2}	M_b/M_a	M_c/M_b
H1: Downstream of Hanamuro Riv.	53%/36.8km^2	40.2	41.9	65.3	1.0	1.6
H2: Middle stream of Hanamuro Riv.	51%/26.8km^2	22.9	76.6	–	3.3	–
H3: Upper stream of Hanamuro Riv.	49%/10.8km^2	36.5	94.3	–	2.6	–
S1: Downstream of Sakura Riv.	16%/325.7km^2	30.0	29.5	32.9	1.0	1.1
S2: Upper stream of Sakura Riv.	16%/298.5km^2	31.7	23.2	–	0.7	–
K1: Upper stream of Koise Riv.	11%/136.2km^2	35.8	37.7	35.8	1.1	1.0
K2: Middle stream of Koise Riv.	9%/212.3km^2	31.5	35.8	31.0	1.1	1.0
B: Bizen Riv.	51%/6.0km^2	(30.0)[a]	–	91.7	–	(3.1)

[a]The measured radioactivity at S1 is shown instead, because the locations are very close to each other

had not been directly monitored. We have measured the dose rate at 1 cm above a paved road and at an uncultivated neighboring paddy field in Fukushima on February 2012 by insulating the scintillation detector with a lead block having a thickness of 30 mm. Only 50% of dose rate was recorded on the road compared to the paddy soil surface. Because the measured dose rates included the effect of the background radiation, more than 50% of fallout onto the paved road was probably scoured by water. Even if the land use is categorized to 'building', not all the ground is covered by buildings or pavement. Thus, the percentage of outflow of radiocesium might have been less. However, it is highly plausible that the radiocesium outflow from urbanized regions should have been much higher than that from the farmlands and forests.

Figure 15.7 shows the vertical profiles of radiocesium measured by the undisturbed soil samples taken at the river beds of Hanamuro Riv. At the measured points of the watershed H1 and H2, the thickness of the contaminated sediment was approximately 10 cm. The total amounts of radiocesium per unit area were estimated to be 47 and 58 kBq m^{-2}, respectively, which were close to the radioactivity per unit area (inventory) measured by the scintillation detector. At the measured

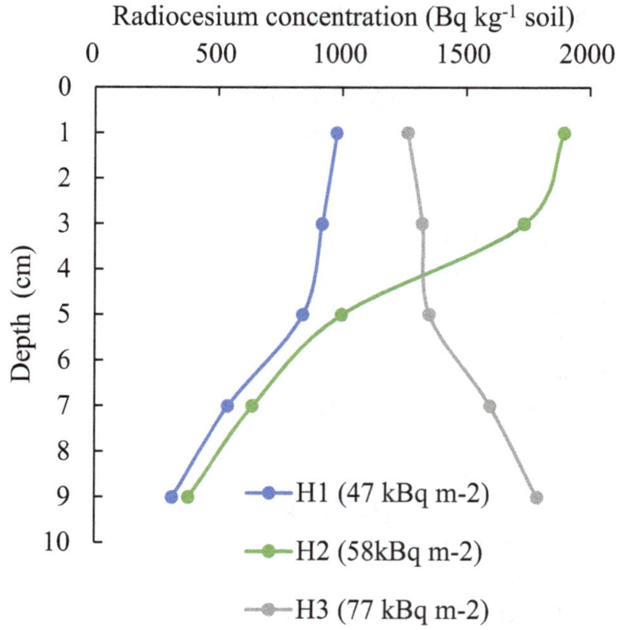

Fig. 15.7 Vertical profile of radiocesium concentration measured in the dry river beds of Hanamuro river

points of the watershed H3, the whole radioactivity per unit area (inventory) should have been higher because radiocesium was distributed deeper than 10 cm; the total amount of radiocesium to 10 cm was estimated to be 77 kBq m^{-2}. The vertical profile of the radiocesium concentration at the measured points of the watershed H3 also suggests that the observed higher radioactivity per unit area (inventory) than the fallout was not caused by the lateral flow from the slope of the bank, but by the accumulation of contaminated mud due to flooding; the vertical profile of radiocesium formed by the percolation of water had hardly extended to deeper layers (e.g. Honda et al. 2015; Koarashi et al. 2012). Finally, the influence of the radiocesium outflow from the basin on the contamination of Kasumigaura Lake was analyzed. Neither the total amount of radiocesium accumulating in Lake Kasumigaura nor the total fallout onto the lake has been accurately estimated because airborne monitoring does not work on the water surface. Tanaka (2015) began surveying the sediment of Kasumigaura Lake in 2011 at a spatial interval of a minute (1′) mesh to reveal the inflow of radiocesium and its horizontal movement within the lake. From a map in that report, the authors estimated the total amount and the mean inventory of radiocesium in Nishiura portion of Kasumigaura Lake. Because only the categorized range of the inventory of each polygon was shown on the map, the mean of the categorized range was used for the representative inventory of the polygon. The inventory on the ground surrounding the lake was also cited from the airborne survey conducted by MEXT in 2011 (JAEA 2014). The domain was determined by forming a 500 m buffer from the lake shore by GIS software. Table 15.2 shows the areal percentage of the inventory in the lake and the lake shore. The comparison suggested that Lake Kasumigaura has almost the same amount of radiocesium from the upstream basin as the direct fallout onto the lake, assuming the amount of direct fallout per unit area of the lake was the average in the surrounding ground. The area of the lake (Nishiura Lake portion of the whole Kasumigaura Lake system) and its

Table 15.2 Estimated radioactivity per unit area of the lake bottom and the surrounding ground of Kasumigaura Lake

Inventory class	Lake Kasumigaura[b] (%)	Surrounding ground[a] (%)
<10 kBq m^{-2}	6%	30%
10–30 kBq m^{-2}	57%	61%
30–60 $kBqm^{-2}$	26%	8%
60–100 $kBqm^{-2}$	11%	0%
Estimates of the mean	32.2 kBq m^{-2}	17.3 $kBqm^{-2}$

[a]Estimated from the data of the airborne survey conducted by MEXT in 2011 (JAEA 2014)
[b]Compiled from the distribution map estimated by Tanaka (2015)

basin is 170 km^2 and 1400 km^2, respectively. If 12% of the fallout on the basin was scoured and flowed out to the rivers, the observed accumulation of radiocesium in the lake can be accounted for. However, since most of the basin was used for agriculture or forestry, only a small proportion of fallout was released to the water system from the soil surface. Thus, the urbanized areas, which are located at the western edge of Lake Kasumigaura, were probably the main additional source of radiocesium to the bottom of the lake.

15.4 Conclusions

On the dry river beds of the rivers flowing from the urbanized area to Kasumigaura Lake, the deposition of radiocesium per unit area was found to be much higher than the direct fallout per unit area at each surveying point. On the other hand, the dry river beds of the rivers flowing from the rural area stored almost the same amount of radiocesium as the direct fallout. These observations revealed that the urbanized areas located at the upstream of Kasumigaura Lake were a major additional source of radiocesium contamination in the lake.

Acknowledgements The map of the sewage system was kindly provided by Tsukuba city. The measurement of the soil samples were performed in Isotope Facility for Agricultural Education and Research, The University of Tokyo.

References

Eguchi S (2017) Behavior of radioactive cesium in agricultural environment. J Jpn Soc Soil Phys 135:9–23 in Japanese with English abstract

Honda M, Matsuzaki H, Miyake Y, Maejima Y, Yamagata T, Nagai H (2015) Depth profile and mobility of 129I and 137Cs in soil originating from the Fukushima Dai-ichi nuclear power plant accident. J Environ Radioact 146:35–43

JAEA (2014) http://emdb.jaea.go.jp/emdb/portals/b224/

Kakamu T, Kanda H, Tsuji M, Kobayashi D, Miyake M, Hayakawa T, Katsuda S, Mori Y, Okouchi T, Hazama A, Fukushima T (2012) Differences in rates of decrease of environmental radiation dose rates by ground surface property in Fukushima City after the Fukushima Daiichi nuclear power plant accident. Health Phys 104(1):102–107

Koarashi J, Atarashi-Andoh M, Matsunaga T, Sato T, Nagao S, Nagai H (2012) Factors affecting vertical distribution of Fukushima accident-derived radiocesium in soil under different land-use conditions. Sci Total Environ 431:392–401

Shiozawa S (2016) Radiocesium migration in soil and outflow into river system. Water Land Environ Eng 84:495–499 (in Japanese)

Tanaka A (2015) Dynamics of radiocesium in the river and lake environment around high air dose rate areas and mapping of radiocesium in the lake bottom sediment, Final research report of Grants-in-Aid for Scientific Research (KAKEN), No. 24510076. https://kaken.nii.ac.jp/ja/file/KAKENHI-PROJECT-24510076/24510076seika.pdf (in Japanese)

Yoshioka K, Sato M, Ohkoshi S, Eguchi S, Yoshikawa S, Mishima S, Itabashi S (2013) Monitoring of radiocesium behavior in farmland. 2013's Annual Report of the project on the establishment of the measures for long-term risk assessment of radionuclides due to the accident of Fukushima-Dai-ichi Nuclear Power Plant. JAEA. 159–166 (in Japanese)

Chapter 16
Challenges of Agricultural Land Remediation and Renewal of Agriculture in Iitate Village by a Collaboration Between Researchers and a Non-profit Organization

Masaru Mizoguchi

Abstract We have tested several ways to revitalize agriculture in Fukushima by developing farmland decontamination methods that farmers can undertake by themselves. As a result, the rice harvested in a test field passed the official inspection of Fukushima Prefecture in 2014. Despite the efforts of local people, we have not yet succeeded to dispel the anxieties in the general public who believe that Fukushima's agricultural crops might contain radioactive cesium. Such "harmful rumors" are hampering the recovery of local agriculture in Fukushima. In this chapter, we review the challenges of agricultural land remediation and renewal of agriculture from a collaboration between researchers and a non-profit organization (NPO) and propose the scenario for the recovery of local agriculture and village life.

Keywords Agricultural land · Decontamination · Agricultural regeneration · Iitate Village · Radiation · Rural planning · Rural reconstruction scenario

16.1 Introduction

The accident at TEPCO's Fukushima Daiichi Nuclear Power Plant, which occurred in March 2011, became an unprecedented nuclear disaster. As a result, the forests, agricultural lands and oceans were contaminated extensively by radioactive cesium. In Iitate Village, Fukushima Prefecture, where the evacuation order was canceled in March 2017, decontamination work has been carried out using thousands of workers in preparation for the return of the Iitate Villagers in the spring of 2017.

M. Mizoguchi (✉)
Graduate school of Agricultural and Life Sciences, The University of Tokyo, Tokyo, Japan
e-mail: amizo@mail.ecc.u-tokyo.ac.jp

T. M. Nakanishi et al. (eds.), *Agricultural Implications of the Fukushima Nuclear Accident (III)*, https://doi.org/10.1007/978-981-13-3218-0_16

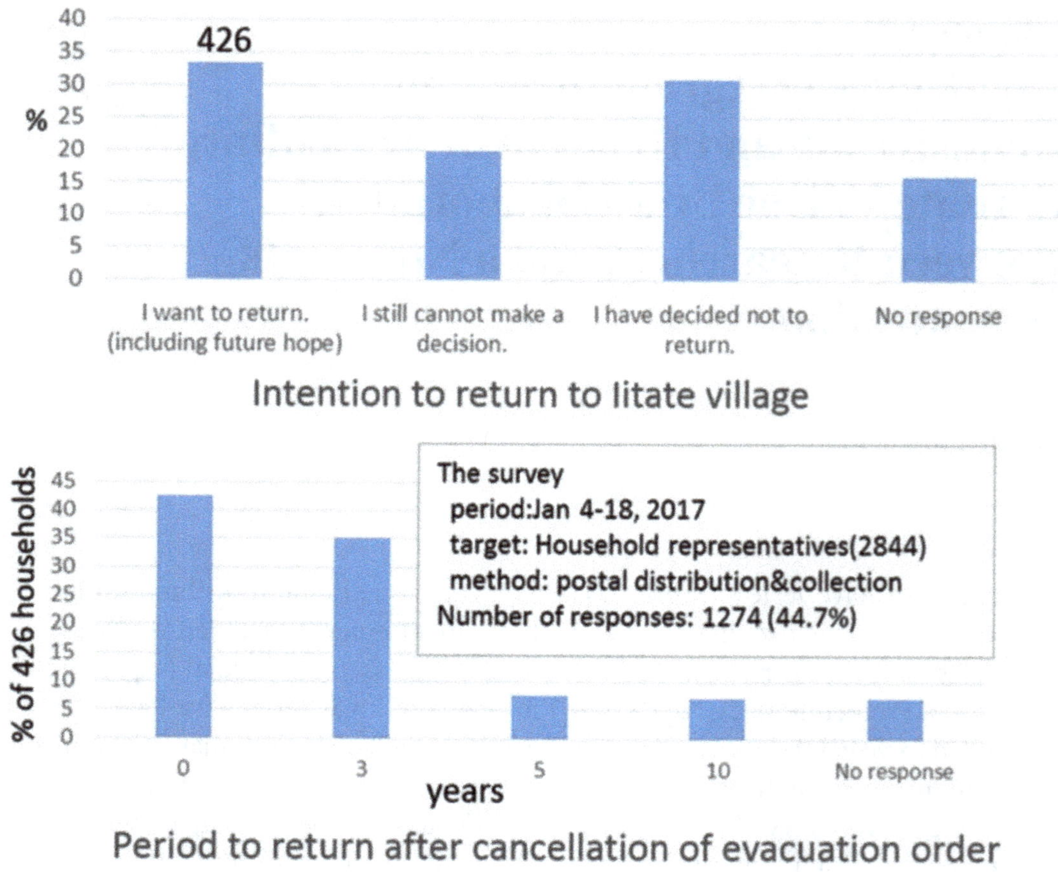

Fig. 16.1 Present situation as seen from the resident's intention survey (Iitate Village 2017) (Extracted from the Public Information "Iitate", issue 1, April 2017)

In January 2017, the reconstruction agency, Fukushima prefecture, and Iitate Village jointly conducted a questionnaire survey. Figure 16.1 presents some of the results from the survey (Iitate Village 2017). Among the 1271 households (44.7% of the respondents), 33.5% of responding households said that "we want to return (including future hope)". Thirty-one percent replied "we have decided not to return", and 19.7% replied that they "still cannot judge". Of the 426 households who replied "they want to return", 42.5% said they want to return immediately, but many residents want to return in 3, 5 or 10 years. Moreover, 59.4% said "Improvement in medical and nursing care welfare facilities" in the village is one of the conditions that needs to be met before returning. Two hundred and fifty-one households responded "I cannot judge yet". For these households, the conditions that need to be met before returning include: "When the infrastructure such as roads, buses, schools and hospitals will be restored (46.2%)", "how many other residents are planning to return (45.4%)", and "prospects for radiation dose reduction (44.2%)". These results mean that many people intend to observe the situation for a while even though the evacuation order was canceled.

I entered Iitate Village 3 months after the nuclear accident and have tested several ways to revitalize agriculture with the collaboration of local farmers, NPO members and other researchers (Mizoguchi 2015a). In this chapter, I review the challenges of agricultural land remediation and the renewal of agriculture and propose the scenario for the recovery of local agriculture and village life.

16.2 Collaboration Between Researchers and NPO

Our activities are based on the collaboration of groups beyond the existing organizational framework (Kanno and Mizoguchi 2014; Yokokawa and Mizoguchi 2013).

16.2.1 Authorized NPO "Resurrection of Fukushima" (Resurrection of Fukushima 2017)

"Resurrection of Fukushima" was founded in June 2011 in Sasu district in Iitate Village, Fukushima Prefecture. As of the end of May 2017, there were 290 members and 6 groups of regular members. Senior members called "Ara Koki" (average age is about 70 years) are the core of the NPO. Volunteers with diverse backgrounds gather in Iitate Village every weekend and discuss various ways to help the recovery of Iitate Village based on the data acquired in collaboration with local farmers. More than 1000 volunteers participated each year in activities such as radiation measurement and analysis of radioactivity, development of decontamination technologies, and health care for the residents, etc. The latest discussion and data are uploaded promptly on the website (http://www.Fukushima-saisei.jp).

16.2.2 Fukushima Reconstruction Agricultural Engineering Group (Fukushima Reconstruction Agricultural Engineering Meeting 2017)

This is a group organized immediately after the nuclear accident by agricultural engineering researchers who are specialists of agricultural land, irrigation, soil physics, hydrology, and agro-informatics in the Graduate School of Agriculture and Life Sciences (GSALS), The University of Tokyo. The group which is a group member of the NPO "Resurrection of Fukushima" plays an important role carrying out field and soil surveys.

16.2.3 Campus Group "Madei"

In September 2012, the mayor of Iitate Village and the Dean of GSALS exchanged a letter of research cooperation between Iitate Village and GSALS. Based on the letter, a volunteer group was organized by employees of GSALS, who want to support the activities of the NPO without going to Iitate Village. The group is named "Madei" which is a dialect of Iitate Village meaning "politely". Members of the group meet after work every Tuesday, and pack samples, such as soil and plants collected by the NPO members in Iitate Village on weekends, into vial tubes, then send the tubes to a researcher at the Radioisotope facility (RI facility) on campus.

16.2.4 Rehabilitation Support Project (University of Tokyo Agricultural Life Science Graduate School of Grants-in-Aids Rehabilitation Support Project 2017) of the Graduate School of Agriculture and Life Sciences (GSALS), The University of Tokyo

This is a project which was initiated by the Dean of the GSALS immediately after the earthquake. Numerous researchers are involved in field surveys and reconstruction assistance in various forms. Under the project, measurement of the radioactive cesium concentration of samples collected in Iitate Village is carried out at the Radioisotope facility. This project regularly holds "Research report meetings on the influence of radioactivity on agricultural and livestock products, etc." The meeting has been held 13 times as of January 2017.

16.3 Development of Agricultural Land Decontamination Method by Farmers Themselves (Mizoguchi 2013)

A technical document for agricultural land decontamination (Ministry of Agriculture, Forestry and Fisheries 2013) (MAFF; Ministry of Agriculture, Forestry and Fisheries, 2013) recommends three decontamination methods according to the level of farmland contamination. They are (1) the topsoil stripping-off method (>10,000 Bq/kg soil), (2) the muddy water method (>5000 Bq/kg soil), and (3) reverse cultivation method (<5000 Bq/kg soil). However, only the topsoil stripping-off method has been used in the field, which has resulted in the contaminated soil being stored in black container bags; these bags are currently being stored on agricultural land that is easily accessible (Resurrection of Fukushima 2017). These storage places are referred to as a "temporary-temporary storage place" because a temporary/final storage place has still not been determined by the government (Photo 16.1).

We focused on the property that cesium is fixed onto clay particles. Based on the idea that "the decontamination of agricultural land is equivalent to the removal of clay particles", we developed several agricultural land decontamination methods that farmers themselves can easily practice (Mizoguchi 2014a), which are different from the methods proposed by MAFF.

Photo 16.1 Contaminated soil bags piled up in a temporary-temporary storage place (Resurrection of Fukushima 2017). (Photo and video provided: NPO Resurrection of Fukushima)

16.3.1 Muddy Water Flushing Out Method with a Hand Weed Machine

In April 2012, we conducted an experiment at a paddy field in Sasu district in Iitate Village, Fukushima Prefecture. We irrigated water into a 5 m × 10 m paddy field and mixed the surface soil of the paddy field with a hand weed machine (this machine was often used after rice planting more than 50 years ago). Next, a tennis court brush was used to flush out the muddy water containing radioactive cesium. Using this method, it was found that 80% of radioactive cesium can be removed.

With this method, muddy water was flushed out into a 1 m-deep drainage channel which was dug around the paddy field. After 3 months, the muddy water which was removed from the paddy field penetrated underground, and only dry clay remained on the bottom of the drainage channel. We measured radioactive cesium concentration in this dry clay and found that radioactive cesium remained in the soil layer to a depth of 6–7 cm from the bottom of the drainage channel (Mizoguchi 2013a). This is because the clay particles which contain radioactive cesium were trapped in the soil pores by the soil filtration function (Yahata 1980). To explain this principle, we video recorded a simple experiment using a plastic bottle and uploaded it to YouTube as teaching material for the public (Kato et al. 2016).

16.3.2 Muddy Water Flushing Out Method with a Tractor (Mizoguchi 2014b)

The muddy water flushing out method with a hand weed machine cannot be used for the decontamination of a large area. In addition, paddy fields have been abandoned without cultivation for 6 years. Therefore, grasses and shrubs extend roots into the soil, and wild boars have been digging the ground surface in some paddy fields. As a result, the radioactive cesium accumulated in the surface layer has been mixed. In May 2013, we tested another method using a tractor at a paddy field in Komiya district in Iitate Village, Fukushima Prefecture. We mixed soil with water in the paddy field using a tractor and flushed out the muddy water into a moat that was dug at a corner of the paddy field (Photo 16.2).

The muddy water flushing out method using either the hand weed machine or tractor corresponded to Method 2 in the MAFF manual (Ministry of Agriculture, Forestry and Fisheries 2013). With both our methods, the settling time of clay particles in water is important to collect clay particles efficiently. Also, as muddy water penetrates into the bottom of the moat, clay particles must be trapped in the soil pores. However, radioactive cesium might be adsorbed to organic matter which sometimes forms a colloid. In that case, there is a possibility that the settling velocity of the particles and the trapping rate of the particles in the soil pores may change. Therefore, we need to elucidate the phenomena of colloid migration and clogging in the soil.

Photo 16.2 Decontamination of farmland by muddy water flushing out method with a tractor

16.3.3 Burial Method of Contaminated Soil-Madei Method (Mizoguchi et al. 2013)

According to a previous study (Miyazaki 2012), if contaminated soil is covered over with a 50 cm layer of non-contaminated soil, the radiation dose can be attenuated from 1/100 to 1/1000. In December 2012, we conducted an in-situ burial experiment of contaminated soil at a paddy field in Sasu district in Iitate Village, Fukushima Prefecture. We stripped 5 cm of contaminated topsoil from a paddy field (10 m × 30 m), and dug a trench (2 m wide, 30 m long, 1 m deep) in the center of the paddy field. Then, we buried the contaminated soil at a depth of 50–80 cm in the trench and covered the trench with non-contaminated soil to a depth of 50 cm above the contaminated soil (Mizoguchi et al. 2013; Photo 16.3).

This method is a combination of Method 1 and 2 in the MAFF manual. We named this method "Madei construction method".

We continued the cultivation test of rice in this paddy field, and confirmed (Ii et al. 2015) that the radioactive cesium concentration of brown rice was below the Japanese standard value of 100 Bq/kg. However, since this method is not an approved decontamination method in Japan, the local government requested us to bury the rice harvested in 2012 and 2013 under the ground. However, after tenacious negotiations between the NPO and the local government, the brown rice harvested in 2014 was permitted to receive the official inspection under the management

Photo 16.3 Burial method of contaminated soil

of the prefectural government and succeeded to pass the inspection. In this way, it was proved that safe rice can be produced by our own decontamination method.

16.3.4 *Monitoring of Buried Contaminated Soil (Mizoguchi et al. 2015)*

Although we proved the safety of rice production in the paddy field decontaminated by the Madei construction method, we also needed to prove that radioactive cesium cannot leak out from the buried contaminated soil. Therefore, we set an observation well using a PVC pipe in a paddy field where contaminated soil was buried under the ground, and periodically measured the soil radiation dose by inserting a radiometer into the well. Figure 16.2 shows the soil radiation dose (cpm) measured by a new 10-unit Geiger–Müller tube radiometer that we developed. Solid lines are

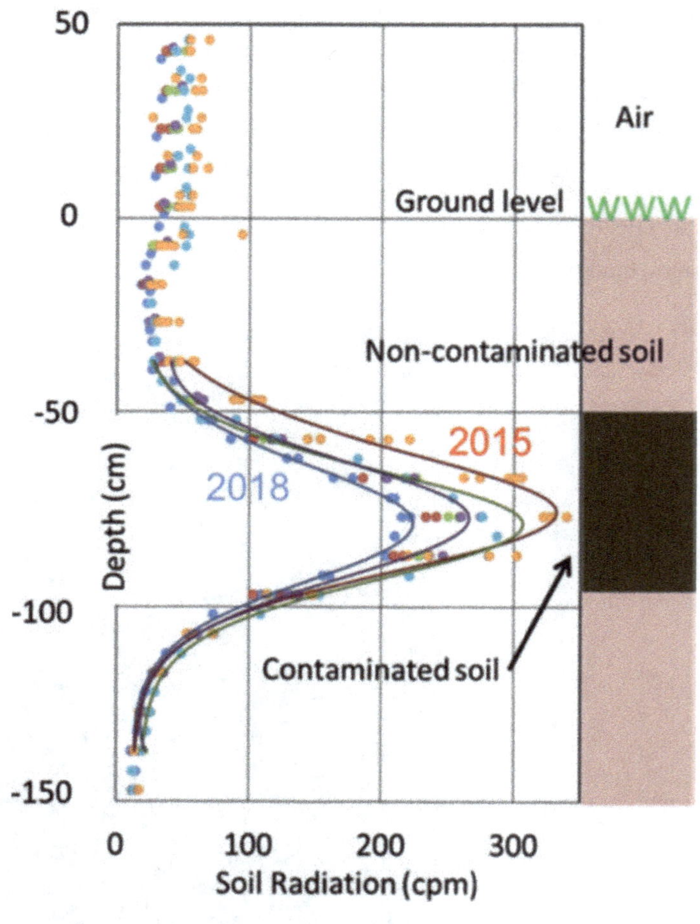

Fig. 16.2 Profiles of radiation dose in the soil. All data measured every 6 months from March 2015 are plotted. Solid lines are fitting curves of the data measured in March 2015, 2016, 2017 and 2018. The Cs-contaminated soil was buried to a depth of about 50–90 cm under the ground

fitting curves of data measured in March 2015, 2016, 2017 and 2018. The radiation dose showed a normal distribution shape with a peak at a depth of about 70 cm. This is because the radiation from the contaminated soil buried in 50–90 cm is attenuated in the soil.

We need to continue monitoring buried contaminated soil over the long-term. On the other hand, a study which predicts long-term cesium migration in the soil using a simulation model is needed because it is difficult to directly demonstrate that radioactive cesium does not move.

16.3.5 Environmental Monitoring in the Iitate Village (Mizoguchi 2013b)

Environmental monitoring of the radiation level in the village is important to evaluate the decontamination effect and to judge when the villagers should return to the village. In particular, it is important to accumulate scientific data such as the relationship between wind direction/wind speed and radiation dose, the relationship between precipitation intensity and the amount of outflow/turbidity of rivers. Based on those data, we must discuss the villagers' lifestyle and agricultural revitalization after returning to the village. We added a radiation meter to our field monitoring system (Mizoguchi and Ito 2015) (FMS) that we developed in our laboratory, and have been conducting environmental monitoring at several points, such as paddy fields, forests and houses, in Iitate Village since October 2011. Some meteorological sensors and monitoring equipment that we use here were donated to the Japanese Society of Irrigation, Drainage and Reclamation Engineering (JSIDRE) by sensor/equipment developers after the 2011 earthquake.

Figure 16.3 is an example of weather and radiation dose data. Because the normal radiation sensor was expensive and difficult to obtain after the nuclear power plant accident, we measured radiation dose using a cheap photodiode-semiconductor sensor at the garden of a residence in Iitate Village. Therefore, the sensor sometimes picked up noise and the data were unstable. However, since the radiation dose was observed automatically at the fixed point, it was effective to see the trend of the change of radiation. By comparing with the on-site image data, we found that the radiation dose decreases due to snow cover (Fig. 16.4). In addition, we found a tendency that the radiation dose will rise on dry and fine days. Currently, the price of the radiation sensor has decreased, so we have upgraded to a more accurate and stable sensor and are continuing to observe the radiation dose at the same site. We also developed a portable device (Suzuki et al. 2015) that measures soil radiation in cooperation with radiation measurement engineers that we became acquainted with after the accident. These devices are now used by villagers to inspect their own farmland by themselves.

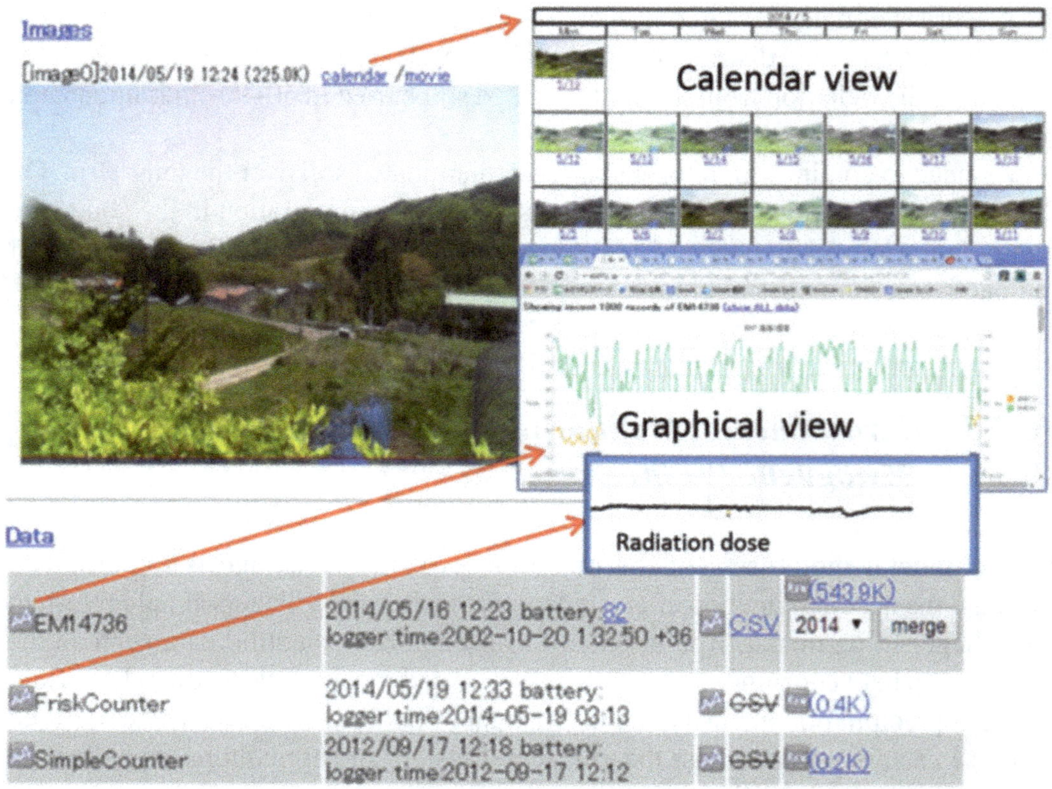

Fig. 16.3 Environment monitoring in Iitate Village using the field monitoring system (FMS). Data such as radiation dose, air temperature, humidity, precipitation, wind direction/wind speed, solar radiation, etc. are automatically transmitted to the cloud server together with the image data every day

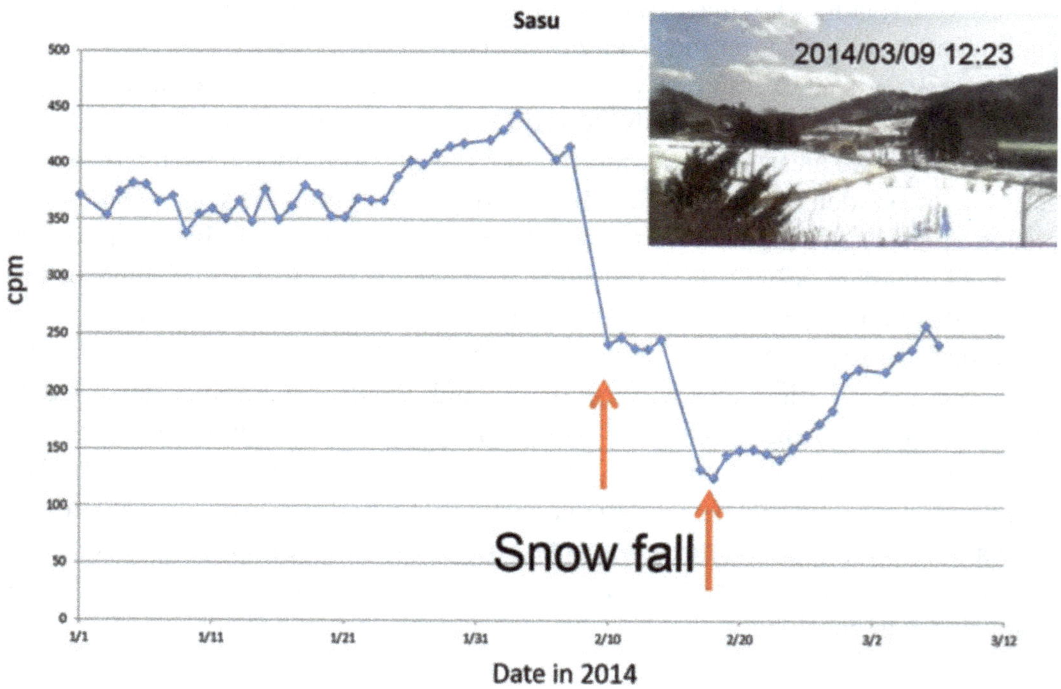

Fig. 16.4 Effect of radiation shielding by snow

16.4 The Current Status of Agricultural Land After Decontamination

After decontamination work, to decrease the radiation dose in the environment the topsoil in paddy fields in Iitate Village was replaced with mountain soil which is not fertile (Photo 16.4).

In 2015, we conducted a survey of paddy fields (Mizoguchi 2017) with regards to (1) radioactive cesium concentration in agricultural soil, (2) the thickness of the covered soil, (3) soil hardness, and (4) drainage characteristics of the soil. The following were our findings:

(1) The degree of decontamination varied depending on the location, even in the same paddy field.
(2) The levee of the paddy field was not decontaminated.
(3) The thickness of the covered soil is not necessarily uniform.
(4) In the paddy field after decontamination, a hard layer was formed at the boundary (5–10 cm) between the covered soil and the original paddy soil in addition to the conventional tiller layer (20–30 cm).
(5) Due to poor drainage, many paddy fields become flooded after heavy rain. The poor drainage is due to a hard layer formed by a heavy machinery used for decontamination work. Agricultural land regeneration based on these findings is an important mission for agricultural engineers.

Photo 16.4 Soil survey after decontamination work. The topsoil in paddy fields in Iitate Village was replaced with mountain soil

16.5 Rural Reconstruction Scenario

16.5.1 Creation of a New Japanese Agriculture Model (Mizoguchi 2015b)

In the past, the ordered beautiful paddy fields in Japan were constructed by public land improvement projects. Immediately after the land improvement, soil fertility was normally low. Therefore, high crop yields were not expected for the first few years after construction. Perhaps, farmland in Fukushima after decontamination will recover after several years. In other words, we must continue to improve the soil for several years considering the usage of the farmland.

The biggest problem is to keep farmers motivated to continue agriculture in spite of the hard conditions. The future of agriculture in Japan is uncertain, not only in Fukushima. There is also the damaging rumours about the safety of agricultural products of Fukushima. However, hard times bring opportunity! We will be able to use Fukushima's world-famous name to sell Fukushima's agricultural and livestock products to the world. Through discussions among all stakeholders, we need to develop novel schemes to encourage companies and new farmers to come to Fukushima. Moreover, we can control the environment inside the greenhouses and cultivate flowers and vegetables while using state-of-the-art ICT technology. And, we need to think positively that we can create a new Japanese agriculture model whose products will be internationally competitive.

16.5.2 Human Resource Development

The important point in thinking about regional reconstruction after the nuclear accident is to nurture young people who overcome Fukushima's handicap. We sometimes take students on field trips to Iitate Village. In a class lecture (Mizoguchi 2017) back on campus, we talk to the students about the mission of researchers to solve on-site problems in Iitate Village. After visiting the Iitate Village, the student's opinions changed greatly.

In January 2016, a student-organized agricultural food exhibition was held at the Faculty of Agriculture, the University of Tokyo. A graduate student whose hometown is Iitate Village planned tasting sales with young people of the same generation with the aim to promote the recovery of Iitate beef. In addition, female college students in Yokohama also developed a cake using vegetables test-cultivated in Iitate Village as an attempt by Community Supported Agriculture (CSA) to connect producers and consumers. In this way, the young generation is also interested in rural reconstruction. The university needs to to develop human resources among the young who can draw a scenario of rural reconstruction from a comprehensive perspective and work on problem-solving with local people.

16.6 Conclusion

In this chapter, I reviewed the challenges of agricultural land remediation and the renewal of agriculture that I conducted with the collaboration of local farmers, NPO members and researchers since I entered Iitate Village immediately after the nuclear accident. In addition, I proposed some of my own ideas regarding rural reconstruction.

I believe we have done almost all that we can to decontaminate the agricultural land affected by radioactive cesium fallout. However, we often heard from evacuated villagers that "What we are most sorry is that the family gathering place is gone". From the viewpoint of rural planning by a sociological approach, we need to still consider more ways to revive village life.

Acknowledgments I would like to thank the NPO members (President: Mr. Yoichi Tao) for their collaboration, and the local farmers (Mr. Muneo Kanno, Mr. Kinichi Okubo, Mr. Takeshi Yamada, Mr. Keiichi Kanno) who provided us land in Iitate Village for the past 7 years to carry out our field experiments.

References

Fukushima Reconstruction Agricultural Engineering Meeting. Tackling agricultural restoration from decontamination of agricultural land – the way to realistic decontamination technology development utilizing the research of "Agricultural Engineering". http://utf.u-tokyo.ac.jp/2013/07/post-43c5.html (As of April 2017)

Ii I, Tanoi K, Uno Y, Nobori T, Hirose A, Kobayashi N, Nihei N, Ogawa T, Tao Y, Kanno M, Nishiwaki J, Mizoguchi M (2015) Radioactive Caesium concentration of lowland rice grown in the decontaminated paddy fields in Iitate village in Fukushima. Radioisotopes 64(5):299–310

Iitate Village. Public information "Iitate", issue 1, April (2017)

Kanno M, Mizoguchi M (2014) Soil decontamination in Iitate by cooperation of village, NPO and academic. Trends Sci 19(7):36–39

Kato C, Sakai M, Nishiwaki J, Tokumoto I, Hirozumi T, Watanabe K, Shiozawa N, Mizoguchi M (2016) Classes on reconstruction agriculture at elementary school in Fukushima Prefecture. Water Land Environ Eng 84(6):15–18

Ministry of Agriculture, Forestry and Fisheries (2013) Technical document on measures against decontamination of agricultural land, February 2013

Tsuyoshi Miyazaki (2012) Attenuation effect of cesium radiation by soil, toward recovery of agriculture, forestry and fishery industry from eastern Japan great earthquake disaster – technical recognition for damage recognition and understanding, reconstruction-, Association of Japanese Agricultural Scientific Society, 21. http://www.ajass.jp/pdf/recom2012.1.13.pdf (As of April 2017)

Mizoguchi M (2013a) Remediation of Paddy soil contaminated by Radiocesium in Iitate Village in Fukushima prefecture. In: Agricultural implications of the Fukushima nuclear. Springer, Tokyo, pp 131–142

Mizoguchi M (2013b) Soil decontamination and radiation monitoring. J Soc Instrum Control Eng 52(8):730–735

Mizoguchi M (2013) Development of agricultural land decontamination method that can be done by farmers themselves, soil science of radiation decontamination – from forests, fields, fields to home garden, Scientific meeting series 20. Japan Science Support Foundation, pp 135–151

Mizoguchi M (2014a) Agricultural restoration by soil physicists – agricultural land decontamination for farmers by farmers, Columbus (March issue), pp 22–24

Mizoguchi M (2014b) Radioactive matter problem – what should soil physics do? J Jpn Soc Soil Phys 126:3–11

Mizoguchi M (2015a) The farmer's spirit will want to decontaminate his own farmland by himself, how to act after the nuclear accident – the trajectory of experts and victims, the University of Tokyo Hospital, pp 45–61

Mizoguchi M (2015b) Finding the possibility of Japanese style agriculture in Iitate Village by agricultural engineering for reconstruction, Columbus (May issue), pp 20–22

Mizoguchi M (2017) Current status and issues of radioactive contamination in Fukushima. J Jpn Soc Soil Phys 135:5–7

Mizoguchi M. Lectures related to Iitate Village, http://www.iai.ga.a.u-tokyo.ac.jp/mizo/edrp/fukushima/Iitate-lec14.html (As of April 2017)

Mizoguchi M, Ito T (2015) Field monitoring technology to change agriculture rural. Water Land Environ Eng 83(2):3–6

Mizoguchi M, Ito T, Tao Y (2013) Burial experiment of soil contaminated by radiocesium at a paddy field in Iidate Village, Fukushima Prefecture. Abstracts of Water, Land and Environmental Engineering meeting

Mizoguchi M, Itakura Y, Kanno M, Tao Y (2015) Radiation measurement in paddy soil layer that was buried contaminated topsoil. Abstracts of Water, Land and Environmental Engineering meeting

Resurrection of Fukushima. http://www.fukushima-saisei.jp (As of April 2017)

Resurrection of Fukushima: Reality of provisional casual storage place. https://www.facebook.com/FukushimaSaisei/videos/1054291244592879/ (As of April 2017)

Suzuki S, Itakura Y, Mizoguchi M (2015) Development and application of the device for measuring the vertical distribution of radiocaesium concentration in soil. Abstracts of Water, Land and Environmental Engineering meeting

University of Tokyo Agricultural Life Science Graduate School of Grants-in-Aids Rehabilitation Support Project. http://www.a.u-tokyo.ac.jp/rpjt/ (As of April 2017)

Yahata T (1980) Matters related to filtration function. Physics of soil (3rd press), The University of Tokyo Press, Tokyo, pp 142–156

Yokokawa H, Mizoguchi M (2013) Collaboration structure aimed at resurrection of Iitate Village. Phys Properties Soil 125:53–54

Chapter 17
Radiocesium Contamination on a University Campus and in Forests in Kashiwa City, Chiba Prefecture, a Suburb of Metropolitan Tokyo

Kenji Fukuda

Abstract Kashiwa, a city in Chiba Prefecture, became the most contaminated suburb of Metropolitan Tokyo after the Fukushima Daiichi nuclear accident. The Kashiwa Campus of the University of Tokyo and nearby urban forests were surveyed to examine the distribution of radiocesium in the aboveground parts of trees, turf grass, and soil. The air dose rate 1 m aboveground in the summer of 2011 was 0.3–0.6 μSv/h and more than 90% of the radiocesium was in the surface soil. A nursery lawn was effectively decontaminated by removing the turf and surface soil using a sod cutter. In the forests, the radiocesium concentration was higher in the leaves of evergreen trees and outer bark of trees, while the total amount of radiocesium in the aboveground parts of trees was less than 10% of the amount in the surface soil. Therefore, decontamination by cutting trees would not be effective. The decrease in the radiocesium concentration in the surface soil could be explained by natural decay, while the effects of cesium movement to deeper soil were not prominent.

Keywords Contamination in a University Campus · Vegetation cover · Cesium distribution in trees · Soil contamination · Decontamination of lawn

17.1 Introduction

Kashiwa is a city located in northwestern Chiba Prefecture. One of the satellite cities around Metropolitan Tokyo, it is located about 30 km northeast of the center of Tokyo. The landscape is a mosaic of urbanized areas around train stations, residential areas, farmland, factories, and secondary forests. Kashiwa and the surrounding

K. Fukuda (✉)
Department of Forest Science, Graduate School of Agricultural and Life Sciences, The University of Tokyo, Tokyo, Japan
e-mail: fukuda@fr.a.u-tokyo.ac.jp

T. M. Nakanishi et al. (eds.), *Agricultural Implications of the Fukushima Nuclear Accident (III)*, https://doi.org/10.1007/978-981-13-3218-0_17

Tokatsu area suffered from the most serious radiocesium contamination in suburban Tokyo after the Fukushima Daiichi accident in March 2011. The estimated contamination by ^{134}Cs and ^{137}Cs totaled 60–100 kBq/m^2 in September 2011 according to a contamination map produced by airplane monitoring (MEXT 2016).

The Kashiwa Campus is one of the three main Campuses of the University of Tokyo (UTokyo Campus hereafter), and it has a children's nursery (Kashiwa Donguri Day Nursery). To ensure safety, the radiocesium contamination on the Campus and in nearby forests was surveyed in the summer of 2011, and decontamination of the nursery lawn was attempted in October 2011 (Fukuda et al. 2013a; Kitoh 2013). To estimate the radiocesium distribution in suburban Satoyama forests (i.e., located within 1 km from Campus), the radiocesium contamination was surveyed in the aboveground parts of thinned trees and in litter and surface soil samples collected from the winter of 2011 to the winter of 2014. The details of Satoyama forests were previously reported in Fukuda et al. (2013b).

17.2 Study Area and Methods

The study was conducted in three areas: the UTokyo Campus, Oaota Forest, and Konbukuro Park, which are 1–3 km distant from one another (Fig. 17.1). On the UTokyo Campus, the air dose rate was measured in a remnant secondary deciduous and evergreen oak forest, Campus green areas (planted deciduous and evergreen trees), lawns and grassland, tree pits planted with *Zelkova serrata* and *Rhododendron indicum*, and paved areas. The air dose rate was measured 1 m and 1 or 5 cm above the ground using a NaI(Tl) scintillation survey meter (Hitachi, Aloka TCS-171 or Clearpulse, Mr. Gamma A2700). Soil cores up to 20 cm in depth were taken from several points on the Campus using a liner hand auger (Daiki, DIK-100C). A line transect was set from a forest patch and bamboo stand to grassland in an unused area on the UTokyo Campus in the summer of 2013, and the air dose rate was measured, and soil cores collected (100 mL: 0–5 cm depths) at 10-m intervals.

In Konbukuro Park, the land cover is a mosaic of conifer plantations, deciduous and evergreen broadleaved forests, and wetland vegetation. There, the air dose rate was measured along a footpath, and leaf, litter, and soil samples were taken.

Oaota is also a mosaic of small patches of private conifer and deciduous broadleaf forests with a total area of about 50 ha. We set study plots in a conifer–deciduous mixed stand, deciduous broadleaf stand, grassland, and bamboo bush. The air dose rate was measured in each plot, and leaf and litter samples were taken. Leaf samples, mushrooms, and insects were collected randomly from these forest plots.

The radiocesium concentration of each sample was measured using a NaI(Tl) scintillation counter (Hitachi, Aloka Auto-well gamma counter ARC-370 M) in the Radioisotope Laboratory of the Graduate School of Frontier Science, UTokyo.

In addition, some trees were felled in the conifer and deciduous stands to estimate the aboveground contamination of forest stands, and 100-mL soil core samples were collected. After thinning in Oaota Forest in the winter of 2011, six trees (two

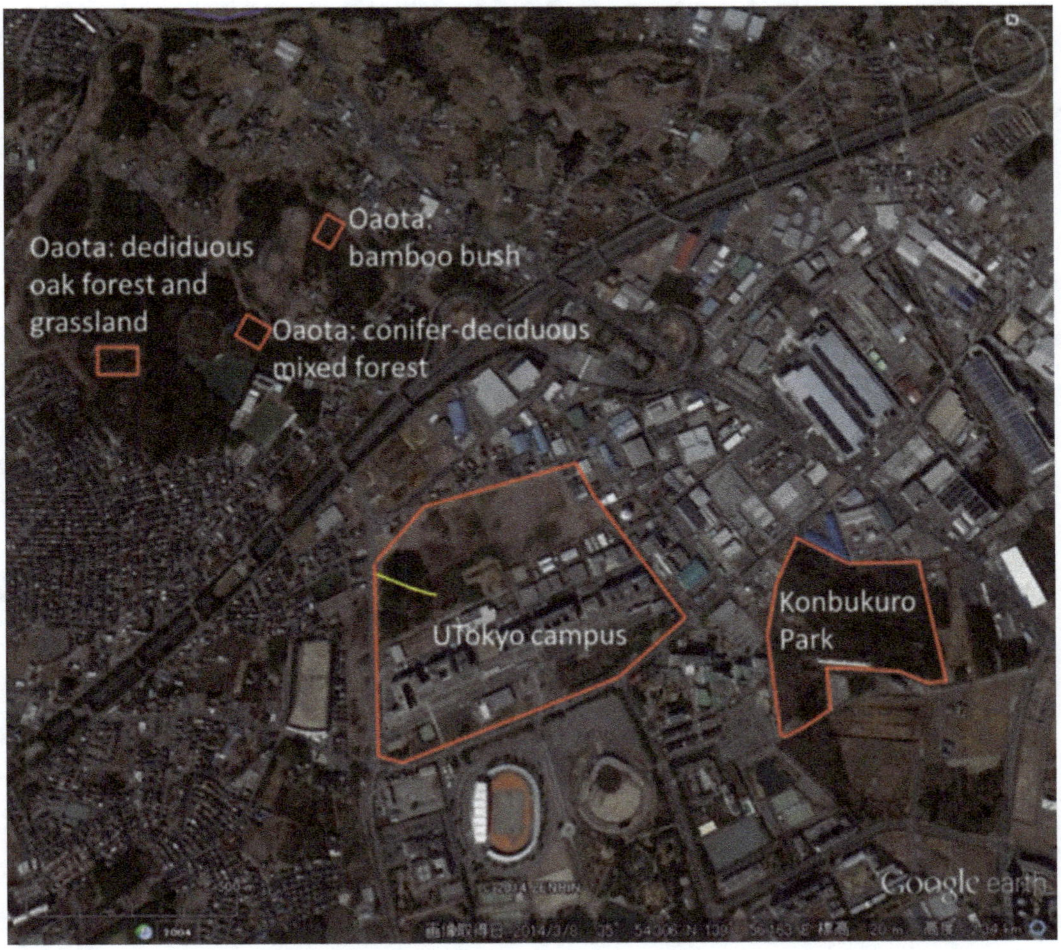

Fig. 17.1 Study sites (plotted on a Google Earth image)
The yellow line in UTokyo Campus shows the line transect

Chamaecyparis obtusa and two *Carpinus tshonoskii* from a mixed conifer–deciduous stand and two *Quercus serrata* from a deciduous stand) were sampled to estimate the distribution of radiocesium in standing trees. Immediately after the trees were felled, wood discs were sampled at 0.3 and 1.3 m aboveground and at 4-m intervals to the tree top, and one to three 2-m-long branches were sampled from the top and bottom of each crown. The disks were oven-dried and divided into outer bark, inner bark, current-year annual ring, sapwood, and heartwood. The branches were divided into branch, old leaves, new leaves, cones, and winter buds and oven-dried. The dried samples were cut into small pieces and milled in a coffee mill, and the radiocesium concentration was measured.

Litter samples of the L- and FH-layers were taken from nine $0.2 \times 0.2\text{-m}^2$ quadrats set in two forest stands (a mixed conifer–deciduous stand and a deciduous oak stand) in Oaota Forest in the winters of 2012, 2013, and 2014. At the same time, a 100-mL soil core was taken from the center of each quadrat, divided into 1-cm depths, and weighed after oven drying. The radiocesium concentrations of the litter and soil samples were measured to estimate the soil contamination by radiocesium concentration per ground area.

17.3 Air Dose Rate and Soil Contamination in 2011 in Relation to the Land Cover

Table 17.1 summarizes the air dose rates measured from August to November of 2011. On the UTokyo Campus, the Campus greenery and lawns showed a roughly uniform contamination level of 0.3~0.4 μSv/h at 1 m aboveground regardless of the land use, such as forest floor, forest edge, lawn, and bare ground. The air dose rate was lowest on the granite benches, where rainwater washed the polished smooth surface (0.28 μSv/h), while asphalt and the concrete pavement had air dose rates similar to or higher than those of the forest floor and lawn. The rough surface asperity of the pavement served as a reservoir for radiocesium-containing mud particles. A drainage pit and the pavement edge receiving rainwater from paved areas had the highest air dose rates, where clay particles and organic compounds in soil concentrated radiocesium.

Table 17.1 Air dose rate (μSv/h) and land cover at the study sites

Land cover	Month	Number of points	1~5 cm	1 m
UTokyo Campus	Aug.–Nov. 2011			
Asphalt & concrete pavement		45	0.54 ± 0.19	0.38 ± 0.10
Natural stone bench		6	0.28 ± 0.09	0.28 ± 0.09
Drainage, Pavement edge		92	0.99 ± 0.58	0.42 ± 0.12
Water canal's edge		4	0.29 ± 0.06	0.26 ± 0.04
Lawn		88	0.52 ± 0.19	0.38 ± 0.08
Bare soil		11	0.48 ± 0.12	0.31 ± 0.05
Tree planting pit		29	0.58 ± 0.34	0.36 ± 0.08
Forest edge		13	0.46 ± 0.09	0.36 ± 0.03
Forest center		37	0.63 ± 0.40	0.40 ± 0.09
Konbukuro Park	Dec. 2011			
Grassland		5	0.39 ± 0.04	0.35 ± 0.03
Pleioblastus chino bamboo bush		2	0.35 ± 0.00	0.33 ± 0.01
Evergreen forest stand		15	0.36 ± 0.05	0.32 ± 0.03
Decidous forest stand		15	0.36 ± 0.06	0.32 ± 0.03
Forest edge		27	0.36 ± 0.10	0.30 ± 0.05
Side of the drainage canal		8	1.16±1.05	0.78 ± 0.31
Oaota forest				
Conifer-deciduous mixed: thinned	July 2013	56	0.28 ± 0.11	0.14 ± 0.01
Deciduous oak: thinned	Nov. 2012	18	0.34 ± 0.13	0.19 ± 0.02
Deciduous oak: control	June 2013	15	0.28 ± 0.03	0.23 ± 0.02

Modified from Fukuda et al. (2013b)

In Konbukuro Park, the air dose rate was 0.30~0.35 μSv/h at 1 m aboveground regardless of forest type. A drainage canal at the northern edge of the park had the highest contamination, 0.78 μSv/h, after collecting rainwater from surrounding areas including a road gutter. Benten Pond, a spring which is supplied by shallow groundwater, had the lowest radiocesium concentration in the sediment (Fukuda et al. 2013b).

In Oaota Forest, a conifer stand and a deciduous stand thinned after the Fukushima accident had lower air dose rates (0.1~0.2 μSv/h) than did non-thinned stands and other forests in Konbukuro Park and UTokyo.

Figure 17.2 shows the radiocesium measured in soil cores sampled on the UTokyo Campus in the summer of 2011. Soil contamination was higher at the edges of the lawn and a planting pit, which receive rainwater from pavement and asphalt roads, than it was in the center of the lawn and pit, respectively. In an area of planted trees, deposition was higher at the forest edge. In the lawn center, radiocesium deposition was concentrated at 0~1-cm depth, while in the forest, where there was almost no herbaceous vegetation or litter layer, radiocesium was distributed to 5 cm or deeper. The turfgrass shoots, thatch, and root mat of the lawn seemed to have intercepted radiocesium-contaminated rain effectively, preventing penetration into the soil. This suggests that the decontamination of lawns would be relatively easy, by removing the turf grass and shallow surface soil.

17.4 Radiocesium Concentrations in Biological Samples

Table 17.2 summarizes the radiocesium concentrations in living leaves, fungi, and litter samples. Old leaves of evergreen conifers and branches of some deciduous trees and shrubs that were contaminated directly after the Fukushima accident had higher concentrations, while the leaves of deciduous trees, which flushed after the accident, had lower values. New leaves of evergreen conifers and broadleaves had higher values than did leaves of deciduous trees, indicating the translocation of radiocesium from contaminated old shoots to newly developed shoots (IAEA 2006; Tagami et al. 2012; Yoshihara et al. 2013; Masumori et al. 2015a, b; Tagami et al. 2012; Takata 2013, 2015). Mushrooms contained a range of radiocesium concentrations, with some wood-rotting species and ectomycorrhizal species having extraordinarily high levels, 36–60 kBq/kg (dry weight), indicating bioconcentration of radiocesium by these fungi (*e.g.*, Yoshida and Muramatsu 1994; Yamada 2013). In Oaota Forest, citizen volunteers help with the management of the forest, as contracted between a volunteer NPO (a non-profit organization) and the landowner. The volunteers used the thinned oak wood as fuel for BBQs and as bed logs for cultivating shiitake mushrooms on the forest floor. The shiitake mushrooms cultivated there had higher radiocesium concentrations than the government reference level (0.1 kBq/kg).

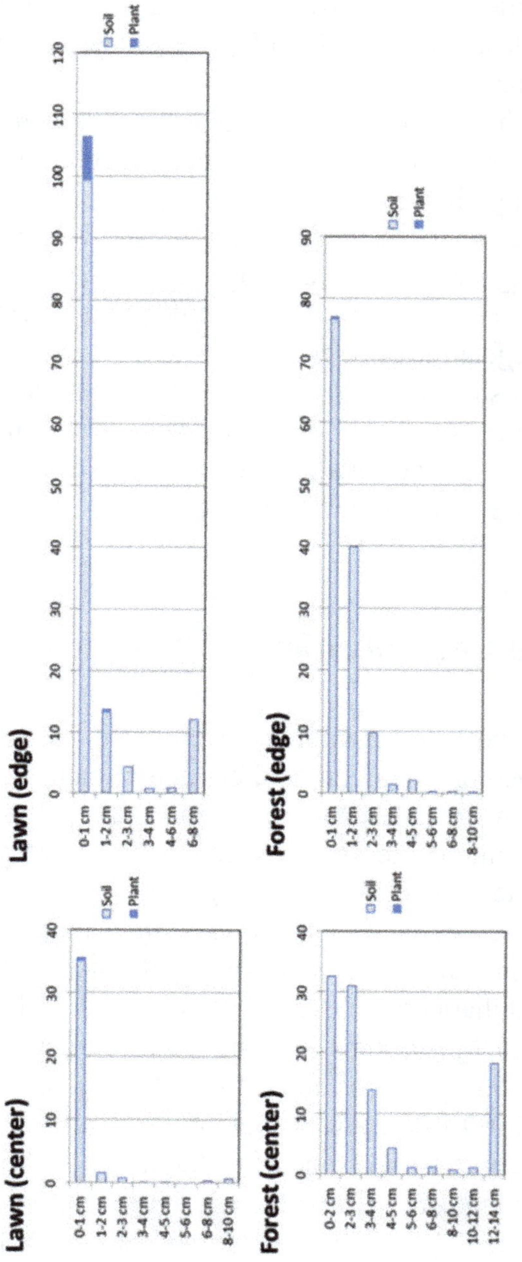

Fig. 17.2 Vertical profile of radiocesium deposition (kBq/m^2) in soil cores taken from lawns and forests on the UTokyo Campus in the summer of 2011. (Modified from Fukuda et al. 2013a)

Table 17.2 Radiocesium concentrations (kBq/kg) in biological samples collected in 2011–2012

Life form	Species		Autumn to winter 2011					Spring to Autumn 2012		
			Oaota			Konbukuro		Oaota		UTokyo
			Mixed	Oak	Bamboo	Conifer	Deciduous	Oak	Bamboo	Campus
Evergreen conifers	*Cryptomeria japonica*	Branch		5.87		3.48	6.70			
		Old needle and twig		4.91		2.42	3.12			
		Current-year needle		7.97		1.99	1.16			
	Chamacyparis obtusa	Branch				1.63				
		Twig				3.55				
		leaf				4.86				
	Chamaecyparis pysifera	Branch				4.11	1.63			
		Twig				8.77	3.55			
		Leaf				6.52	4.86			
Evergreen broadleaves	*Quercus myrsinaefolia*	Old branch								2.53
		1-year branch								1.68
		1-year leaf				3.31				1.60
		Current twig								1.23
		Current leaf								1.34
	Neolitsea sericea	1-year branch				2.57				
		1-year leaf				2.45				
		Current twig				0.68				
		Current leaf				2.34				
		Leaf gall				13.34				
	Aucuba japonica	Branch		1.42						
		Leaf		1.40						
		Fruit		1.75						

(continued)

Table 17.2 (continued)

Life form	Species		Autumn to winter 2011					Spring to Autumn 2012		
			Oaota			Konbukuro		Oaota		UTokyo Campus
			Mixed	Oak	Bamboo	Conifer	Deciduous	Oak	Bamboo	
Deciduous broadleaves	*Carpinus tschonoskii*	Branch		0.57						
		Current twig		0.50						
		Leaf		0.33						
	Castanea crenata	Branch		0.49						1.11
		Leaf		0.56						2.76
		Acorn	0.38	0.84						
	Quercus serrata	Branch		0.63						
		Leaf		1.10						
	Robinia pseudoacacia	Branch								ND
		Leaf								0.84
	Celcis sinensis	Branch								2.80
	Prunus buergeriana	Branch								0.80
		Current twig								0.54
		Leaf								0.84
	Prunus grayana	Branch								3.68
		Leaf								2.16
	Viburnum dilatatum	Branch								5.65
		Current twig								1.21
	Zanthoxylum piperitum	Branch								4.41
		Leaf								1.62
	Rosa multiflora	Branch								ND
		Leaf								ND
Dwarf bamboo	*Pleioblastus chino*	Stem			1.53				1.71	
		Leaf			2.10		1.54		0.58	

Leaf litter	*Cryptomeria japonica*	Litter	4.99	9.60		7.67	21.00	2.27		
	Chamaecyparis obtusa	Litter	5.93	8.19				3.52		
	Carpinus tschonoskii	Litter	0.85	3.19						
	Quercus serrata	Litter	1.13	1.21		0.74	2.81	0.37		
	Quercus acutissima	Litter		1.00				0.20		
	Castanea crenata	Litter	1.98							
	Aphananthe aspera	Litter	0.32				2.79			
	Swida controversa	Litter	0.67					0.19		
	Magnolia kobus	Litter					1.94			
	Parthenocissus tricuspidata	Litter		2.44						
	Pleioblastus chino	Litter							0.24	
		Chip compost							1.62	
Saprophytic fungi	*Lentinula edodes* (Shiitake) on *Quercus serrata* wood	Pileus		6.07						
		Pileus						2.28		
		Stipe						0.65		
		Outer bark		1.01						
		Inner bark		0.29						
		Wood		0.40						
	Mycena galericulata on unidentified wood	Fruit body						31.41		
	Hypholoma fasciculare on *Cryptoemeria japonica* wood	Fruit body				38.14				
		Bark				8.42				
		Wood				3.34				
	Antrodiella gypsea on *Cryptomeria japonica* wood	Fruit body				5.76				
		Bark				3.38				
		Wood				0.49				

(continued)

Table 17.2 (continued)

Life form	Species		Autumn to winter 2011					Spring to Autumn 2012		
			Oaota			Konbukuro		Oaota		UTokyo
			Mixed	Oak	Bamboo	Conifer	Deciduous	Oak	Bamboo	Campus
	Trametes hirsuta on *Qercus serrata* wood	Fruit body						1.97		
	Trametes versicolor on *Q. serrata* wood	Fruit body						1.85		
Ectomycorrhizal fungi	*Tylopilus rigens*	Fruit body						1.02		
		Soil						2.39		
	Entoloma sp.	Fruit body								3.40
	Amanita sp.1	Fruit body								ND
	Amanita sp.2	Fruit body								ND
	Amanita sp3.	Pileus								8.94
		Stipe								3.26
		Pileus								1.64
		Stipe								1.41
		Immature volva								0.80
		Immature pileus								0.72
	Lactarius subindigo	Fruit body								18.61
		Litter								4.55
	Russula bella	Fruit body								4.40
	Cortinarius sp.	Fruit body								60.88
	Inocybe umbratica	Fruit body								4.39

Modified from Fukuda et al. (2013b)

17.5 Radiocesium Contamination in Forest Trees and Soil in the Winter of 2011

Figure 17.3 shows the radiocesium distributions in the felled trees. The lower branches of *Chamaecyparis obtusa* (Hinoki) had the highest contamination load, and the outer bark of the upper stem had the highest load in the stem. For *Carpinus tshonoskii* (Inushide) trees, the outer bark at a height of 0.8 m and above had a high contamination load. The outer bark of *Quercus serrata* (Konara) at a height of 10 m had the highest level of contamination. In all tree species, the sapwood and some heartwood samples showed radiocesium contamination. The general trend of the radiocesium distribution in the aboveground parts of standing trees was similar to that in Fukushima Prefecture (Kaneko and Tsuboyama 2012; Kato et al. 2012; Koarashi et al. 2012; Kuroda et al. 2013; Masumori et al. 2015a, b; Miura 2015; Ohashi et al. 2014; Ohte et al. 2015) (Figs. 17.4 and 17.5).

Tables 17.3 and 17.4 summarize the total deposition of radiocesium in the aboveground parts of forest trees in these stands. Each tree contained 40–113 kBq, and the total estimated deposition was 3.7–5.7 kBq/m^2. The radiocesium deposition in soil was roughly estimated to be 50–90 kBq/m^2 (Fukuda et al. 2013b). Therefore, more than 90% of the radiocesium contamination was deposited on the soil surface in the mixed and deciduous forests in this area, concurring with some low-level contaminated sites in Fukushima Prefecture (Kaneko and Tsuboyama 2012; Kuroda et al. 2013). Therefore, decontamination by cutting trees or collecting litter would not be effective in these forests.

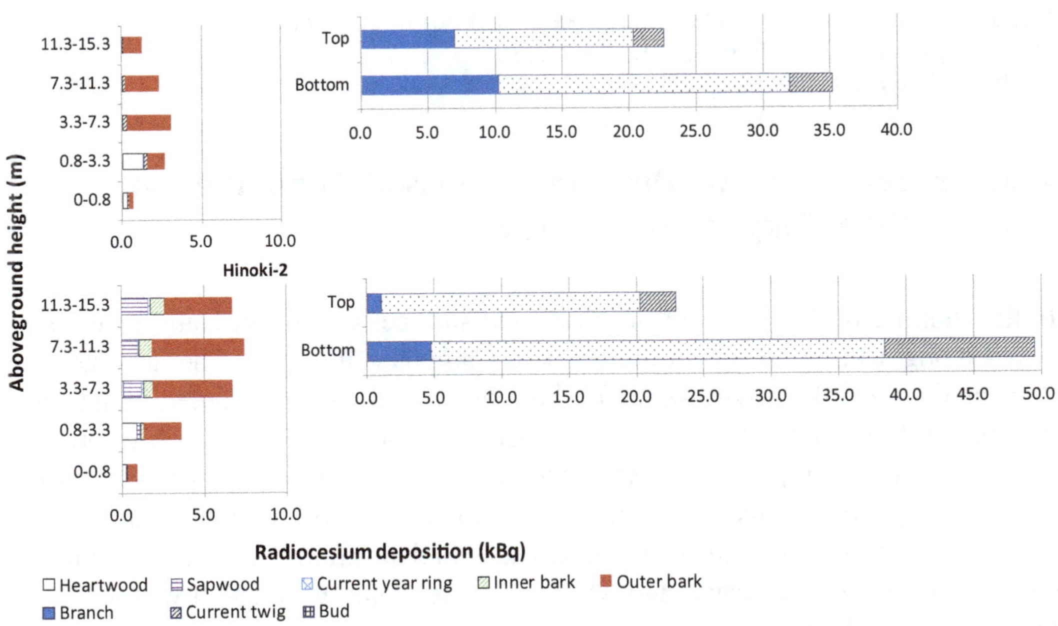

Fig. 17.3 Aboveground deposition of radiocesium (kBq) in trees felled in Oaota Forest in the winter of 2011. (Modified from Fukuda et al. 2013b)
Hinoki: *Chamaecyparis obtusa*

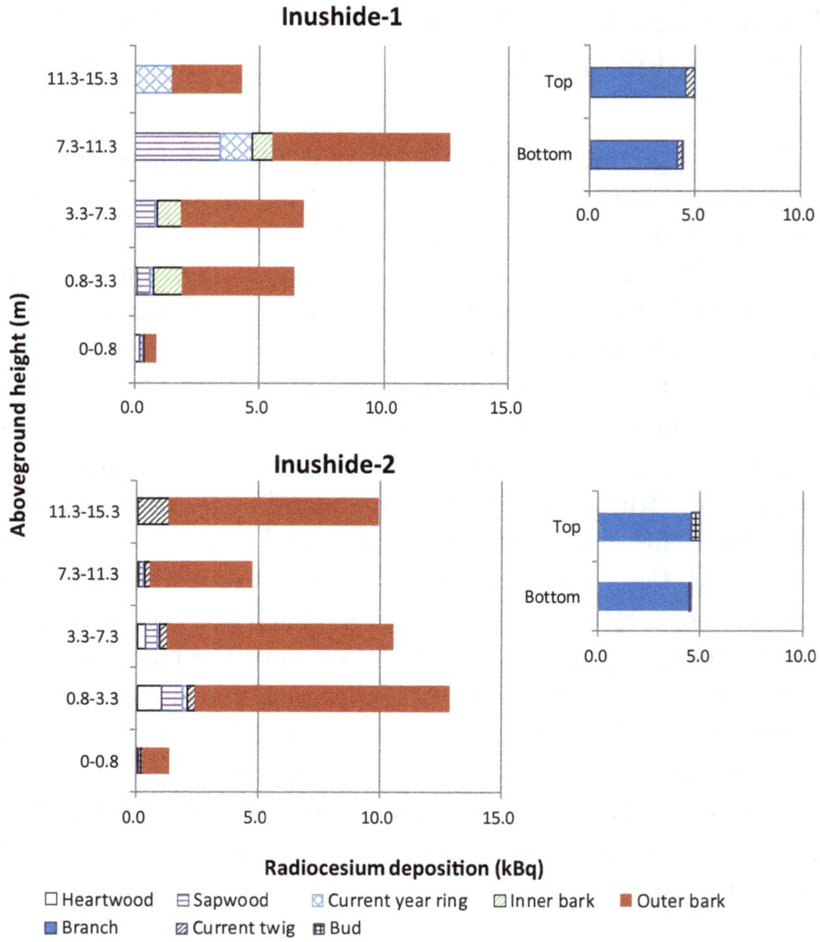

Fig. 17.4 Aboveground deposition of radiocesium (kBq) in trees felled in Oaota Forest in the winter of 2011. (Modified from Fukuda et al. 2013b)
Inushide: *Carpinus tschonoskii*

17.6 Forest Type, Air Dose Rate, and Soil Contamination in the UTokyo Campus Forest in 2013

In the summer of 2013, the air dose rate and surface soil contamination were surveyed along a line transect on the UTokyo Campus (Fig. 17.6). The air dose rate fluctuated around 0.2 μSv/h, which is about one-half the level recorded in the summer of 2011. In this unused area of the Campus, no decontamination effort was undertaken; therefore, the decrease in the air dose rate was solely attributed to the natural decay and movement of radiocesium into the soil. Soil contamination differed significantly among the land cover types. The total deposition of radiocesium in 0~5-cm soil was highest at a canal into which rainwater could flow, and it was relatively high on bare land (unpaved road) and under a deciduous tree canopy. The contamination level was low under the dense tall bamboo stand and in the soil under the evergreen tree canopy. Although the radiocesium deposition in the aboveground parts of bamboo and evergreen trees was not estimated in this forest, we postulate that the evergreen canopy intercepted the radiocesium deposition (Hashimoto et al. 2012; Kaneko and Tsuboyama 2012), and decontamination by cutting trees and bamboo

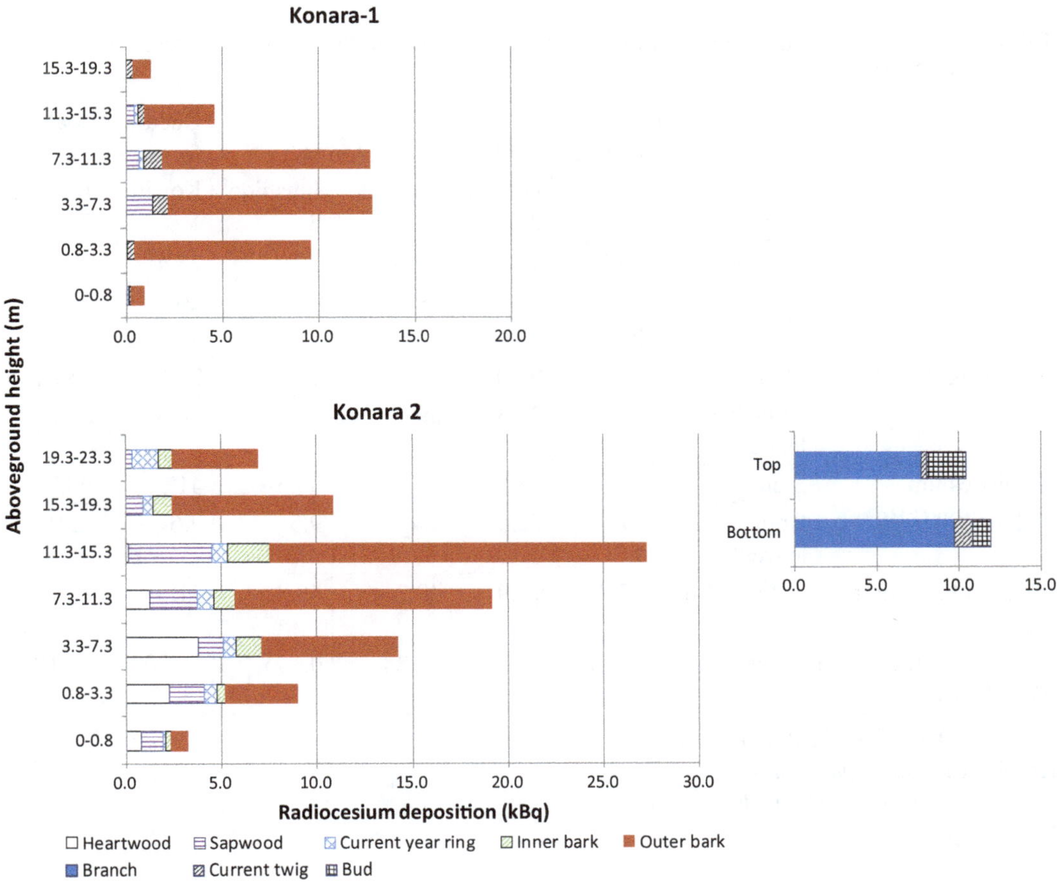

Fig. 17.5 Aboveground deposition of radiocesium (kBq) in trees felled in Oaota Forest in the winter of 2011. (Modified from Fukuda et al. 2013b)
Konara: *Quercus serrata*

might be effective in bamboo stands and evergreen forests. In deciduous forests, radiocesium was thought to be concentrated in surface soil, so decontamination would be much more difficult.

17.7 Decontamination Experiment in a Nursery Lawn

In the autumn of 2011, a decontamination experiment was conducted in the lawn of the nursery on the UTokyo Campus (Fukuda et al. 2013a). Based on the vertical distribution of radiocesium contamination in the lawn soil (Fig. 17.2, upper), removing the turfgrass and surface soil was thought to be effective. Therefore, the area within 10 m of the nursery building was treated with a sod cutter (Iwamoto TS-1F) which removed 2 cm of surface soil with the turf root mat (Fig. 17.7, left). A second decontamination method recommended in Fukushima Prefecture, which involved removing turfgrass shoots and thatch with a reel (Kyoei LM22) and rotary mowers (Kyoei GM530C) and preserving the turf roots to reduce the cost of returfing, was

Table 17.3 Aboveground deposition of radiocesium in trees felled in Oaota Forest estimated in the winter of 2011

	Plot	Conifer-deciduous mixed forest				Deciduous oak forest	
	Tree	Hinoki 1	Hinoki 2	Inushide 1	Inushide 2	Konara 1	Konara 2
	DBH (cm)	19.5	17.4	20.0	20.0	23.4	34.9
	H (m)	15	16.9	19.0	16.5	20.1	21.1
	BA (cm$^{3)}$	299	238	314	314	430	957
Dry weight (kg)	Stem	86.0	92.0	150.6	132.2	228.9	659.1
	Branches & leaves	25.8	23.3	39.9	39.9	38.4	89.6
Radiocesium deposition (kBq)	Stem	10.0	25.5	30.9	39.3	41.9	90.7
	Branches & leaves	57.7	72.5	9.4	9.5	9.6	22.3
Total radiocesium depositon (kBq/tree)		67.7	97.9	40.3	48.8	51.5	113.0
Radiocesium deposition per BA (kBq/cm^2)		0.227	0.412	0.128	0.155	0.120	0.118

Modified from Fukuda et al. (2013b)
Hinoki: *Chamaecyparis obtusa*
Inushide: *Carpinus tschonoskii*
Konara: *Quercus serrata*

Table 17.4 Stand-level deposition of radiocesium in Oaota Forest

Land cover		Conifer-deciduous mixed forest	Deciduous forest
Dominant species		*Chamaecyparis obtusa*	*Querus serrata*
		Carpinus tschonoskii	*Q. accutissima*
Number of species		9	5
Max DBH (cm)		35.0	36.6
Max H (m)		22.0	22.1
Stem density (/ha)		1175	725
Total BA (m^2/ha)		25.3	30.9
BA (m^2/ha)	Conifers	11.5	0.37
	Evergreen broadleaves	0.21	0.00
	Deciduous broadleaves	13.5	30.5
Radiocesium deposition in trees(kBq/m^2)	Conifers	3.69	0.12
	Evergreen broadleaves	0.07	0.00
	Deciduous broadleaves	1.91	3.63
	Aboveground total	5.67	3.74
Radiocesium deposition in soil (kBq/m^2)		60	85
Total deposition (kBq/m^2)		70	90

Modified from Fukuda et al. (2013b)

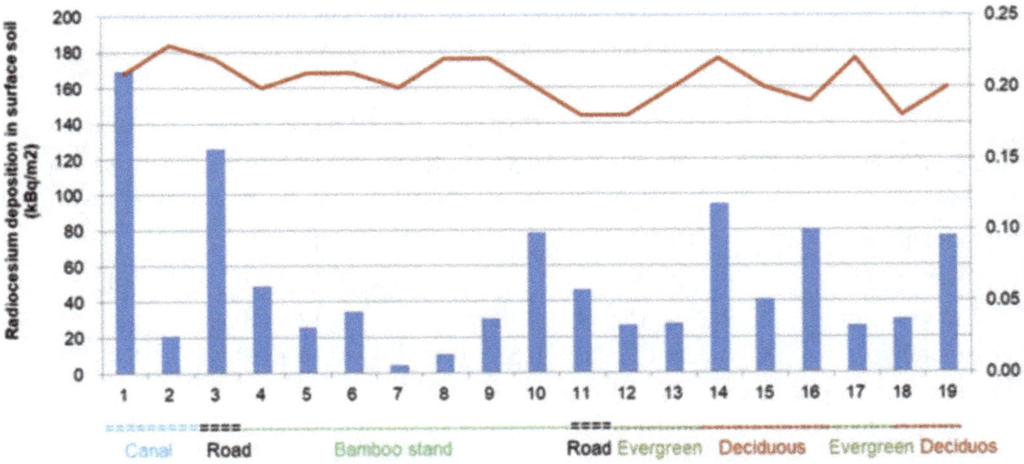

Fig. 17.6 Air dose rate at 1 m above the ground (red line) and radiocesium deposition in 0–5 cm surface soil (blue bars) along a transect on the UTokyo Campus measured in the summer of 2013

Fig. 17.7 The decontamination experiment conducted on a nursery lawn on the UTokyo Campus (Fukuda et al. 2013a)
Left: A sod-cutter removing the root mat of turf with 2 cm of surface soil. Right: A rotary mower collecting thatch and the cover soil preparation (the bare area in front of the nursery building had been treated by sod-cutting method)

also tested in the outer area (Fig. 17.7, right). The results are shown in Table 17.5. Removing soil with the root mat effectively decreased the air dose rate at 5 cm aboveground to 0.13 μSv/h, while the low-cost method failed to reduce the air dose rate sufficiently. This clearly demonstrated that to reduce the air dose rate of the lawn, it is important to remove the contaminated surface soil.

17.8 Change in Radiocesium Distribution in Deciduous Forest Soil in Oaota in 2013–2015

The vertical distribution of radiocesium in the surface soil samples collected in the deciduous forest of Oaota over a 3 year period is shown in Fig. 17.8. These observations showed that the movement of radiocesium into deeper soil was very slow, as

Table 17.5 Effect of decontamination of a lawn by sod-cutting (removing the root mat with 2-cm of surface soil) and mowing (low-cost method) on air dose rate (μSv/h) at 5 cm above the ground in 2011

	Date	Control	Sod-cutting	Mowing
Before decontamination	Sept. 13	0.54 ± 0.16	0.67 ± 0.24	0.53 ± 0.03
After mowing	Oct. 29	–	–	0.49 ± 0.10
After sod-cutting	Oct. 29	–	0.16 ± 0.02	–
After covering soil	Oct. 29	–	0.13 ± 0.02	0.42 ± 0.02
After 2 weeks	Nov. 15	0.49 ± 0.11	0.12 ± 0.05	0.42 ± 0.03
After 2 months	Jan. 17	0.52 ± 0.14	0.16 ± 0.04	0.43 ± 0.04

Modified from Fukuda et al. (2013a)

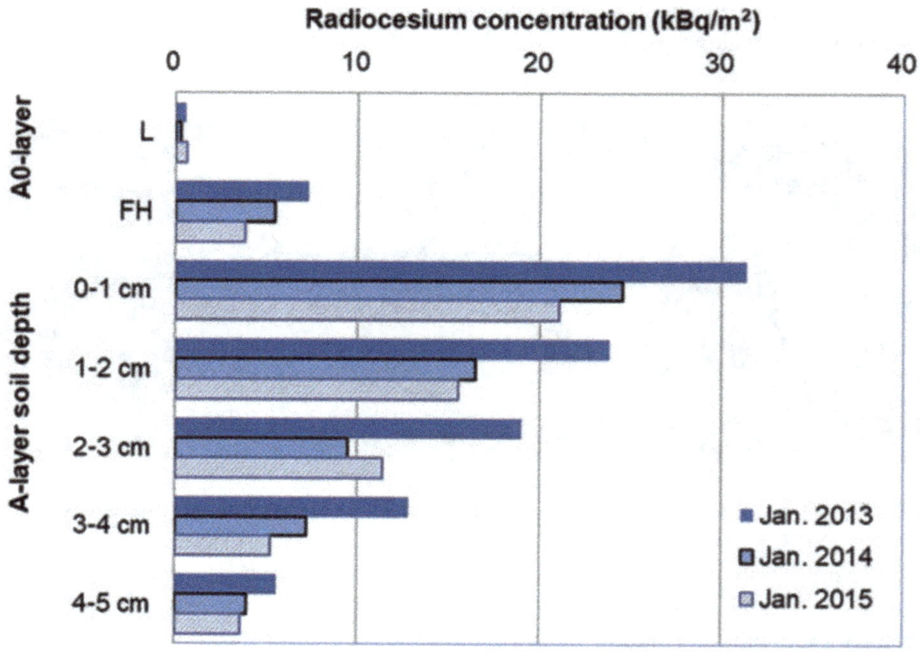

Fig. 17.8 Vertical distribution of radiocesium deposition in the deciduous oak stand in Oaota Forest

has been suggested in many other studies (IAEA 2006; Shiozawa 2013; Yamaguchi et al. 2012). The decrease in radiocesium concentration could be explained by the natural decay of ^{134}Cs, which accounted for about half the level in 2011. Therefore, the subsequent decrease in radiocesium concentration will slow down, because ^{137}Cs with its longer half-life remains in the soil. The absorption of radiocesium by tree roots and ectomycorrhizal fungi should be monitored carefully to predict the dynamics of radiocesium in Satoyama forest ecosystems.

17.9 Conclusion

The total contamination level in the UTokyo Campus and nearby forests matched the deposition map (MEXT 2011). In the lawn on the UTokyo Campus, most of the radiocesium contamination was restricted to the very shallow soil surface (0–1 cm), and removing soil with the root mat was effective for reducing the air dose rate in the lawn. In evergreen forests and bamboo stands, the soil contamination was lower than that in bare land and deciduous forests. Deciduous forests are the most common vegetation of Satoyama forests, where many citizen volunteers work, and most of the radiocesium deposition in such forests was in the surface soil, making decontamination of deciduous forests difficult. The movement of radiocesium into deeper soil was negligible until the winter of 2014, and only natural decay seemed to decrease the air dose rate in these forests. Both the sapwood and heartwood of the standing trees were slightly contaminated, suggesting translocation of radiocesium into the tree stem. The use of cut logs for fuelwood and for the cultivation of shiitake mushrooms was impossible in these forests in the winter of 2011. Continued monitoring of the radiocesium dynamics in trees is necessary for the safety of Satoyama citizen volunteers in the hot spot areas of suburban Tokyo.

Acknowledgment I thank Dr. Natsumaro Kutsuna of the Radioisotope Laboratory of the Graduate School of Frontier Sciences, the University of Tokyo, for measuring the radiocesium concentrations in the samples. I also thank Prof. Shuichi Kitoh and the members of the "Donguri Day Nursery Decontamination Project," as well as Prof. Makoto Yokohari, Prof. Hirokazu Yamamoto, Prof. Toru Terada, and the members of Agriculture–Landscape Group of the "JST Urban Reformation Program for the Realization of a Bright Low-Carbon Society Project" for their financial support, discussion, and assistance in the field survey and experiments.

I also thank the members of the Department of Natural Environmental Studies, Graduate School of Frontier Sciences, the University of Tokyo, for their assistance in the line transect survey in 2013, and the volunteers from the NPOs "Chiba Satoyama Trust" and "Konbukuroike Shizen-no-Mori" for their kind cooperation with the field survey and sample collection.

References

Fukuda K, Kutsuna N, Terada T, Mansounia MR, Uddin MN, Jimbo K, Shibuya S, Fujieda J, Yamamoto H, Yokohari M (2013a) Radiocesium contamination in suburban forests in Kashiwa city, Chiba Prefecture. Jpn J For Environ 55:83–98 (in Japanese with English abstract)

Fukuda K, Kutsuna N, Kitoh S, Yamaji E, Saito K, Onuki M, Koibuchi Y, Mitani H, Yoshida Z, Jimbo K, Matsuo Y, Sueyoshi K (2013b) Radioactive cesium contamination in Kashiwa Campus of the University of Tokyo, Chiba prefecture and decontamination experiment for lawn. J Jpn Soc Turfgrass Sci 42:20–30 (in Japanese with English abstract)

Hashimoto S, Ugawa S, Nanko K, Shichi K (2012) The total amounts of radioactively contaminated materials in forests in Fukushima. Jpn Sci Rep 2:416

IAEA (International Atomic Energy Agency) (2006) Environmental consequences of the chernobyl accident and their remediation: twenty years of experience. http://www-pub.iaea.org/mtcd/publications/pdf/pub1239_web.pdf

Kaneko S, Tsuboyama Y (2012) Radioactive contamination in forests and its decontamination. Gakujutsu no Doko (in Japanese)

Kato H, Onda Y, Teramage M (2012) Depth distribution of ^{137}Cs, ^{134}Cs, and ^{131}I in soil profile after Fukushima Dai-ichi Nuclear Power Plant accident. J Environ Radioact 11: 59–64.

Kitoh S (2013) Ethics of scientists facing the low-level contamination caused by Fukushima nuclear accident. In: FGF, TGF (eds) Nuclear disaster and academism: question and action from Fukushima University and the University of Tokyo. Godo Shuppan, Tokyo, pp 80–104 (in Japanese)

Koarashi J, Atarashi-Andoh M, Matsunaga T, Sato T, Nagao S, Nagai H (2012) Factors affecting vertical distribution of Fukushima accident-derived radiocesium in soil under different land-use conditions. Sci Total Environ 431:392–401

Kuroda K, Kagawa A, Tonosaki M (2013) Radiocesium concentrations in the bark, sapwood and heartwood of three tree species collected at Fukushima forests half a year after the Fukushima Dai-ichi nuclear accident. J Environ Radioact 122:37–42

Masumori M, Nogawa N, Sugiura S, Tange T (2015a) Radiocesium in stem, branch and leaf of *Cryptemria japonica* and *Pinus densiflora* trees: case of forests in Minamisoma in 2012 and 2013. J Jpn For Soc 97:51–56

Masumori M, Nogawa N, Sugiura S, Tange T (2015b) Radiocesium in tnimber of Japanese cedar and Japanese red pine, in the forests of Minamisoma, Fukushima. In: Nakanishi TM, Tanoi K (eds) Agricultural implications of the Fukushima nuclear accident. Springer, Tokyo, pp 161–174. https://doi.org/10.1007/978-4-431-55,828-6_13

MEXT (Japanese Ministry of Education, Culture, Sports, Science, and Technology) (2011–2016) Extension site of distribution map of radiation dose, etc. http://ramap.jmc.or.jp/map/eng/. Accessed 5 Dec 2016

Miura S (2015) The effects of radioactive contamination on the forestry industry and commercial mushroom-log production in Fukushima, Japan. In: Nakanishi TM, Tanoi K (eds) Agricultural implications of the Fukushima nuclear accident. Springer, Tokyo, pp 145–160. https://doi.org/10.1007/978-4-431-55,828-6_12

Ohashi S, Okada N, Tanaka A, Nakai W, Takano S (2014) Radial and vertical distributions of radiocesium in tree stems of *P. densiflora* and *Quercus serrata* 1.5 y after the Fukushima nuclear disaster. J Environ Radioact 134:54–60

Ohte N, Murakami M, Endo I, Ohashi M, Iseda K, Suzuki T, Oda T, Hotta N, Tanoi K, Kobayashi NI, Ishii N (2015) Ecosystem monitoring of radiocesium redistribution dynamics in a forested catchment in Fukushima after the Nuclear Power Plant accident in March 2011. In: Nakanishi TM, Tanoi K (eds) Agricultural implications of the Fukushima nuclear accident. Springer, Tokyo, pp 175–188. https://doi.org/10.1007/978-4-431-55,828-6_14

Shiozawa S (2013) Vertical migration of radiocesium fallout in soil in Fukushima. In: Nakanishi TM, Tanoi K (eds) Agricultural implications of the Fukushima nuclear accident. Springer, Tokyo, pp 49–60. https://doi.org/10.1007/978-4-431-54,328-2_6

Tagami K, Uchida S, Ishii N, Kagiya S (2012) Translocation of radiocesium from stems and leaves of plants and the effect on radiocesium concentrations in newly emerged plant tissues. J Environ Radioact 111:65–69

Takata D (2013) Distribution of radiocesium from the radioactive fallout in fruit trees. In: Nakanishi TM, Tanoi K (eds) Agricultural implications of the Fukushima nuclear accident. Springer, Tokyo, pp 143–162. https://doi.org/10.1007/978-4-431-54,328-2_14

Takata D (2015) Translocation of radiocesium in fruit trees. In: Nakanishi TM, Tanoi K (eds) Agricultural implications of the Fukushima nuclear accident. Springer, Tokyo, pp 119–144. https://doi.org/10.1007/978-4-431-55,828-6_11

Yamada T (2013) Mushrooms: radioactive contamination of widespread mushrooms in Japan. In: Nakanishi TM, Tanoi K (eds) Agricultural implications of the Fukushima nuclear accident. Springer, Tokyo, pp 163–176. https://doi.org/10.1007/978-4-431-54,328-2_15

Yamaguchi N, Takada Y, Hayashi K, Ishikawa S, Kuramata M, Eguchi S, Yoshikawa S, Sakaguchi A, Asada K, Wagai R, Makino T, Akahane I, Hiradate S (2012) Behavior of radiocesium in soil-plant systems and its controlling factor. Bull Natl Inst Agro Environ Sci 31:75–129 (in Japanese with English abstract)

Yoshida S, Muramatsu Y (1994) Accumulation of radiocesium in basidiomycetes collected from Japanese forests. Sci Total Environ 157:197–205

Yoshihara T, Matsumura T, Hashida S, Nagaoka T (2013) Radiocesium contaminations of 20 wood species and the corresponding gamma-ray dose rates around the canopies at 5 months after the Fukushima Nuclear Power Plant accident. J Environ Radioact 115:60–68

Chapter 18
The State of Fisheries and Marine Species in Fukushima: Six Years After the 2011 Disaster

Nobuyuki Yagi

Abstract The recovery of Fukushima fisheries remains sluggish 6 years after the disaster. The Fukushima Fisheries Cooperative Association (FCA) decided to allow limited fishing in June 2012 (known as the trial operation). Total landing value of fish and fishery products from the trial operation has been gradually increasing due to the increased number of catchable target species and increased fishing areas. But the landing value in 2016 was only 5% of the value recorded in the pre-disaster years. Safety of the products has been demonstrated by various surveys conducted by the government authorities and independent researchers. Several studies indicated that the population of key fish species in Fukushima waters showed a tangible increase after the 2011 disaster reflecting low fishing pressures in this period. Weak consumer confidence would have contributed to the extremely slow recovery of Fukushima fisheries. In addition to the consumers' attitudes, fish distributors' risk-averse attitudes could have brought additional adverse effects against the recovery of Fukushima fisheries. This situation could continue for several more years. Continued support for fishers in Fukushima is needed for the foreseeable future to sustain the livelihood of small fishing households as well as maintain societies, traditional knowledge, and other human or social capital in the region.

Keywords Consumer confidence · Fishery · Fukushima · Radiocesium · Tsunami

18.1 Introduction

The tsunami triggered by the Great East Japan Earthquake on March 11th, 2011 damaged around 29,000 fishing boats and 319 fishing ports in Japan (Fisheries Agency of Japan 2017). These numbers account for some 10% of the respective national totals. Periodical updates on the recovery status of fisheries in the Tsunami-damaged area have been provided by the Fisheries Agency of the Japanese Ministry

N. Yagi (✉)
Graduate School of Agricultural and Life Sciences, The University of Tokyo, Tokyo, Japan
e-mail: yagi@fs.a.u-tokyo.ac.jp

T. M. Nakanishi et al. (eds.), *Agricultural Implications of the Fukushima Nuclear Accident (III)*, https://doi.org/10.1007/978-981-13-3218-0_18

of Agriculture, Forestry and Fisheries (MAFF). According to the most recent report, approximately 18,486 of those boats and all ports (i.e., 319 ports) had returned to operation as of March 31st, 2017 (Fishery Agency of Japan 2017).

Despite the unprecedented scale of the disaster, the rehabilitation of the fisheries in the tsunami-damaged areas other than Fukushima has been relatively expeditious in terms of the fishing capacity as measured in the number of boats and ports. However, Fukushima fisheries are still suffering from the large-scale release of radioactive substances from the Fukushima Dai-ichi Nuclear Power Plant operated by the Tokyo Electric Power Company (TEPCO) even 6 years after the disaster.

18.2 Declining Level of Radiocesium Contained in Fish and Fishery Products

On a weekly basis, the Government agencies, including the Fisheries Agency of MAFF, Fukushima prefectural government and relevant municipal authorities, examine the level of radiocesium in fish and fisheries products taken from Fukushima and its surrounding waters. Samples are collected from scientific research vessels of the Fukushima prefectural government and other research boats (http://www.pref.fukushima.lg.jp/uploaded/attachment/231539.pdf). This is not random sampling across species occurring in Fukushima waters. Rather, a large number of samples are collected and species that have past records of high concentrations of radiocesium are analyzed (http://www.jfa.maff.go.jp/j/housyanou/attach/pdf/kekka-79.pdf). Time-series changes of radioactive cesium levels contained in these samples from April 2011 to June 2017 are shown in Fig. 18.1. More than 100,000 samples were examined during this period, and the proportion of samples with over 100 Bq/Kg radiocesium decreased relatively quickly in 2011 and 2012. In and after 2015, the proportion of samples with over 100 Bq/Kg of radiocesium was almost zero. 100 Bq/Kg is regarded as an important threshold because the allowable level of radioactive cesium established by the Japanese government was 100 Bq/kg for all fisheries products after April 2012 (Fisheries Agency of Japan 2014).

The declining trend of radiocesium in fish and fisheries products is consistent with independent research results published after the Fukushima disaster. Wada et al. (2013), for instance, discussed that, although the time-series trends of radioactive cesium concentrations differ greatly among taxa, habitats, and spatial distributions, higher concentrations have been observed in shallower waters, and that radioactive cesium concentrations decreased quickly or were below detection limits in pelagic fish and some invertebrates. However, in some demersal fish,[1] the declining trend was much more gradual, and concentrations above the regulatory limit (100 Bq/kg-wet weight) were frequently found, indicating continued uptake of radioactive cesium through the benthic food web (Wada et al. 2013).

[1] The term demersal fish means fish species living near the bottom of the sea

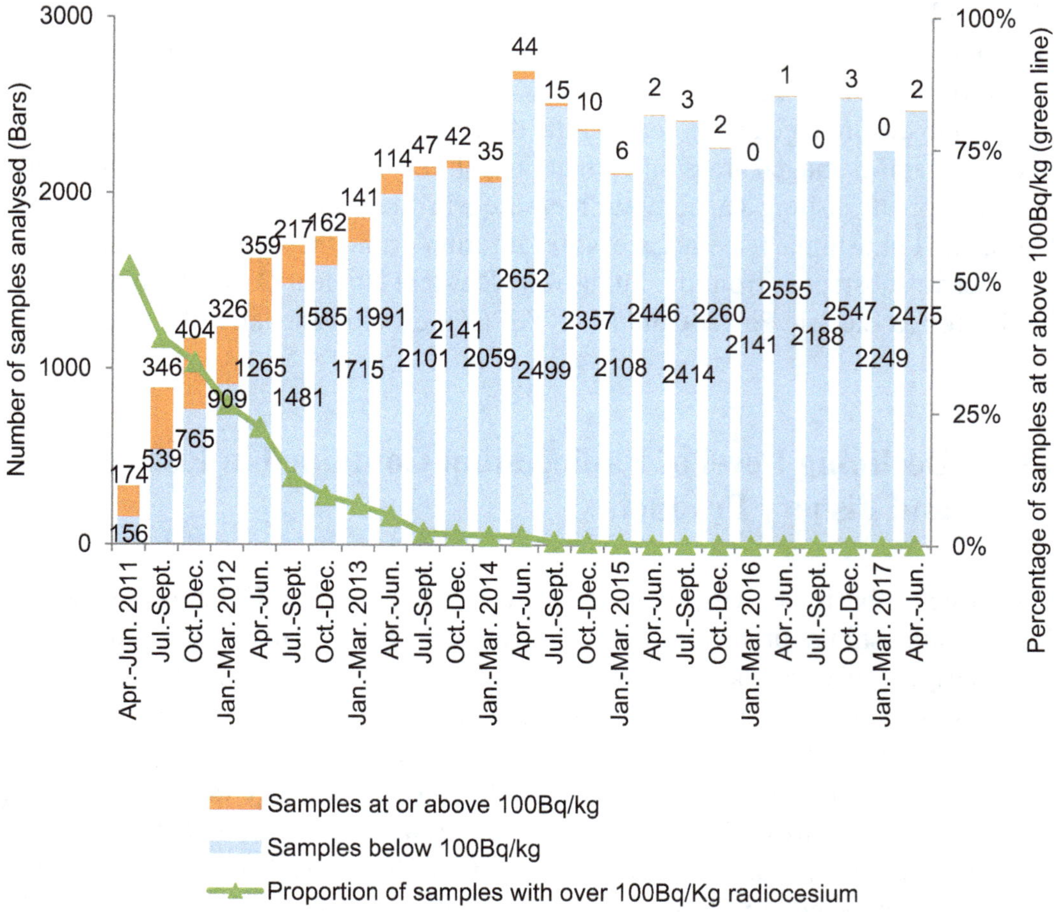

Fig. 18.1 Level of radiocesium in sampled marine organisms in Fukushima waters. (Data source: Fisheries Agency of Japan http://www.jfa.maff.go.jp/j/housyanou/kekka.html. Accessed in October 2017)

18.3 Development of Biological Studies on Fish and Radioactive Substances

Even before the disaster, it is widely known that the level of radiocesium concentration is different by species. For instance, according to an IAEA report (IAEA 2004), saltwater fish tend not to accumulate cesium at as high a level as freshwater fish. This mechanism can be explained as follows: in freshwater, the osmotic pressure of a fish's body fluids is higher than that of the surrounding water, and the fish actively cycle water out of their bodies while keeping salt and minerals to maintain normal levels of osmotic pressure. During this process, cesium is accumulated within the collected salt and minerals. On the other hand, the osmotic pressure in the bodies of saltwater fish is lower than that of the surrounding seawater. To prevent the loss of too much water from their bodies and to maintain balance, saltwater fish actively cycle salt and minerals – along with cesium – out of their bodies.

A more detailed mechanism of excretion of cesium from the body of fish was reported by Furukawa et al. (2012) and Kaneko et al. (2013). They reported that the gills of fish eliminate unnecessary cesium (Cs^+) from body fluid, presumably through the same pathway as potassium (K^+) excretion (Furukawa et al. 2012; Kaneko et al. 2013). Cesium is known to be a biochemical analog of K, and through the K transporting pathway Cs can be released from the gills of fish (Furukawa et al. 2012; Kaneko et al. 2013).

Results of empirical studies on radiocesium concentration in various fish species are reported through a number of independent research teams after the Fukushima disaster. Kikkawa et al. (2014), using published data of 8683 samples obtained off the coast of Fukushima during the 2-year period after the disaster, identified that fish and fisheries products in Fukushima can be categorized into four groups in terms of concentration of radioactive cesium (134Cs + 137Cs). They are (1) low concentrations of radiocesium both in 2011 and 2012, (2) some decline of concentration in 2012 but still high, (3) high initial concentrations in 2011 but levels quickly dropped and became almost undetectable in 2012, and (4) high concentration of radiocesium both in 2011 and 2012 (Kikkawa et al. 2014).

Specifically, out of 97 fish and fishery items (a total of 95 species including two species that are marketed separately for the adult and immature stages), 60 items fall in the first group (low concentration) which include invertebrate animals as well as a wide variety of vertebrate fish such as pelagic species (Kikkawa et al. 2014). The second group (some decline but still high in 2012) had 21 items and many of them are demersal species (Kikkawa et al. 2014). The third group (extremely high in 2011 but quickly reduced to undetectable levels in 2012) had 1 item, namely whitebait-size Japanese sand lance (*Ammodytes personatus*) (Kikkawa et al. 2014). The fourth group (high concentration both in 2011 and 2012) had 15 items and most of them were coastal demersal vertebrate fish species such as greenling (*Hexagrammos otakii*) (Kikkawa et al. 2014). Consequently, Kikkawa et al. (2014) discussed that almost all the products of groups (1) and (3) satisfied the government food-safety standard as of 2012, while products in groups (2) and (4) did not satisfy the standard and must be closely monitored.

Takagi et al. (2015) reported their analyses on small epipelagic fishes (such as sardine and Japanese anchovy). Sixty-three samples were collected by commercial fishery boats off the Kashima-Boso area (approximately 100 km south of Fukushima waters) from March 24th, 2011 to March 21st, 2013. The radiocesium concentration in fish muscle reached a maximum of 31 Bq/kg-wet weight in July 2011, and after that, it declined gradually. From 2012 to 2013, the radiocesium concentration in fish muscle was low (0.58–0.68 Bq/kg-wet weight).

It is possible that radiocesium concentration in fish depends on the size and the age of fish. Year-class related difference in radiocesium concentrations in the body of fish was examined by Narimatsu et al. (2015). Pacific cod (*Gadus macrocephalus*) captured from April 2011 to March 2014 off Fukushima Prefecture were analyzed and they found the radiocesium concentration of cod in the year-class of 2009 (i.e., cod born in the year 2009) and earlier was higher than that in the 2010 year-class. In addition, radiocesium was rarely detected or detected at very low levels in the 2011 year-class. Regression analyses showed that the estimated ecological half-life of radiocesium in Pacific cod was from 258 to 309 days; this value is consistent with values in other demersal fish caught off Fukushima Prefecture as reported by Wada et al. (2013). Narimatsu et al. (2015) further pointed out that the half-life was longer in old and larger individuals than in young and smaller individuals, probably a result of differences in metabolic rate and growth rates between age and body size classes as proposed in the study of Doi et al. (2012). Radiocesium was rarely detected in the 2011 year-class, most likely because the fish were only exposed to very low levels of radiocesium when they started their life in the ocean bottom and after that radiocesium in the fish body was diluted by growth (Narimatsu et al. 2015).

18.4 Limited Resumption of Fishing in Fukushima Waters

In Fukushima, a total of 873 fishing vessels were damaged by the tsunami on March 11th, 2011 (Ministry of Agriculture, Forestry and Fisheries, Japan 2012). After March 12th, the large-scale release of radioactive substances from the Fukushima Dai-ichi Nuclear Power Plant occurred. On March 15th, 2011, the Fukushima Prefectural Federation of Fisheries Cooperative Association (hereafter referred to as the Fukushima FCA) voluntarily stopped fishing operations in the waters inside of Fukushima Prefecture (Yagi 2014, 2016). Some fishing activities in the prefectures neighboring Fukushima (namely Miyagi and Ibaraki) were also suspended after the Fukushima disaster, but most of these were subsequently lifted within 2 years (Ibaraki Prefecture 2014; Miyagi Prefecture 2014). Neither the national nor the prefectural governments revoked fishing licenses in Fukushima (Yagi 2016). However, the national government did provide legally binding sales prohibitions on certain marine products caught in the waters off Fukushima Prefecture based on food safety requirements (Yagi 2016).

After a cessation of fishing for more than 1 year, the Fukushima FCA decided to resume fishing activities in June 2012, hereafter referred to as the ‘trial operation’, for three species (two octopus species and one shellfish species) living at depths of more than 150 meters in ocean areas approximately 60–90 km from the damaged nuclear power plant (Yagi 2014, 2016). The trial operation has several limitations, and therefore it is not regarded as a full resumption of commercial fishing. The limitations include: (i) days of operation (usually fewer than 5 days a month); (ii) landing ports (only two ports have been designated: one in Soma City and the other in Iwaki City); (iii) the amount of landed fish (usually less than 10 tons a day); and (iv) the number of vessels involved in fishing operations (Yagi 2014). Operation rules are set to maintain a high frequency of monitoring for radioactive substances and to ensure traceability following the landing of marine products (Yagi 2016). Most of these products were sold at local supermarkets in Fukushima, and sold out very quickly, most likely due to the small number of available items and consumers wishing to help their local fishers by purchasing their products (Yagi 2014).

The Fukushima FCA monitored for radiocesium at the two landing sites (namely, Soma-Futaba and Iwaki) from the start of the trial operation. Analyses were carried out for every species landed during the day of each trial operation. 0.5–1.0 kg of edible parts is randomly taken as a representative sample of each species on the day, and radiocesium (^{134}Cs and ^{137}Cs combined) is measured using NaI scintillation counters installed at the two landing sites. Landed fish have been sold with labels indicating Fukushima as their point of origin (Yagi 2014).

As of May 2017, the trial operation has caught approximately 20–40 species and, therefore, all these species have been measured for radiocesium. The level of radiocesium in samples are reported on the website of the Fukushima FCA, and all records from 2012 until today are available (http://www.fsgyoren.jf-net.ne.jp/kensakekka201209.pdf for Soma-Futaba landing site and http://www.fsgyoren.jf-net.ne.jp/kensakekka-iwaki.pdf for Iwaki landing sites). In almost all samples, the level of radiocesium was below 12.5 Bq/kg and such results are reported as "not detected" on the websites.

The legally binding sales prohibition by the government has been periodically revised. Since December 2013, the prohibition has been in effect for 40 marine species living in the waters of Fukushima (Fukushima Prefecture 2014), and since January 2015, it has been reduced to 35 species (Fukushima Prefecture 2015). As of October 11th, 2017, the number of species subject to the sales prohibition was reduced to 10 marine species (Fukushima Prefecture 2017a). These species are not allowed to land by the trial operation.

Total landing value of fish and fishery products from the trial operation has gradually increased due to the increased number of catchable species and increased fishing areas. But its landing value in 2016 was still at the level of 1/20 of the predisaster years. The total value of fish and fishery products landed in Fukushima from the trial operation was 461 million JPY (approximately 4 million USD) in 2016, while the value of landed fish and fishery products in 2009 and 2010 was 11,280 million JPY and 10,959 million JPY (both values are equivalent to approximately 100 million USD), respectively (Fukushima Prefecture 2017b, for predisaster data, and post-disaster data through personal communications with Mr. Noguchi of the Fukushima FCA).

18.5 Weak Consumer Confidence and Risk-Averse Distributers

Weak consumer confidence, along with the limited fishing (i.e., limited number of fishing boats and fishing days), could have contributed to the extremely slow pace of recovery of Fukushima fisheries. Consumer attitudes in Japan toward agricultural and fishery products have been reported by various researchers since the 2011 disaster. Aruga (2017), for instance, examined consumer attitudes in Japan for seven agricultural products (rice, apples, cucumbers, beef, pork, eggs, and shiitake mushrooms) from regions near the Fukushima Daiichi nuclear power plant. It was found that consumers with children under the age of 15 required a higher discount rate to accept agricultural products from regions near the power plant (Aruga 2017). Conversely, consumers who trusted the current safety standards for radioactive material concentrations in food were more likely to purchase products from regions near the plant (Aruga 2017). As for fishery products, Wakamatsu and Miyata (2017) studied consumers' evaluation of Pacific cod (*Gadus macrocephalus*) and whitebait (*Engraulis japonicus*) and found that the calculated values of consumers' marginal willingness to pay dropped for products originating from Fukushima compared with products originating from other parts of Japan.

In addition to consumers' attitudes, fish distributors' risk-averse attitudes could have brought additional adverse effects against the recovery of Fukushima fisheries. Even though some consumer segments are interested in Fukushima products, such products cannot be delivered to the consumers if distributors avoid trading such products. Fishers have no choice but to accept risk-averse attitudes of distributors or retailors, because fishers are in a weaker position. In the case of Japan's fishery distribution, the market power rests with retail stores rather than producers even before the Tsunami in 2011 (Sakai et al. 2012; Nakajima et al. 2014). In Fukushima after 2011, fishers' market power was weakened further. While fishing activities are geographically linked to specific marine areas designated by fishing licenses, processing or distribution industries are under no such administrative requirements. These industries can move from Fukushima and still continue their business in fish distribution and processing. In fact, some large fish processing facilities in the disaster-affected region decided to relocate their factories outside of these areas, and they still have not returned to the region. Although no systematic study has been conducted on the attitudes of fish distributors and retailers against Fukushima food products, various news reports suggested that risk-averse food distributors could exist and they may have adversely affected the recovery of agriculture and fisheries in Fukushima (for instance, NHK TV broadcast on May 24th, 2017. https://www.nhk.or.jp/gendai/kiji/fukushima/. Accessed October 2017).

18.6 Increased Abundance of Key Target Fish Species in Fukushima

The population of key fish species in Fukushima waters showed a tangible increase after the 2011 disaster reflecting low fishing pressures from 2011 to 2014. Narimatsu et al. (2017) reported that the population of Pacific cod (*Gadus macrocephalus*) inhabiting waters off northeastern Honshu, Japan, which includes all Fukushima waters, has increased in 2012, 2013, and 2014 after the Great East Japan Earthquake in 2011. Their research suggests that Pacific cod increased post-2011 not because of the occurrence of strong year classes followed by good recruitments but because of the lower mortality after recruitment owing to reduced fishing mortality in the Fukushima area.

Shibata et al. (2015) also reported that the result of their analysis based on a fish population model suggested that abundance of sea ravens (*Hemitripterus villosus*) off Fukushima increased after the 2011 accident. Shibata et al. (2017) further reported that, based on the analysis of their fish population model, the abundance of Japanese flounder (*Paralichthys olivaceus*) on and after 2012 increased owing to the decreased fishing effort during this period. These study outcomes are consistent with oral reports received from Fukushima fishers. The author is a member of the "Fukushima Prefectural Fisheries Reconstruction Committee" which was established by Fukushima FCA in 2012 with the aim of reconstructing the fisheries industry and restarting fisheries operations (Yagi 2014). The committee has repeatedly discussed conditions of fish stocks and fishers recognized increased populations of Pacific cod or some other species during their trial operation after mid-2012.

18.7 Conclusion

Fukushima fisheries are slow to recover after 6 years since the disaster occurred. The Fukushima FCA had stopped fishing since March 15th, 2011 and, after a series of internal discussions, it decided to reopen limited fishing in June 2012 (i.e., trial operation). Total landing value of fish and fishery products from the trial operation has been gradually increasing due to the increased number of catchable target species and increased fishing areas. But its landing value in 2016 was still only 5% of the value recorded in the pre-disaster years.

Safety of the products has been demonstrated by various surveys. More than 100,000 samples were examined from April 2011 to June 2017 by government authorities, and the proportion of samples with over 100 Bq/Kg radiocesium decreased relatively quickly in 2011 and 2012. In and after 2015, the proportion of samples with over 100 Bq/Kg of radiocesium decreased to almost zero. Several studies indicated that the population of key fish species in Fukushima waters showed a tangible increase after the 2011 disaster reflecting low fishing pressures from 2011 to 2014.

But the reality is that the recovery of fishery productions has been sluggish. Weak consumer confidence may have contributed to the extremely slow pace of recovery of Fukushima fisheries. Fish distributors' risk-averse attitudes could have brought additional adverse effects against the recovery of Fukushima fisheries. This situation could continue for several more years judging from the fact that production shortfalls in Fukushima fisheries can be easily offset with products from other Japanese regions or foreign countries. Before the disaster in 2011, Fukushima fisheries produced only 1% of Japan's marine products.[2] Fukushima's share of the Japanese market is even lower when we consider that about half of the marine products consumed in Japan are imported from overseas. In other words, meeting the requirements of the Japanese consumer can be fulfilled without using products from Fukushima. The export of seafood from Japan, in particular products from Fukushima and the surrounding waters, is subject to strict import restrictions or

[2] Author's calculation using the data in Fisheries Agency of Japan (2017) and Fukushima Prefecture (2017b).

prohibitions by various countries. Under these circumstances, continued help for Fukushima fishers is required for several more years to sustain the livelihood of small fishing households as well as maintain societies, traditional knowledge, and other human or social capital in the region.

Acknowledgement The author acknowledges Mr. Kazunobu Noguchi of the Fukushima FCA for his support in data compilation of the trial operation.

References

Aruga K (2017) Consumer responses to food produced near the Fukushima nuclear plant. Environ Econ Policy Stud 19:677–690. https://doi.org/10.1007/s10018-016-0169-y

Doi H, Takahara T, Tanaka K (2012) Trophic position and metabolic rate predict the long-term decay process of radiocesium in fish: a meta-analysis. PLoS One 7:e29295. https://doi.org/10.1371/journal.pone.0029295

Fisheries Agency of Japan (2014) Questions and Answers on fisheries. Website accessed in February 2014. http://www.jfa.maff.go.jp/j/kakou/Q_A/index.html (in Japanese)

Fisheries Agency of Japan (2017) The Great East Japan Earthquake's impact on fisheries and future measures. July 2017 (in Japanese)

Fishery Agency of Japan (2017) White paper on fisheries (in Japanese)

Fukushima Prefecture (2014) Website of fisheries division, Fukushima Prefecture. http://www-cms.pref.fukushima.jp/pcp_portal/PortalServlet. Accessed Feb 2014 (in Japanese)

Fukushima Prefecture (2015) Discussion material for "Fukushima Prefectural Fisheries Reconstruction Committee" Materials prepared by Fukushima Prefecture and distributed to the press and committee members on January 22, 2015

Fukushima Prefecture (2017a) Website of Fisheries Division, Fukushima Prefecture. http://www.pref.fukushima.lg.jp/uploaded/attachment/222247.pdf. Accessed Oct 2017 (in Japanese)

Fukushima Prefecture (2017b) Catch Statistics of Fukushima marine capture Fisheries https://www.pref.fukushima.lg.jp/sec/36035e/suisanka-toukei-top.html. Accessed Oct 2017 (in Japanese)

Furukawa F, Watanabe S, Kaneko T (2012) Excretion of cesium and rubidium via the branchial potassium-transporting pathway in Mozambique tilapia. Fish Sci 78:597–602

Ibaraki Prefecture (2014) Restricted items for marketing and distributions as of February 21, 2014. http://www.pref.ibaraki.jp/nourin/gyosei/housyanou_jyouhou.html#4. Accessed Feb 2014 (in Japanese)

International Atomic Energy Agency (2004) Sediment distribution coefficients and concentration factors for biota in the marine environment, Technical report 422. IAEA, Vienna

Kaneko T, Furukawa F, Watanabe S (2013) Excretion of cesium through potassium transport pathway in the gills of a marine teleost. In: Nakanishi TM, Tanoi K (eds), Agricultural implications 105 of the Fukushima nuclear accident. https://doi.org/10.1007/978-4-431-54328-2_11

Kikkawa T, Yagi N, Kurokura H (2014) The state of concentration of radioactive cesium in marine organisms collected from the Fukushima coastal area: a species by species evaluation. Bull Jpn Soc Fish Sci 80:27–33 (in Japanese)

Ministry of Agriculture, Forestry and Fisheries, Japan (2012) Basic statistical data on damages in agriculture, forestry and fisheries caused by the Great East Japan Earthquake. http://www.maff.go.jp/j/tokei/joho/zusetu/pdf/00_2406all.pdf. Accessed Jan 2015 (in Japanese)

Miyagi Prefecture (2014) Restricted items for marketing and distributions as of February 18, 2014. http://www.pref.miyagi.jp/uploaded/attachment/245220.pdf. Accessed Feb 2014 (in Japanese)

Nakajima T, Matsui T, Sakai Y, Yagi N (2014) Structural changes and imperfect competition in the supply chain of Japanese fisheries product markets. Fish Sci 80:1337–1345

Narimatsu Y, Sohtome T, Yamada M, Shigenobu Y, Kurita Y, Hattori T, Inagawa R (2015) Why do the radionuclide concentrations of Pacific cod depend on the body size? In: Nakata K, Sugisaki H (eds) Impacts of the Fukushima nuclear accident on fish and fishing grounds. Springer Open, Tokyo

Narimatsu Y, Shibata Y, Hattori T, Yano T, Nagao J (2017) Effects of a marine-protected area occurred incidentally after the great East Japan earthquake on the Pacific cod (*Gadus macrocephalus*) population off northeastern Honshu Japan. Fish Oceanogr 26:181–192

Sakai Y, Nakajima T, Matsui T, Yagi N (2012) Asymmetric price transmission in Japanese seafood market. Nippon Suisan Gakkaishi 78:468–478 (in Japanese)

Shibata Y, Yamada M, Wada T et al (2015) A surplus production model considering movements between two areas using spatiotemporal differences in CPUE: application to sea ravens *Hemitripterus villosus* off Fukushima as a practical marine protected area after the nuclear accident. Mar Coast Fish 7:325–337

Shibata Y, Sakuma T, Wada T, Kurita Y, Tomiyama T, Yamada M, Iwasaki T, Mizuno T, Yamanobe A (2017) Effect of decreased fishing effort off Fukushima on abundance of Japanese flounder (*Paralichthys olivaceus*) using an age-structured population model incorporating seasonal coastal-offshore migrations. Fish Oceanogr 26:193–207

Takagi K, Fujimoto K, Watanabe T, Kaeriyama H, Shigenobu Y, Miki S, Ono T, Morinaga K, Nakata K, Morita T (2015) Radiocesium concentration of small epipelagic fishes (sardine and Japanese anchovy) off the Kashima-Boso area. In: Nakata K, Sugisaki H (eds) Impacts of the Fukushima nuclear accident on fish and fishing grounds. Springer Open, Tokyo

Wada T, Nemoto Y, Shimamura S, Fujita T, Mizuno T, Sohtome T, Kamiyama K, Morita T, Igarashi S (2013) Effects of the nuclear disaster on marine products in Fukushima. J Environ Radioact 124:246–254

Wakamatsu H, Miyata T (2017) Reputational damage and the Fukushima disaster: an analysis of seafood in Japan. Fish Sci 83:1049–1057

Yagi N (2014) The state of fishing industry in Fukushima after the nuclear power-plant accident. Glob Environ Res 18:65–72

Yagi N (2016) Impacts of the nuclear power plant accident and the start of trial operations in Fukushima fisheries. In: Nakanishi T, Tanoi K (eds) Agricultural implications of the Fukushima nuclear accident. Springer, Tokyo, pp 217–227

Chapter 19
Visualization of Ion Transport in Plants

Ryohei Sugita, Natsuko I. Kobayashi, Atsushi Hirose, Keitaro Tanoi, and Tomoko M. Nakanishi

Abstract We have developed a real-time radioisotope imaging system (RRIS) to visualize ion transport in plants, and to measure radioactivity in living plants. To know the mechanisms of ion transport in plants, the use of living plants allows us to visualize ion movement in real time. In addition, the RRIS can analyze how a change to the plant environment affects ion transport. In this chapter, we will introduce some of the applications of the RRIS. We analyzed the effect of light on cesium, potassium, magnesium, phosphate, and calcium transport in plants using the RRIS. The results show that magnesium, potassium, and calcium transport in plants were not influenced by light. On the other hand, the amount of cesium and phosphate absorption in roots decreased after light-off. Moreover, the amount of phosphate transport from root to shoot also decreased after light-off.

Keywords Live imaging · Radiocesium · Rice · Xylem flow · Real-time radioisotope imaging

Abbreviations

Ca	Calcium
Cs	Cesium
K	Potassium
Mg	Magnesium
P	Phosphate
RRIS	real-time radioisotope imaging system

R. Sugita (✉) · N. I. Kobayashi · A. Hirose · K. Tanoi · T. M. Nakanishi
Graduate School of Agricultural and Life Sciences, The University of Tokyo, Bunkyo-ku, Tokyo, Japan
e-mail: asugita@g.ecc.u-tokyo.ac.jp

T. M. Nakanishi et al. (eds.), *Agricultural Implications of the Fukushima Nuclear Accident (III)*, https://doi.org/10.1007/978-981-13-3218-0_19

19.1 Introduction

After the accident of Fukushima Daiichi nuclear power plant, it is necessary to reduce radiocesium (^{137}Cs) in crops. To solve this problem, it is important to know how much ^{137}Cs is accumulated in each tissue and to understand in detail the mechanism of ^{137}Cs transport in plants. There are two pathways responsible for elemental transport in plants, namely, the symplastic and apoplastic pathways. The symplastic pathway transports water and ions via the plasmodesma, whereas transport by the apoplastic pathway is via cell walls. In addition, transporters within membranes also affect elemental transport. Elements within the vascular bundle are distributed between tissues through vessels and sieve tubes. Water and nutrients are mainly transported via vessels and the driving force is root pressure and transpiration. Water, nutrients, and photosynthetic products are transported via sieve tubes and the driving force depends on the concentration gradient of sucrose. The environmental changes, such as weather, affect ion transport because the rate of transpiration and photosynthesis is influenced by light. Various techniques are available to researchers to analyze transport and distribution of ^{137}Cs in plants. In particular, we often use imaging techniques. One such imaging technique is the real-time radioisotope imaging system (RRIS) which was developed in our laboratory (Nakanishi et al. 2009). The RRIS can visualize transport and distribution of various radioisotopes including ^{137}Cs in living plants. In addition, we can analyze the amount of cesium using the images obtained by the RRIS. To analyze ion transport in plants using the RRIS with living plants is a big advantage because ions in plants are continuously moving. Some of the findings related to Cs transport include the following: Kobayashi visualized ^{137}Cs in rice plants grown in water culture and soil, and found minimal ^{137}Cs translocation to rice grain when grown in soil because of ^{137}Cs adsorption to soil particles (Kobayashi 2013). In addition, Sugita et al. reported that the transport manner of ^{137}Cs and ^{42}K from roots to above-ground parts of Arabidopsis is similar for ^{137}Cs and ^{42}K (Sugita et al. 2016). Kobayashi et al. demonstrated that ^{137}Cs was retained in the root tissues with high efficiency, while ^{42}K was easily exchanged and transported towards the shoots (Kobayashi et al. 2016). Moreover, RRIS can analyze how changes in the growth environment of living plants affect ion transport, such as temperature or nourishment. In this chapter, we will report the characteristics of the Cs transport compared to other elements in roots of rice plants when environmental conditions change from light to darkness.

19.2 Imaging System and Imaging Methods

19.2.1 Obtaining Images

The characteristics of the Real-time radioisotope imaging system (RRIS) are as follows:

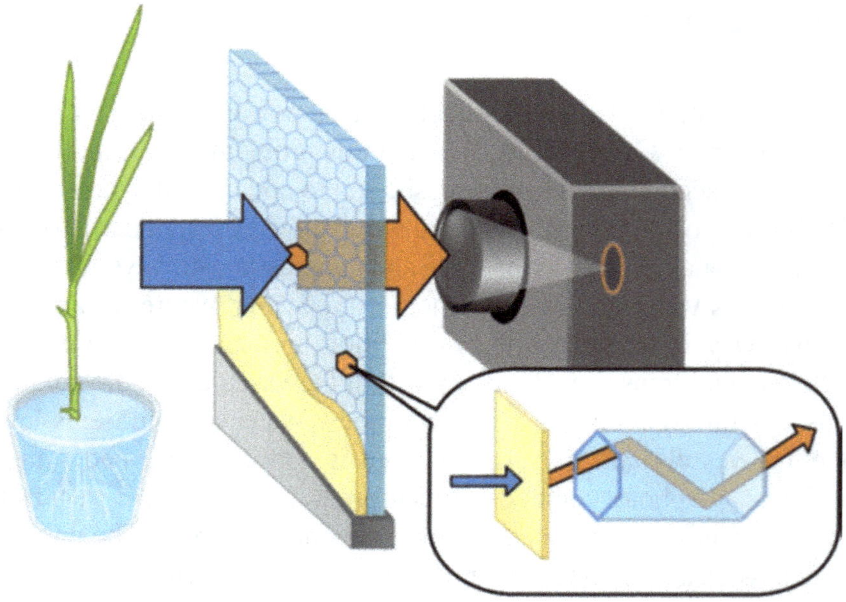

Fig. 19.1 Overview of the mechanism of real-time radioisotope imaging system
The radiation from nuclide is converted to visible light by CsI (Tl) scintillator, and the visible light is then captured by CCD camera
This figure is quoted from Sugita et al. (2017)

- It is possible to visualize and quantify ion movements in living plants using radioisotopes;
- It is possible to detect many types of radiations such as beta-, X-, gamma-rays, and positron (Sugita et al. 2014)

There are two steps to visualize radiation. (1) Radiation is converted to visible light by CsI (Tl) scintillator, and (2) the visible light is captured by CCD camera (Fig. 19.1). RRIS can use living plants for experimentation, so we can get sequential images. The light converted from radiation is very weak, therefore in order to take images, it is necessary to use a dark conditions box.

19.2.2 Two Types of View Areas

The RRIS has two types of view areas. One is 10 × 20 cm in size (w × h) and known as the macro-RRIS. The other, the micro-RRIS, has the ability to take images within the μm range using a microscope. The view area is approximately 600 × 600 μm (optical lens: x20 magnification) to 5.2 × 5.2 mm (x2.5 magnification) in size (W × L).

19.2.2.1 macro-RRIS

The macro-RRIS mainly visualizes ion movements in whole plants. Test plants are placed on the CsI (Tl) scintillators (Fig. 19.2a). The dark box's size is approximately 120 × 150 × 80 cm (W × L × D). We have taken images of rice, Arabidopsis, soy beans, poplar, for example.

19.2.2.2 micro-RRIS

The mechanism of how to take images by micro-RRIS is identical to how images were taken by macro-RRIS. The CsI (Tl) scintillator (1.5 × 1.5 cm) is placed on the plant, and the light converted from radiation by the scintillator is captured by CCD camera through a microscope (Fig. 19.2b).

19.2.3 Radioisotopes

Commercially available radioisotopes are predominantly used for visualizing elements within plants. On the other hand, radioisotopes with short half-lives (e.g., ^{42}K and ^{28}Mg) have to be produced by our laboratory as they are not available commercially. For example, ^{42}K has a half-life of 12 h and ^{28}Mg has a half-life of 20 h.

19.2.3.1 ^{42}K

^{42}K is produced by ^{42}Ar-^{42}K generators (Aramaki et al. 2015). The half-life of ^{42}Ar and ^{42}K is about 33 y and 12 h, respectively. The method of producing ^{42}K is described as follows (Fig. 19.3a): (1) a steel electrode is inserted into a generator

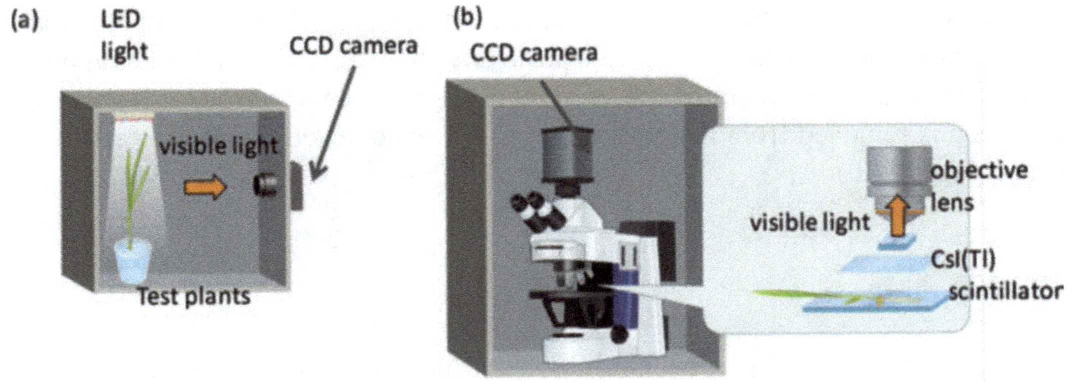

Fig. 19.2 Schematic drawing of the two types of real-time radioisotope imaging system (RRIS)
(**a**) Macro-RRIS; the view area is approximately 10 × 20 cm
(**b**) Micro-RRIS; the view area is approximately 600 × 600 μm (x20 magnification)

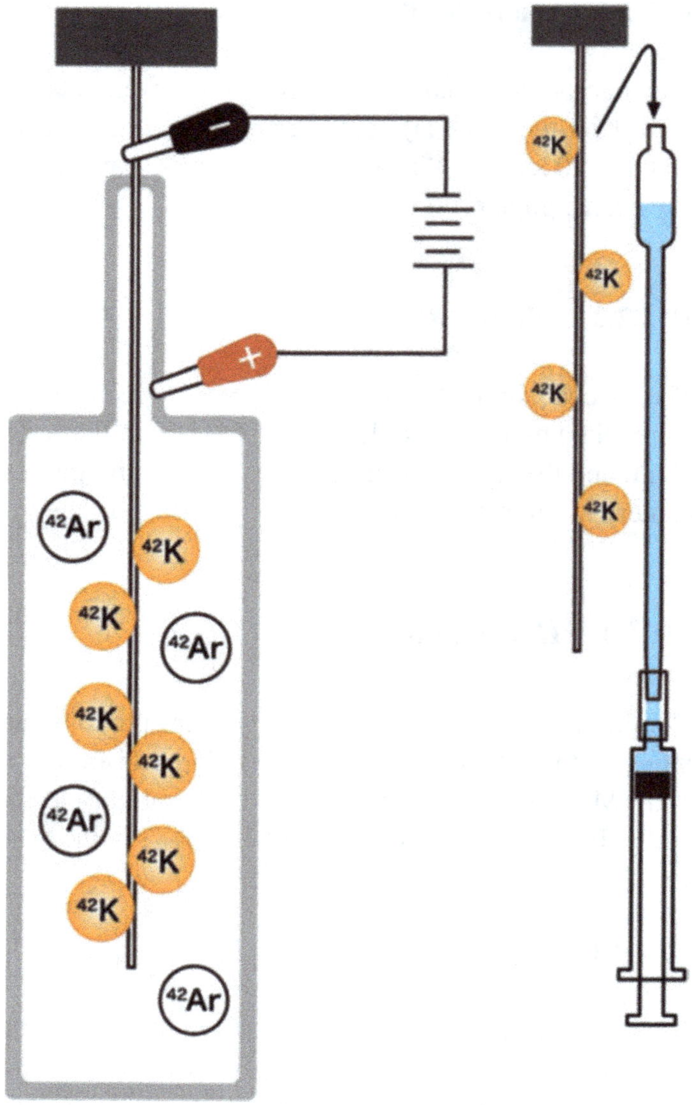

Fig. 19.3 Schematic drawing of the ^{42}Ar-^{42}K generator
(a) ^{42}K is attracted to the cathode when voltage is applied
(b) ^{42}K is extracted from the cathode using water

with a cock which has ^{42}Ar. (2) ^{42}K is produced from ^{42}Ar when the steel electrode has a minus charge and the generator is positively charged. (3) When 60 V is applied to the electrode and generator, ^{42}K is attracted to the steel electrode. (4) the steel electrode is removed from the generator and ^{42}K is collected inside a pipette with water (Fig. 19.3b).

19.2.3.2 ^{28}Mg

^{28}Mg was produced using a cyclotron by ^{27}Al (α, 3p) ^{28}Mg reaction (Tanoi et al. 2013).

19.2.4 Applying Radioisotopes to Plants

Radioisotopes can be applied to plants either in gas or liquid form. Gas radioisotopes such as ^{14}C-labeled CO_2 (Sugita et al. 2013) can be applied to the whole plant or to only specific tissues (e.g., leaves, shoots, stems). Liquid radioisotopes such as ^{137}Cs and ^{42}K are applied to roots when test plants are grown in culture solutions, gels or soils. In addition, it is possible to visualize ion movement from a leaf to other tissues (Sugita et al. 2016).

19.2.5 Applying Light to Plants

The CCD camera in the RRIS has a high sensitivity, but its detector can become damaged if exposed to strong light. In addition, visible light converted from radiation is very weak, and therefore it is necessary to have complete dark conditions while taking images. Because plants need light for growth, an intermittent lighting system was developed (Hirose et al. 2013). This lighting system ensures dark conditions when images are being taken and light conditions for the remaining time. For example, one dark/light cycle is 3 min/7 min and the RRIS gets an image for 3 min in the dark condition. Moreover, the cycle can be set for 24 h, allowing the RRIS to visualize ion movements for 24 h. The micro-system has the same type of system.

19.3 Impact of Light for Ion Transport in Plants

19.3.1 Experimental

19.3.1.1 Test Plants

Rice seeds (*Oryza sativa*. L. 'Nipponbare') were germinated in 0.5 mM $CaCl_2$ solution under dark conditions for 2 days. The plants were transplanted into half-strength Kimura-B nutrient solution (270 mM K^+) for 6 days. Next, four plants were placed in a root chamber with half-strength Kimura-B nutrient solution. For ^{137}Cs imaging, 0.1 μM $CsCl_2$ was added to the Kimura-B nutrient solution. The root chamber was made of a polyethylene bag and polyurethane sheet (Fig. 19.4a). The plant roots were fixed inside the bag using the polyurethane sheet, and the bag was fixed onto the scintillator (Fig. 19.4b).

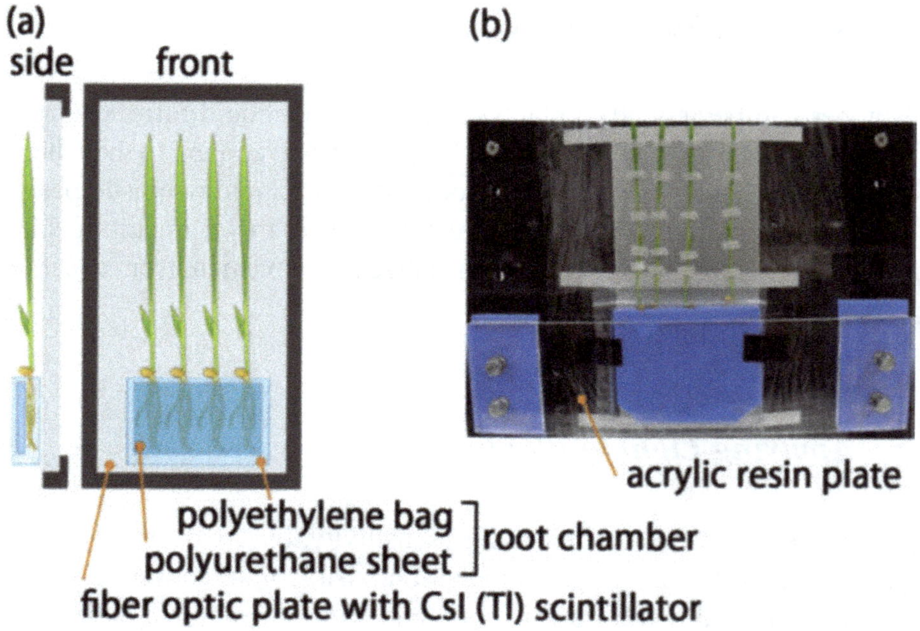

Fig. 19.4 The mounting arrangement of test plants in the real-time radioisotope imaging system (**a**) Four plants were placed in a root chamber. The roots were fixed using the polyurethane sheet (**b**) The root chamber was pressed onto the scintillator using acrylic resin plates
This figure is quoted from Sugita et al. (2017)

Table 19.1 General overview of the radionuclides used in this study
The value of each radiation is the maximum energy (keV)

Nuclide	Mode of decay	Half-life	β-ray	γ-ray
^{28}Mg	β⁻	20.9 h	860	1589
^{28}Al	β⁻	2.24 m	2863	1779
^{32}P	β⁻	14.3 d	1711	–
^{42}K	β⁻	12.4 h	3525	1525
^{45}Ca	β⁻	163 d	257	–
^{137}Cs	β⁻	30.2 y	1176	662
^{137m}Ba	IT	2.55 m	–	662

The ^{28}Al is daughter nuclide of ^{28}Mg
The ^{137m}Ba is daughter nuclide of ^{137}Cs

19.3.1.2 Visualization of Element Movement Using RRIS

The following radioisotopes were applied to the roots: ^{28}Mg: 17 kBq mL^{-1}; ^{32}P: 30 kBq mL^{-1}; ^{42}K: 7 kBq mL^{-1}; ^{45}Ca: 830 kBq mL^{-1}; and ^{137}Cs: 25 kBq mL^{-1}. Their characteristics are shown in Table 19.1. The total length of time for taking images was 10 h. During the first 5 h, the dark/light cycle of 3 min was set up using LED lights (100 μmols^{-1} m^{-2}), and then the RRIS took images using 3 min of exposure time for each 3 min dark condition. During the next 5 h, the RRIS took images using 3 min of exposure time for each 10 min in the dark condition.

19.3.1.3 Transpiration Rate

Transpiration rate was calculated as the decreased level of the culture water under light for 10 h or under light condition for 5 h followed by 5 h of dark condition in the dark box of the RRIS.

19.3.2 Results and Discussion

RRIS was used to visualize the response of ion movement between light and dark conditions. At first, each radioisotope (^{137}Cs, ^{42}K, ^{28}Mg, ^{32}P, and ^{45}Ca) was supplied to roots and visualized for 5 h under light (0–5 h) and dark conditions (5–10 h) (Fig. 19.5a-19.5d). To analyze the changes during uptake in roots and translocation from root to shoot, the radioisotope activities were measured in the region of interest (ROI) of the shoot and root based on the RRIS images (Fig. 19.6a); the time course of signal intensity within the ROI was then obtained. The signal intensity value was shown relative to the 5 h after the radioisotope absorption (Fig. 19.6b–e). In addition, the slopes of the calibration curves were calculated for 3–5 h and 5–7 h to compare the radioisotope accumulation velocities (Fig. 19.6f). The results show that the ^{137}Cs activity in roots only increased slightly after turning the light off (Fig. 19.6b) and the slope of the calibration curve showed a statistically significant decrease (Fig. 19.6f). On the other hand, the activities in the shoots of ^{137}Cs were not influenced by light and continuously increased after turning the light off. These results suggest that Cs uptake from the root is largely influenced by light and Cs transport from root to shoot is not influenced by light. In regards to K, there was no influence of light on either shoot or root (Fig. 19.6b). These results suggest that in Cs absorption from roots, it is possible that transporters which are different to K transporter greatly contributes to Cs transport and the activity of those transporters are affected by light. Both ^{45}Ca and ^{28}Mg accumulation in shoot and root were not influenced by the change from light to dark conditions (Fig. 19.6c, e, f). For ^{32}P, the reaction from light to dark condition was prompt, with ^{32}P accumulation rate in both shoot and root decreasing (Fig. 19.6d). Considering the response time of transpiration from light to dark condition is several minutes (Ishikawa et al. 2011), it was assumed that transpiration greatly influenced ^{32}P transport. After 5 h from light to dark condition, transpiration decreased 57% in comparison with continuous light (Fig. 19.7). This result suggests that P was absorbed and transported with water, while Mg, Ca, and K were not affected by the flow of water. On the other hand, the

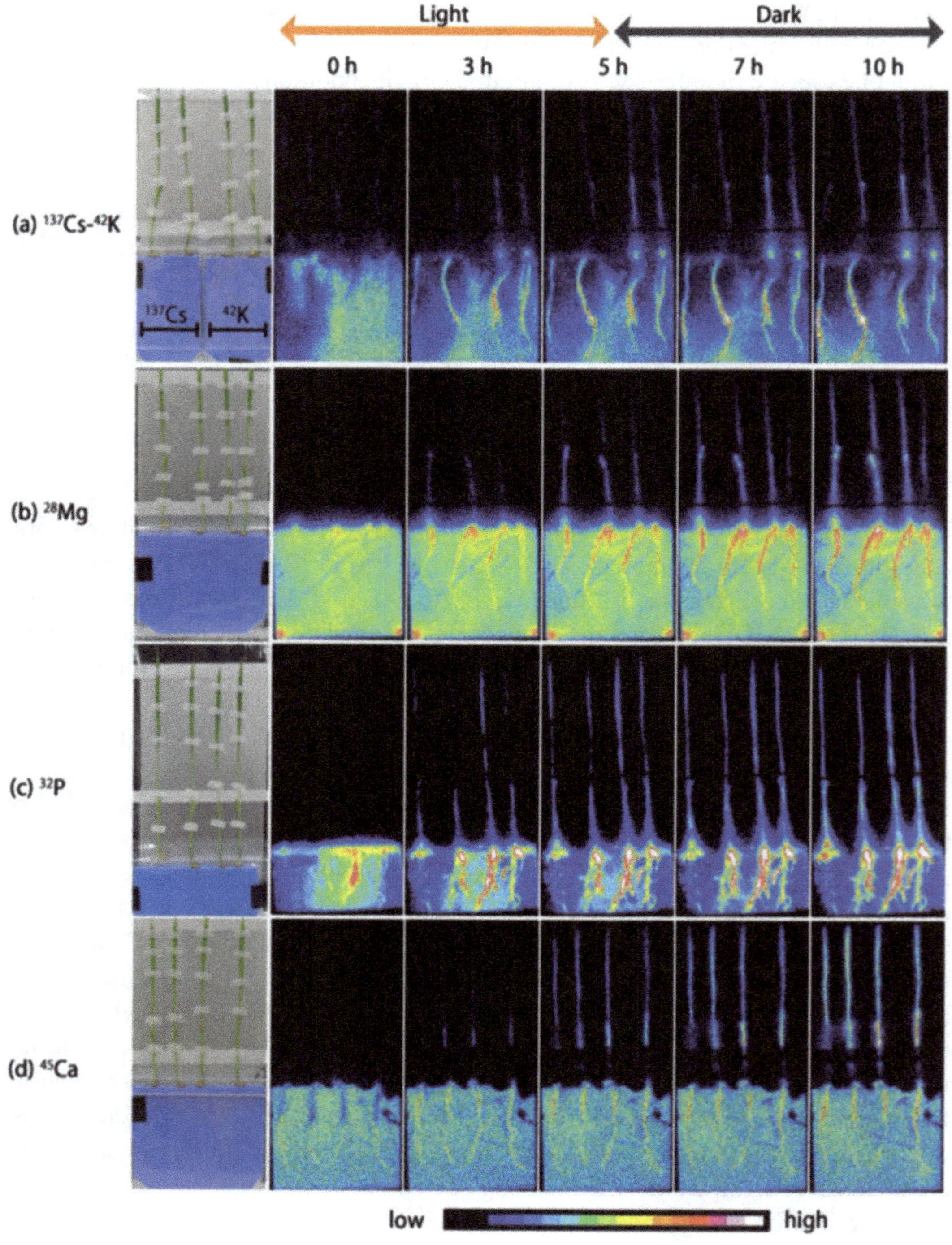

Fig. 19.5 Serial images under light/dark conditions in rice, taken by RRIS
After the radioisotope was added, the serial images were taken under light conditions for the first 5 h and dark conditions for the following 5 h
(**a**) ^{42}K-^{137}Cs, (**b**) ^{28}Mg, (**c**) ^{32}P, and (**d**) ^{45}Ca
This figure is quoted from Sugita et al. (2017)

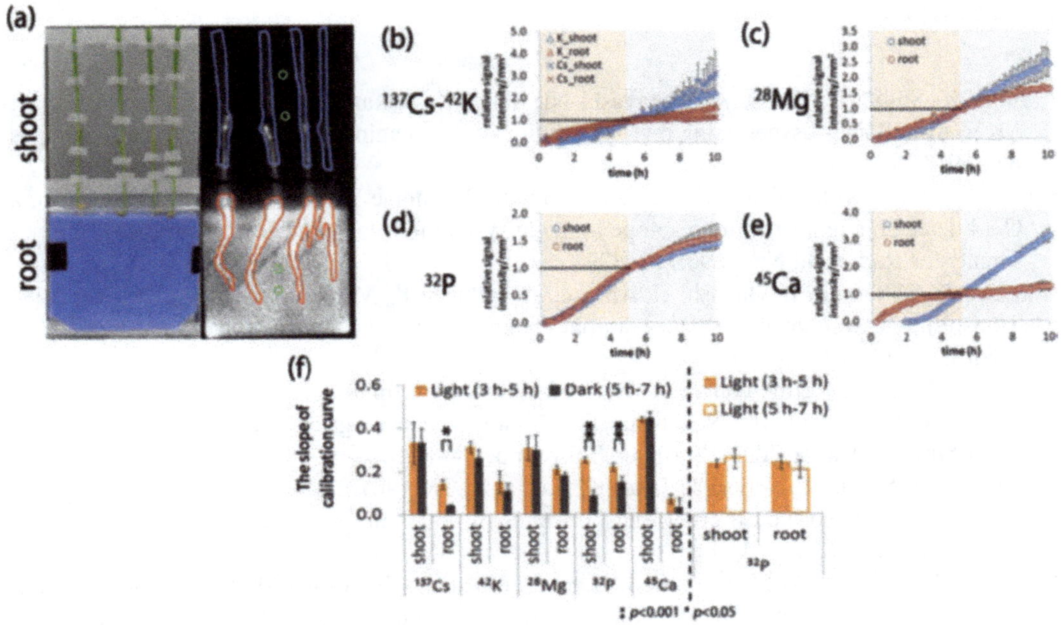

Fig. 19.6 The influence of ion movement under light/dark conditions
(**a**) Photograph of test plants. The blue, red, and green lines indicate the region of interest (ROI) of shoot, root, and background, respectively. Time-course analysis of the radioactivity of (**b**) ^{42}K-^{137}Cs, (**c**) ^{28}Mg, (**d**) ^{32}P, and (**e**) ^{45}Ca. The relative signal intensity was normalized at 5 h. (**f**) The slope of the calibration curve between 3–5 h (light) and 5–7 h (dark) based on Fig. **b-e**. Data represent means ± standard deviation (n = 4 plants)
This figure is quoted from Sugita et al. (2017)

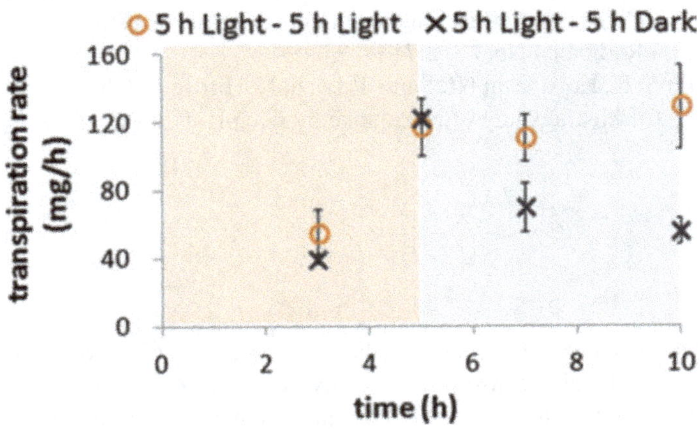

Fig. 19.7 Transpiration rate was calculated under the light condition for 10 h (circles), or under light condition for 5 h followed by 5 h of dark condition (crosses) at 3, 5, 7 and 10 h. Data represent means ± standard deviation (n = 4 plants)
This figure is quoted from Sugita et al. (2017)

signal level of ^{45}Ca in the root only increased slightly after 3 h after its addition (Fig. 19.6e). This result shows that the replacement of Ca^{2+} with ^{45}Ca in roots is quick. In addition, ^{45}Ca accumulation in shoot continuously increased, suggesting that surplus ^{45}Ca is transported to the shoot.

As shown in the examples above, RRIS is a powerful tool to analyze responses in plants to their environment. To decrease ^{137}Cs accumulation in plants, we will continue to research the uptake and translocation of ^{137}Cs in plants using RRIS.

References

Aramaki T, Sugita R, Hirose A, Kobayashi NI, Tanoi K, Nakanishi TM (2015) Application of ^{42}K to Arabidopsis tissues using real-time radioisotope imaging system (RRIS). Radioisotopes 64:169–176

Hirose A, Yamawaki M, Kanno S, Igarashi S, Sugita R, Ohmae Y et al (2013) Development of a C-14 detectable real-time radioisotope imaging system for plants under intermittent light environment. J Radioanal Nucl 296:417–422

Ishikawa JS, Hatano MM, Hayashi H, Ahamed A, Fukushi K, Matsumoto T, Kitagawa Y (2011) Transpiration from shoots triggers diurnal changes in root aquaporin expression. Plant, Cell Environ 34:1150-1163

Kobayashi NI (2013) Time-course analysis of radiocesium uptake and translocation in rice by radioisotope imaging. In: Nakanishi TM, Tanoi K (eds) Agricultural implications of the Fukushima nuclear accident. Springer, Tokyo, pp 37–48

Kobayashi NI, Sugita R, Nobori T, Tanoi K, Nakanishi TM (2016) Tracer experiment using $^{42}K^+$ and $^{137}Cs+$ revealed the different transport rates of potassium and caesium within rice roots. Funct Plant Biol 43:151–160

Nakanishi TM, Yamawaki M, Kannno S, Nihei N, Masuda S, Tanoi K (2009) Real-time imaging of ion uptake from root to above-ground part of the plant using conventional beta-ray emitters. J Radioanal Nucl 282:265–269

Sugita R, Kobayashi NI, Hirose A, Ohmae Y, Tanoi K, Nakanishi TM (2013) Nondestructive real-time radioisotope imaging system for visualizing C-14-labeled chemicals supplied as CO_2 in plants using Arabidopsis thaliana. J Radioanal Nucl 298:1411–1416

Sugita R, Kobayashi NI, Hirose A, Tanoi K, Nakanishi TM (2014) Evaluation of in vivo detection properties of 22 Na, 65 Zn, 86 Rb, 109 Cd and 137 Cs in plant tissues using real-time radioisotope imaging system. Phys Med Biol 59:837–851

Sugita R, Kobayashi NI, Hirose A, Saito T, Iwata R, Tanoi K et al (2016) Visualization of uptake of mineral elements and the dynamics of photosynthates in arabidopsis by a newly developed real-time radioisotope imaging system (RRIS). Plant Cell Physiol 57:743–753

Sugita R, Kobayashi NI, Hirose A, Iwata R, Suzuki H, Tanoi K et al (2017) Visualization of how light changes affect ion movement in rice plants using a real-time radioisotope imaging system. J Radioanal Nucl 312:717–723

Tanoi K, Kobayashi NI, Saito T, Iwata N, Hirose A, Ohmae Y et al (2013) Application of 28Mg to the kinetic study of Mg uptake by rice plants. J Radioanal Nucl 296:749–751

Chapter 20
^{90}Sr Analysis Using Inductively Coupled Plasma Mass Spectrometry with Split-Flow Injection and Online Solid-Phase Extraction for Multiple Concentration and Separation Steps

Makoto Furukawa and Yoshitaka Takagai

Abstract We aimed to develop a rapid and sensitive method to analyze the radioactivity of ^{90}Sr by combining multiple techniques, including online solid-phase extraction (SPE) and inductively coupled plasma mass spectrometry (ICP-MS). An automatic analytical system was designed to execute the proposed process from sample injection to measurement. The analysis time is approximately 20 min and the limit of detection is 0.3 Bq/L (equivalent to 0.06 pg/L) with 50 mL of the sample. Although several challenges were encountered with the ICP-MS measurements of ^{90}Sr, several techniques were leveraged to overcome them. Online solid-phase extraction (SPE) was used to concentrate the sample automatically; the interference from polyatomic ions and isobars was removed by an oxidation, and the extraction and recovery ratio of solid phase were measured by split-flow injection with internal standard correction during the transient signal measurement. These improvements were shown to allow measurements of ^{90}Sr in various kinds of samples to be conducted more quickly than by alternative conventional radiometric methods.

Keywords Strontium-90 · ICP-MS · Online solid phase extraction · Split-flow injection

---

M. Furukawa (✉)
PerkinElmer Japan, Yokohama, Kanagawa, Japan

Faculty of Symbiotic Systems Science, Fukushima University, Fukushima, Japan

Graduate School of Agricultural and Life Sciences, The University of Tokyo, Bunkyo-ku, Tokyo, Japan
e-mail: makoto.furukawa@perkinelmer.com

Y. Takagai
Faculty of Symbiotic Systems Science, Fukushima University, Fukushima, Japan

Institute of Environmental Radioactivity, Fukushima University, Fukushima, Japan
e-mail: takagai@sss.fukushima-u.ac.jp

T. M. Nakanishi et al. (eds.), *Agricultural Implications of the Fukushima Nuclear Accident (III)*, https://doi.org/10.1007/978-981-13-3218-0_20

20.1 Introduction

Measurements of pure-beta-emitting radioactive ^{90}Sr require that it be isolated from other beta nuclides. The standard analysis process is milking, and low-back-gas-flow counting requires multi-step chemical separation; this process is complex, and takes a lot of time and human handling. Moreover, radioactive ^{90}Y production is required to conduct highly sensitive ^{90}Sr radiometric measurements and the entire analysis takes 1–2 weeks. However, there is a need for a simple and rapid analysis method, particularly for emergency situations such as the Fukushima Daiichi nuclear accident. In this chapter, we introduce a rapid method conducted in a fully automated analysis system for ^{90}Sr analysis using inductively coupled plasma mass spectrometry (ICP-MS) and online solid-phase extraction (online SPE) with a flow-injection system (Takagai et al. 2014, 2017). This method requires only nitric acid and water as reagents. The use of automation mitigates the radioactive hazards to human health and improves the precision of the analysis. The mass spectrometer includes a system to separate the elements in aqueous samples by mass (more precisely, the ratio between an element's mass and its electric charge (m/z)). The mass resolution depends on the type of device used; the quadrupole-type mass spectrometer used in ICP-MS has an integral resolution. Therefore, this technique suffers from poor sensitivity and interference from isobars since there are several elements with masses of 90 such as the stable isotope ^{90}Zn, naturally occurring in the environment. Though ICP-MS can be used to analyze inorganic elements with very high sensitivity, this analysis alone cannot be used to detect low concentrations of ^{90}Sr on the order of a few Bq/L such as the environmental ^{90}Sr level. The half-life of ^{90}Sr is 28.9 years and 1 Bq/L of ^{90}Sr is equivalent to a mass concentration of almost 0.2 pg/L. The limit of detection of the stable isotope, ^{88}Sr, by ICP-MS in an ordinary laboratory environment is approximately 500 pg/L because of contamination from the environment and the insufficient sensitivity. Our research team has previously developed a method of concentration and separation to improve the sensitivity and prevent the interference from other nuclides. In addition, as explained herein, several additional techniques are leveraged for this analysis. The proposed system measures ^{90}Sr based on a working curve determined using a standard solution of a stable isotope of Sr based on the correlation between the detection intensity of the stable isotope of Sr and ^{90}Sr. This feature removes the need for a radioactive standard solution, which is difficult to obtain and handle.

20.2 Materials and Methods

Figure 20.1 explains the online the SPE/ICP-MS method. This system combines several techniques. A FIAS400 flow injection device, which has an eight-direction switching valve and two peristaltic pumps, was used. Either a NexION300S or ELAN DRC II (PerkinElmer) ICP-MS instrument, which have collisions and

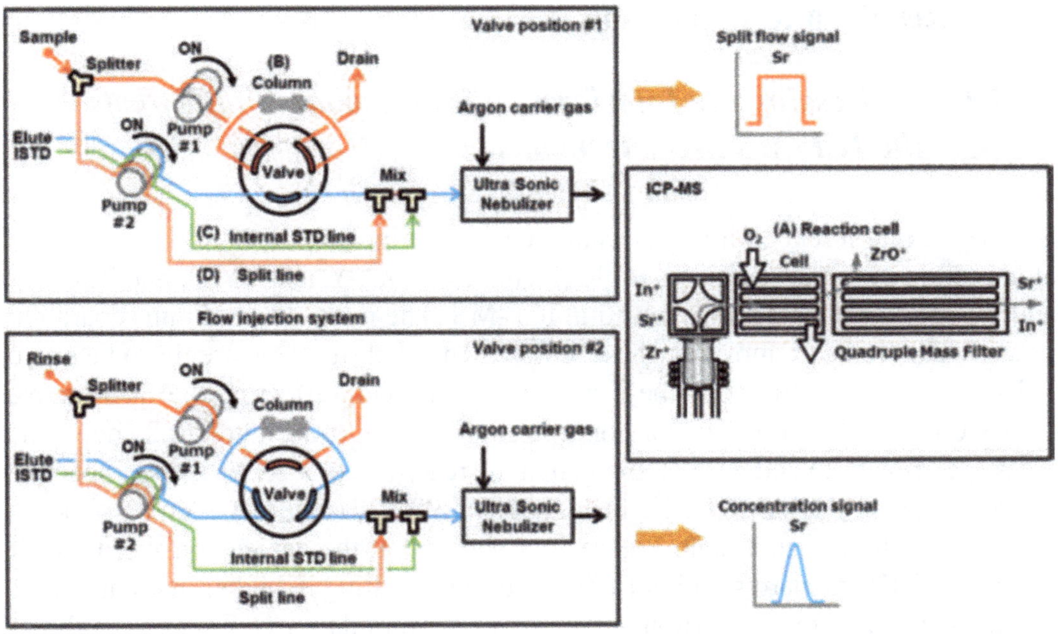

Fig. 20.1 Illustration of the online SPE/ICP-MS system with split flow. The injected sample solution is divided into two flows. One flow is measured by direct injection of the sample and the other flow is measured after preconcentration following online column separation. The measurements from both the direct injection mode and the online preconcentration mode are conducted continuously and in parallel in this automated system. In this way, the relative recovery percentage (related to the SPE efficiency) can be calculated and the targets can be quantified simultaneously. This system can rapidly and very sensitively determine the amount of ^{90}Sr by combining (**a**) inference removal with the oxygen reaction, (**b**) column preconcentration and separation of ^{90}Sr, (**c**) online correction with an internal standard, and (**d**) measurement of the relative recovery

reaction cells to remove the spectrum interference, were also used. As Fig. 20.1 shows, the system incorporates several techniques: (A) a mechanism to remove the interference from isobars with the ^{90}Sr measurements using the collisions and reaction cells in the ICP-MS instrument. (B) PEEK columns (PerkinElmer Japan, φ4 × L50 mm) filled with Sr resin (Eichrom Technology, Particle size 50–100 μm) were used with the flow-injection system for online concentration and separation. Because the adsorption capacity of the Sr resin is limited, one or two columns may be connected alternatively based on the Sr concentration of the sample (not only ^{90}Sr but also stable Sr). (C) To monitor the fluctuation in the ICP-MS measurement sensitivity and the measurement intensity due to the sample matrix (several interference matters) over the long term, internal standard elements are introduced at a fixed speed at a point below the columns and before the nebulizer; this is expected to minimize the error in the measurements. (D) To measure the relative recovery ratio, a novel technique based on split-flow injection was developed; this mechanism allows the sample to be split online and analyzed in parallel both before and after the column concentration step. These strategies are seamlessly integrated into a single system to enable rapid analysis.

20.3 Results and Discussion

20.3.1 Addressing the Interference from Isobars that Affects the ICP-MS Measurements

Sources of interference for the measurement of ^{90}Sr include other isobars and polyatomic ions with mass numbers of 90 (including ^{90}Zr, ^{90}Y, and $^{89}Y^{1}H$, and $^{54}Fe^{36}Ar$, and $^{74}Ge^{16}O$) or the plasma involved in ICP-MS. The multi-step column separation process can remove most of these components before they are introduced to ICP-MS. However, traces of these components remain, affecting the quantitative values obtained because the ^{90}Sr concentration is also low. Figure 20.2 shows the mass spectrum with and without oxygen in the reaction cells. A certified radioactive reference material (DAMRI, 51.7 kBq/g, 01.04.1993) was diluted as an RI standard solution for the measurement. Figure 20.2 shows that the device converted the Zr and Y in the RI standard solution to ZrO and YO, resulting in higher mass numbers. On the other hand, only a portion of the Sr was converted to SrO. Figure 20.3 shows the relationship between the introduced oxygen gas flow causing a reaction with oxygen and the differences between the intensities of the detected ions. Sr, Zr, and Y were converted to plasma ions and exposed to collisions in the reaction cells, resulting in oxidation. The susceptibility of an ion to oxidization depends on the

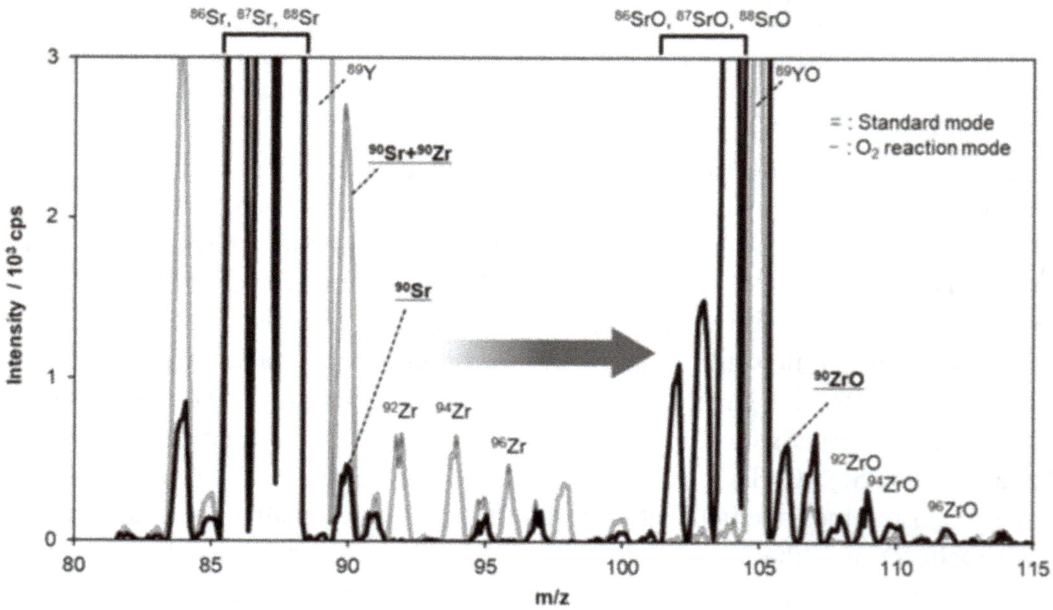

Fig. 20.2 Mass spectra of a radioactively certified ^{90}Sr solution containing stable Sr, Y, and, Zr without the oxygen reaction (white line) and with the oxygen reaction (solid black line) in the dynamic reaction cell (Takagai et al. 2014). In the spectrum in the absence of oxygen, the peak attributed to ^{90}Zr overlapped with that associated with ^{90}Sr. In the presence of oxygen, the Zr peak was shifted to a higher mass number as it existed as ZrO while some of the Sr remained as ^{90}Sr in the position of m/z = 90. In the oxygen reaction in the dynamic reaction cell, only a small portion of the Sr was converted to SrO. The spectra showed that 25 ppt was present in the ^{90}Sr certificated solution (equivalent to 127 Bq/g) when the sample was directly injected through a concentric nebulizer into the ICP-MS instrument without online column preconcentration

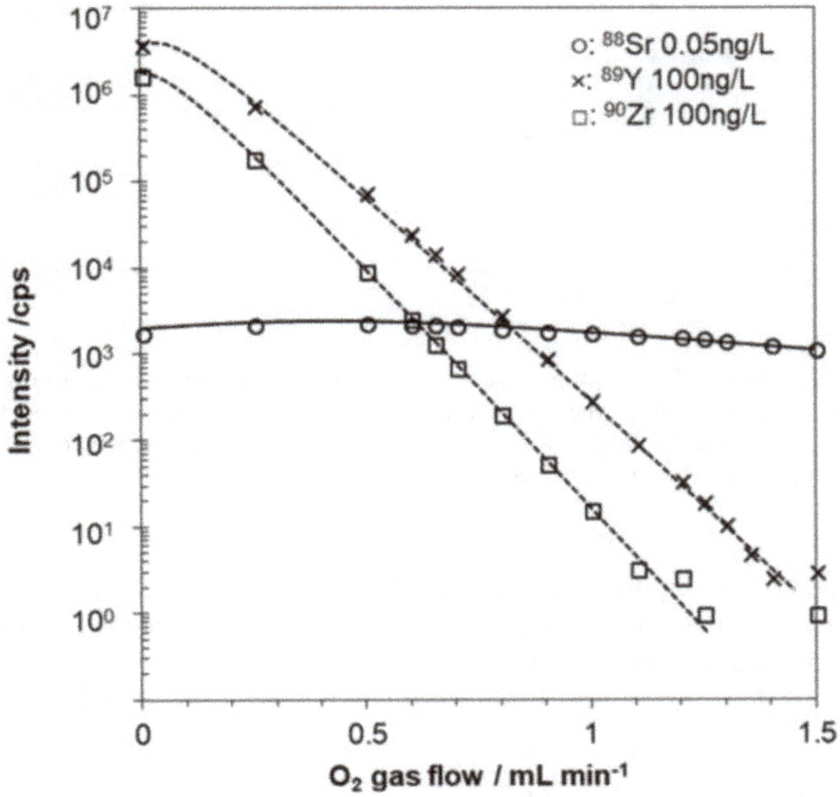

Fig. 20.3 Intensities of the Sr (circles), Y (crosses), and Zr (squares) peaks as functions of the flow rate of O_2 into the dynamic reaction cell (Takagai et al. 2014). The initial ^{88}Sr, ^{89}Y and ^{90}Zr concentrations were 0.05, 100, and 100 ng/L, respectively. The interference of isobars (^{90}Zr and $^{89}Y^1H$ etc.) could be removed effectively by the introduction of oxygen into the dynamic reaction cell in ICP-MS

element. The detected ion Zr and Y intensities decreased proportionally as the oxygen gas flow increased, indicating that most of the Zr and Y were oxidized. However, the detected Sr intensity remained approximately fixed, meaning that Sr is relatively resistant to oxidization. This confirms that Sr can be physically separated from Zr and Y on the basis of the susceptibility to oxidization. Thus, combining this technique with column separation removes the interference from isobars and polyatomic ions completely such that the detection counts of m/z = 90 is approximately the same as background level in the blank sample. Thus, it is possible to measure lower concentrations over the relatively low background signal.

20.3.2 Addressing the Sr Selectivity of the Resin Based on Its Adsorption Characteristics

As shown in Fig. 20.4, the adsorption characteristics of the Sr resin were evaluated in terms of the recovery ratios for 53 kinds of elements. The results show that Sr was adsorbed selectively by the resin. However, Ba and Pb, whose mass numbers are 138 and 208 respectively, were also adsorbed. Therefore, for the measurement of

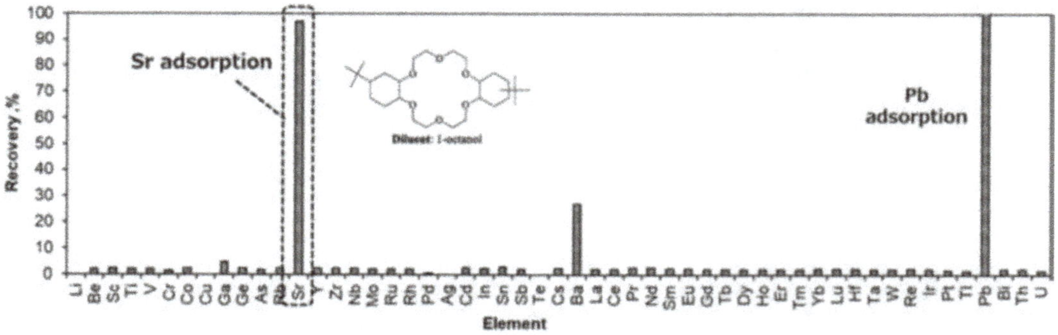

Fig. 20.4 Evaluation of adsorption capacity of the Sr resin for several metal ions. Multiple elements were adsorbed under the optimum adsorption conditions (20 vol% HNO_3) and, then, eluted in ultra-pure water from the Sr resin and the recovery percentages were compared. Large amounts of Sr, Ba and Pb were recovered

beta rays, it is necessary to separate Sr and Pb. However, ICP-MS separates nuclides based on their masses for detection and, thus, Pb in an eluate from the columns does not affect the detection of m/z = 90.

20.3.3 Radioactive ^{90}Sr Measurements Using a Standard Solution of the Stable Isotope, ^{88}Sr

In general, ICP-MS measures analytical targets by comparing the detected intensity of a sample with that of a standard solution with a known concentration. Therefore, it is necessary to measure an ^{90}Sr solution with a known concentration for the subsequent analysis of ^{90}Sr. However, radioactive ^{90}Sr has to be handled carefully in a specialized facility for radioactive materials. Thus, to allow easier and safer measurement, it is better to avoid the use of radioactive standard materials. We hypothesized that a ^{90}Sr calibration curve of a stable isotope of Sr could be used to measure Sr because the detected intensities of ^{90}Sr and the stable Sr isotope are correlated. However, ICP-MS exhibits a mass bias, a phenomenon where the efficiency varies for different ions depending on the mass. Several methods can be used to correct the mass bias such as the relative standardization method to derive a correction factor by measuring the certified standard materials whose isotopic ratio is guaranteed before and after the sample measurement and the internal correction method to correct the results using the measured values of two stable isotopes when an element has multiple stable isotopes. Both of the methods are used to correct the mass bias in the measured values and to calculate their true isotopic ratios.

In this study, we conducted ^{90}Sr measurements using ^{88}Sr as the target as the ratio of ^{88}Sr/^{86}Sr is the same as the natural isotope abundance ratio. This ratio can be controlled by changing the cell entrance voltage, one of the voltage parameters related to the ion vitrification in ICP-MS. Figure 20.5 shows the changes in the measured ratio of ^{88}Sr/^{86}Sr as the cell-entrance voltage was varied. In one condition,

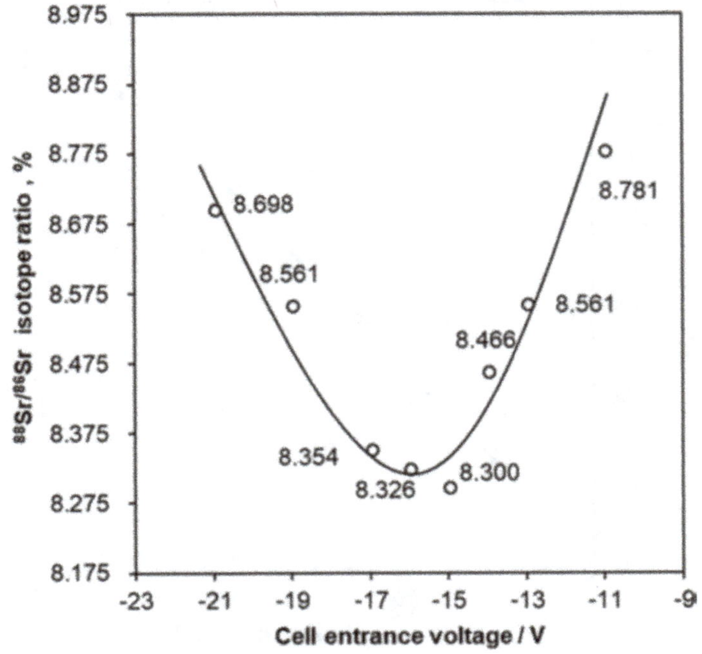

Fig. 20.5 Cell-entrance voltage profiles of different Sr isotope ratios obtained using the MS detector and adjustment of the cell-entrance voltages for Sr detection. The observed isotope ratio (i.e., the ratio between the intensities of ^{88}Sr and ^{86}Sr) as measured by QMS was gradually varied. The natural isotope ratio of ^{88}Sr/^{86}Sr is 8.375; a similar value was observed (8.326, −0.247% difference) when the cell-entrance voltage was −17 V. Thus, this voltage is suitable to quantify Sr because the determination error is sufficiently smaller than the measured values under these conditions. Here, MS was used as quadrupole MS (QMS)

the ratio was close to the natural isotope abundance ratio of 8.375. In Fig. 20.6, the measured intensities of ^{88}Sr and ^{90}Sr at each concentration are plotted on the same graph. The abundance of ^{88}Sr (82.58%) was converted to 100% and the detected intensity at each concentration was plotted; the resulting data demonstrated a good correlation coefficient of 0.9996. This suggests that it is possible to measure ^{90}Sr by using a calibration curve of ^{88}Sr. Although the measurement of the isotopic ratio using ICP-MS contains errors due to the calculation of the ^{90}Sr values using a calibration curve for ^{88}Sr, these errors are smaller than the measured signal errors.

20.3.4 Analysis of ^{90}Sr Using Online SPE/ICP-MS

Online SPE can be used to automatically concentrate Sr in a sample in the column, elute it in an eluent, and introduce it to ICP-MS. The process from concentration to measurement is described as follows. The column was first conditioned in 20 vol% nitric acid. Then, the sample was introduced to the column. It took about 10 min to concentrate a 50 mL sample. The split-flow method was used to measure the recovery ratio at the columns during this process (explained below). Other substances present in the columns were washed out with the 20 vol% nitric acid. The

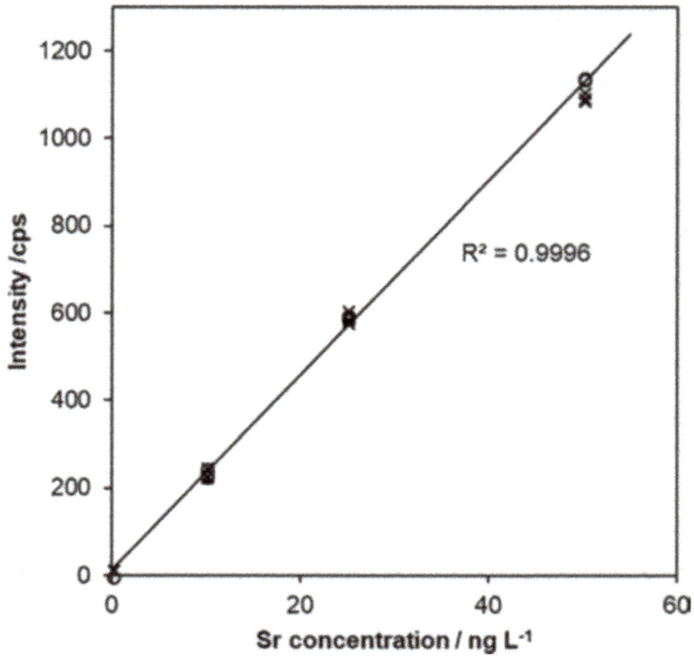

Fig. 20.6 Correlation between the intensity of stable isotope, ^{88}Sr, and the radioactive isotope, ^{90}Sr, at different concentrations. ^{90}Sr (n = 3) measurements are shown as white circles (O) and ^{88}Sr (n = 3) measurements are shown as black crosses (×). When the abundances of ^{88}Sr and ^{90}Sr are the same, the slopes of the calibration corresponded closely. In addition, the background of the artificial ^{90}Sr isotope was almost zero. Therefore, the ^{90}Sr can be indirectly derived from the measurements of the ^{88}Sr stable isotope using the calibration curve. The detection limit of standard ICP-QMS (with a concentric nebulizer and without any accessories or attachments) was 0.055 ng/L (equivalent to 280 Bq/L)

concentrated Sr in the resin was eluted by switching a value position (in flow injection system). Figure 20.7 shows an eluted chromatogram when 5 Bq/L of ^{90}Sr was introduced and measured. Moreover, in the proposed online SPE method, transient signal (= peak) can be observed. It takes about 20 min to analyze a 50 mL sample and about 10 min to analyze a 10 mL sample. It takes only 20 s to identify the peak. Figure 20.8 shows the linearity of the measured ^{90}Sr intensity as a function of its concentration between 5 and 20 Bq/L. The measured ^{90}Sr concentrations were compared with the values measured using an accepted method for radiation measurement—the pico-beta-measurement method. Figure 20.9 shows that the results obtained by the proposed method correlated closely with those obtained by the pico-beta-measurement method in the concentration range from 2 to 100 Bq/L.

The precision of the measurements obtained by ICP-MS analysis also depends on the sample concentration. Figure 20.10 shows the analysis results for samples with 2 and 10 Bq/L (0.4 and 2 pg/L, respectively) of ^{90}Sr in 10 replicates. Even though the 0.4 pg/L concentration is relatively low for ICP-MS, an RSD of 12.6% was achieved, demonstrating the reproducibility of this technique. The measurement of 2 pg/L also showed good reproducibility (RSD of 5.2%). Thus, it can be concluded that, in terms of the method's sensitivity, reproducibility, and consistency with the conventional method, it is a practical analysis technique.

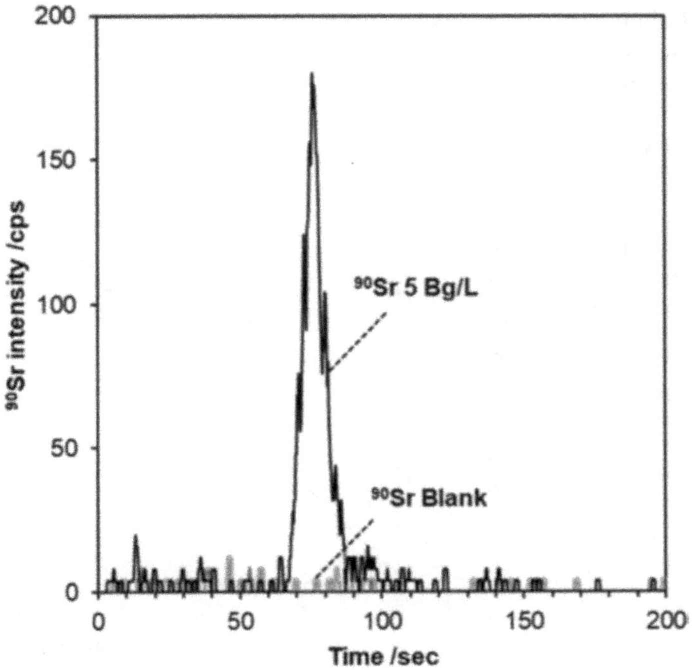

Fig. 20.7 Chromatograms of ^{90}Sr and a blank sample (pure water) purified by online solid extraction. When a blank sample (50 mL of pure water) was injected, no peaks were detected on the chromatogram. When 50 mL of ^{90}Sr was injected, an obvious peak with a width of 20 s was detected. When the elution flow rate was 5 mL/min, the preconcentrated phase volume was approximately 0.8 mL. When 50 mL of the sample was injected, the preconcentration factor (volume ratio) was approximately 63 times

Fig. 20.8 Linear calibration curve for ^{90}Sr. The peak area is used for the quantification of ^{90}Sr. The detection limit was approximately 0.5 Bq/L when 50 mL of the sample was injected. The measurement time was 20 min

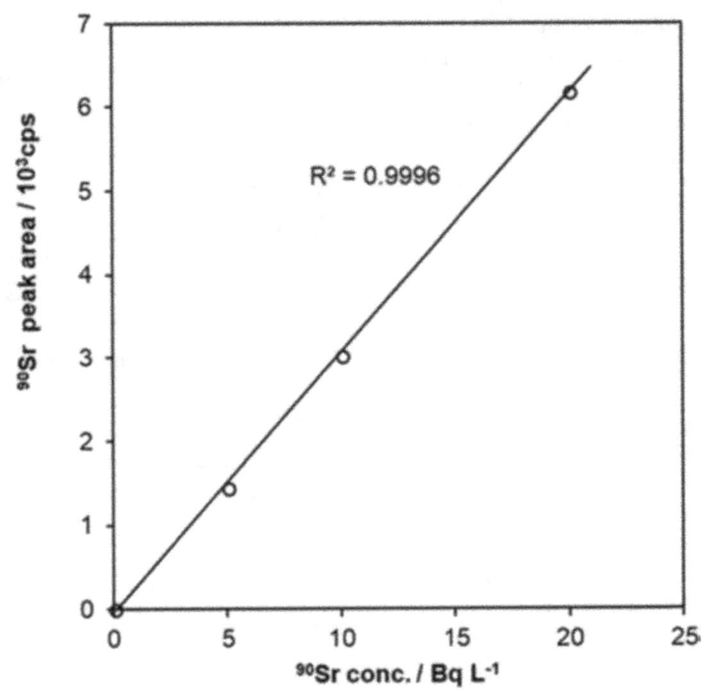

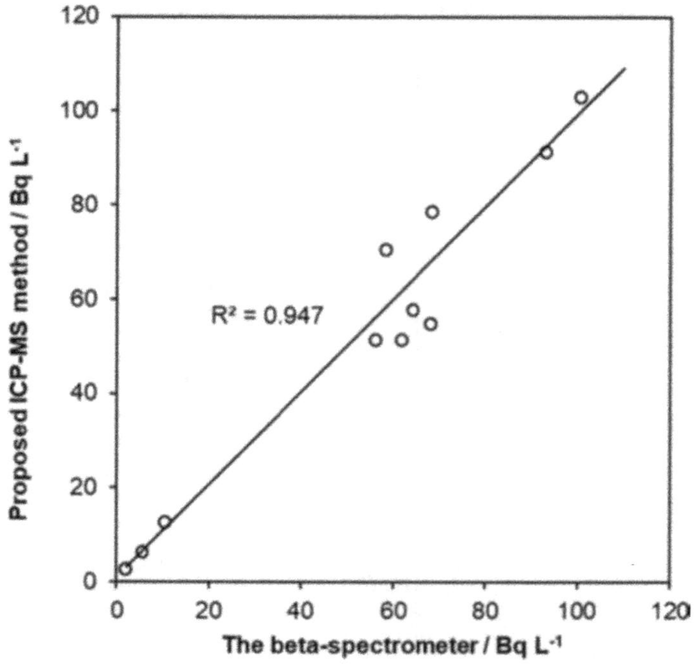

Fig. 20.9 Correlation between the quantitative values obtained by the beta-spectrometer (radiometry) and the proposed ICP-MS method. Sufficient linearity was obtained in the concentration range of 2–100 Bq/L

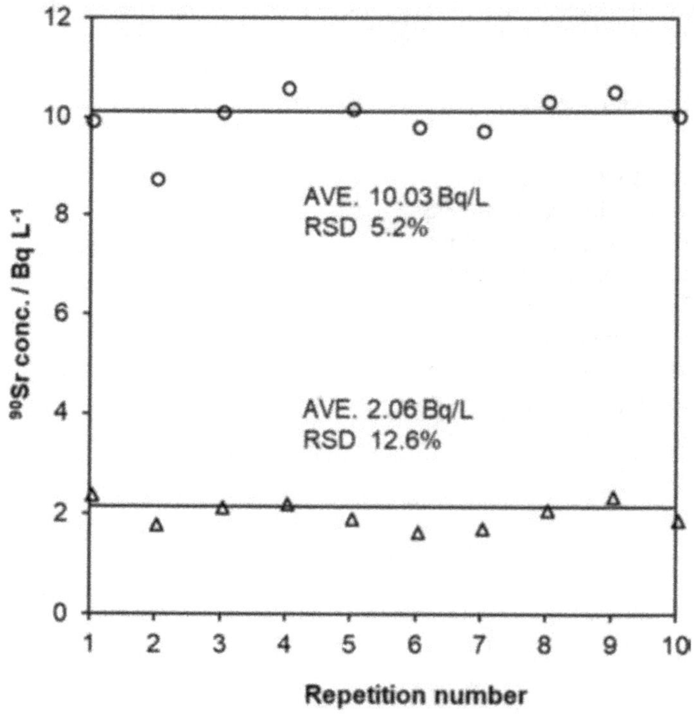

Fig. 20.10 Repeatability of the ^{90}Sr measurements. The measurement repeatability was confirmed over ten replicates of the samples with ^{90}Sr concentrations of 2 and 10 Bq/L (50 mL of the sample were injected each time)

20.3.5 *Addressing the Peaks Associated with the Enriched Stable Sr Isotope*

It is known that the measured intensity of the detected elements decreases when a large quantity of the ion is introduced with ICP-MS. The decrease in the temperature of the plasma and the space-charge effect are considered to be the causes of this phenomenon. Basically, coexisting elements are separated while being concentrated and separated in the columns; therefore, the decrease in the measured intensity does not matter when judging the concentrations of minor elements. However, the coexisting stable isotope of Sr is highly concentrated and eluted when measuring ^{90}Sr. This may cause the decrease in the measured intensity. We established an internal standard correction and a signal integration method to solve this problem (Furukawa et al. 2017). Internal standard elements were introduced after the column to obtain a fixed signal intensity. Figure 20.11 shows a variation of the internal standard signals when a high concentration of the stable isotope of Sr was introduced. As the peak of the eluted Sr became larger, the internal standard signal decreased. The ratio of two things between the increased intensity of Sr peak and the decreased intensity of the internal standard signal was calculated as a correction factor for the measured Sr intensity. ICP-MS detects each ion and records its intensity per unit time because each element number is measured continuously using the instrument's peak-hopping mode. The internal standard correction and signal integration can be used to correct the intensity of the Sr measurements over time to control for the decrease in the whole peak intensity.

20.3.6 *Split-Flow Injection System to Simultaneously Measure the Concentration and Recovery Ratio from a Single Sample (Furukawa and Takagai 2016)*

The recovery ratio is defined as the proportion of the elements that are eluted and measured to those that are introduced and concentrated in the resin. The recovery ratio of the column is affected by the other elements coexisting in the sample, physical obstructions (such as the velocity of flow and viscosity), volume differences between samples, and the deterioration of the resin. Thus, to monitor the changes in the recovery ratio, an experiment was designed to introduce a sample and measure its concentration before and after passing through the columns. In other words, more than two measurements must be conducted. The split-flow injection system developed in this study splits samples online before they are injected into the resin such that one portion can be measured directly and the other portion is first concentrated before being analyzed. Thus, the intensity before and after passing through

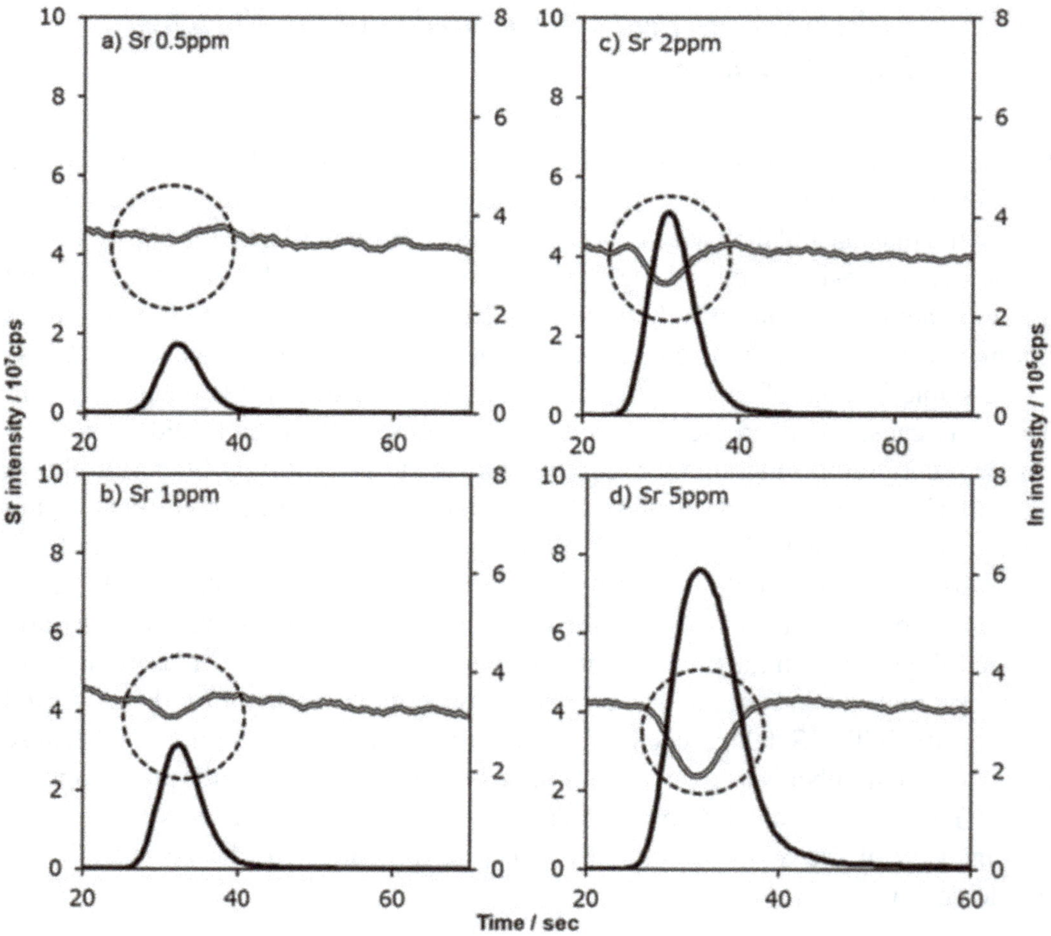

Fig. 20.11 Relationship between the peaks of the internal standard (In) and the preconcentrated stable Sr isotope. In the presence of higher concentrations of stable Sr isotope, the sensitivity of the preconcentrated Sr was decreased. As the Sr peak intensity decreased, the peak of internal standard (In) was seen as a negative chromatographic peak

the resin is measured automatically. Because the measured intensities before and after passing through the resin are proportional to the amount of the substance present, the recovery ratio relative to its absolute quantity can be calculated by integrating the measured intensity. In addition, the split-flow injection method can provide a relative recovery ratio without concentrating the sample based on the correlation between the intensities measured before and after passing through the columns.

Figure 20.1 shows a diagram of the split-flow injection. After the sample is input at valve #1, the splitter divides it into two column pathways. The rotational frequency of the peristaltic pump and the inside diameter of the pump tube control the division ratio. When the division ratio is 1:1, the volume of the sample that is concentrated is half that of the sample that is not. To attain greater preconcentration of the sample solution, the split ratio for the direct injection can be increased such that a larger amount of the sample is introduced into the column. Figure 20.12 shows the correlation between the measured intensity before passing through the columns and

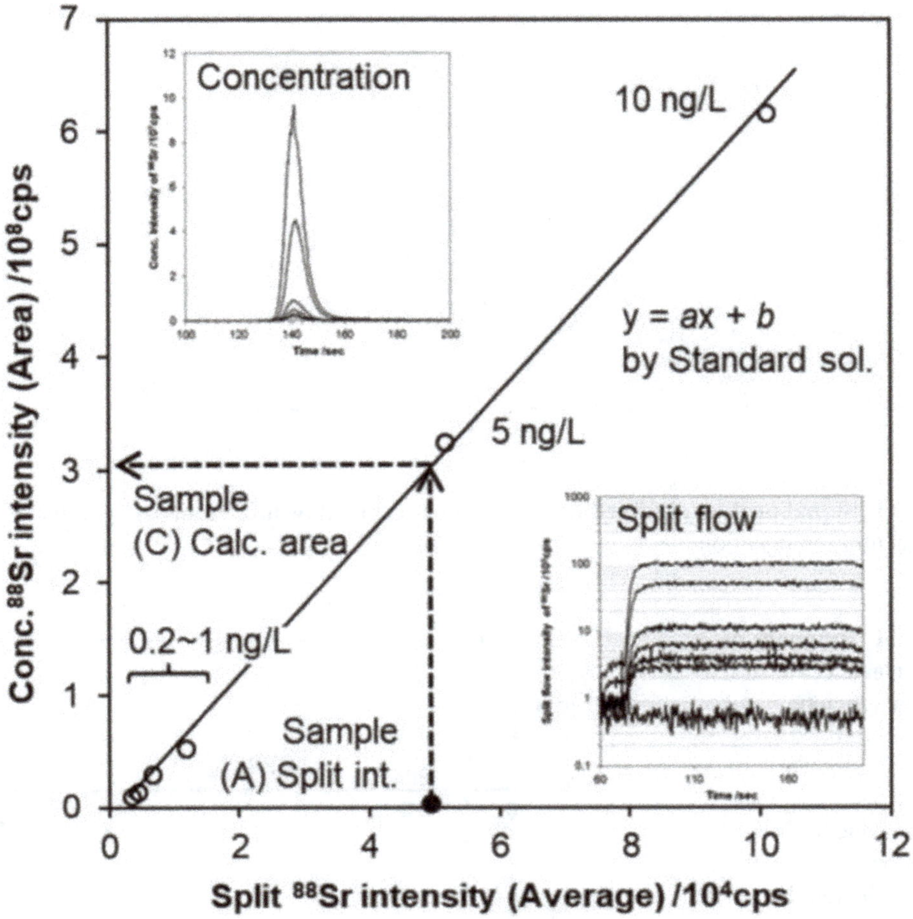

Fig. 20.12 Relationship between the intensities of the split stable Sr isotope and the preconcentrated Sr used to calculate the relative recovery. For the calculation, a linear calibration curve for the split Sr intensity versus the preconcentrated Sr intensity was prepared for a stable Sr standard solution. Based on the line, the relative recovery, R, can be calculated by the following equation given that the split Sr intensity and the preconcentrated Sr peak area have an approximately linear relationship of the form $y = aX + b$

$R = (B)/(C) \times 100 = (B)/[a(A) + b] \times 100$

where *a* and *b* were the slope and intercept of the linear line, respectively, A represents the intensity of the split Sr intensity, B is the detected Sr area, and C is the calculated Sr area

after concentrating the sample for a standard solution when the division ratio (column injection volume: split volume) was 100:4.

The degree of correlation between the detected intensity of the split sample and the Sr measurement after concentration shows the difference between the recovery ratio of a sample compared to that in the standard solution. Taking the recovery ratio of a standard solution as 100%, the relative recovery ratio of each sample was measured. Figure 20.13 shows the relative recovery ratios of the stable isotope of Sr for which m/z = 88, 86, or 84 (as the signal of ^{90}Sr is hardly detected in the split measurements). The relative recovery ratio of the stable Sr isotope was calculated as a correction factor for the ^{90}Sr measurement to obtain the corrected ^{90}Sr value.

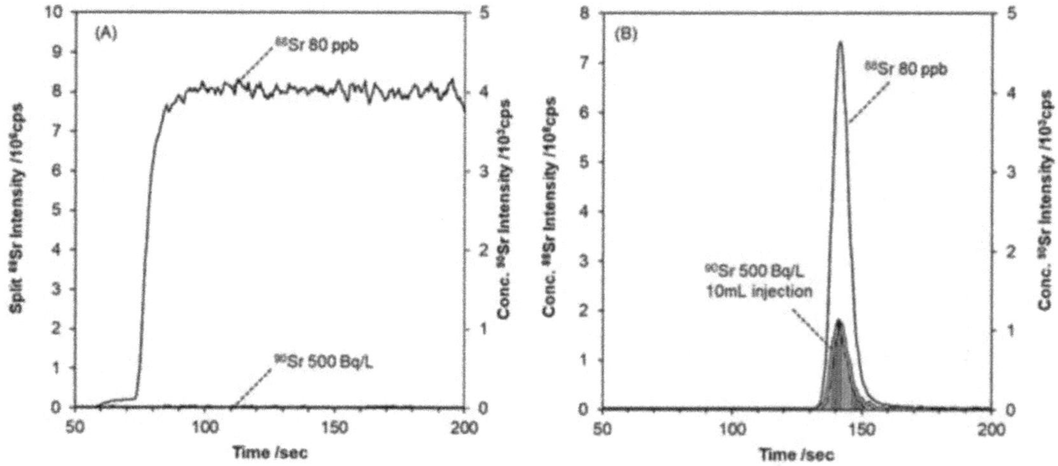

Fig. 20.13 Signal profiles obtained by ICP-MS with split-flow injection. (**a**) Split injection and (**b**) concentration injection

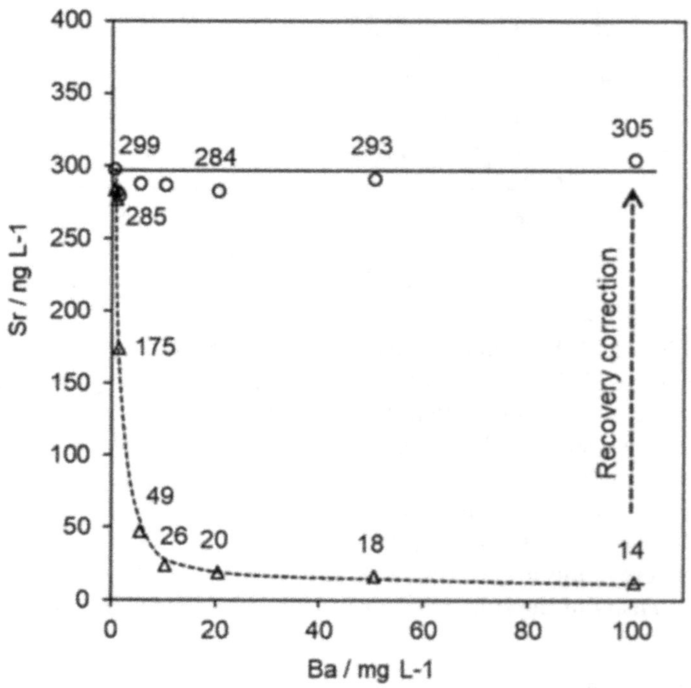

Fig. 20.14 Influence of Ba interference on the Sr recovery efficiency and the corrected concentration. The recovery percentage was decreased due to the interference of Ba with the Sr resin. The relative recovery was determined and the corrected concentration was calculated using the split-flow injection. Δ: observed Sr concentration before the correction using the relative recovery efficiency. O: corrected Sr concentration after the calculation using the relative recovery efficiency

To evaluate the method of split-flow injection, an experiment was conducted with the addition of Ba, known to decrease the Sr recovery ratio. Figure 20.14 shows the measured Sr concentration and the concentration with the corrected relative recovery ratio in the presence of various amounts of Ba and a fixed concentration of Sr. Although the relative recovery ratio of Sr decreased by the addition of Ba, it was confirmed that for the correction of the relative recovery ratio, the use of the split-flow injection method allows the correct concentration (i.e., that without the influence of Ba) to be derived. Therefore, the proposed method is an effective way to correct the recovery ratio. The split measurement does not affect the total measurement time and affects the measurement sensitivity only slightly as the division ratio between the columns and the direct introduction is changed.

20.4 Conclusion

Here, we demonstrated ^{90}Sr analysis using online SPE/ICP-MS with a split-flow injection method. Several techniques were combined in a single automated system. By using split-flow injection to measure the recovery ratio, the drawback of SPE being affected by changes to the matrix over time was mitigated. This approach can complement the radiation measurement method as an alternative technique for ^{90}Sr measurements and the user may select the proper method depending on the specific application (i.e. rapid measurement and the concentration of ^{90}Sr and stable isotopes *etc.*). It takes 10–20 min only to complete the ^{90}Sr analysis, requiring more than 2 weeks with the alternative technique. This analysis requires 50 mL of sample and the limit of detection is 0.3 Bq/L (equivalent to about 0.06 pg/L) with an argon-nitrogen mixed gas effect (Furukawa et al. 2018). This ability to measure concentrations as small as a few Bq/L in a small sample makes this technique suitable for a wide-area, multi-point sampling and analysis. Thus, this method can be readily used not only for environmental water analysis but also for applications requiring prompt measurements, such as the analysis of perishable foods.

Acknowledgment The authors would like to thank Dr. Yutaka Kameo, Dr. Kennichiro Ishimori, Mr. Kiwamu Tanaka, and Mr. Makoto Matsueda (Japan Atomic Energy Agency) and Dr. Katsuhiko Suzuki (Japan Agency Marine-Earth Science and Technology). The work was supported by the *Ministry of Education, Culture, Sports, Science & Technology in Japan (MEXT), Human Resource Development and Research Program for Decommissioning of Fukushima Daiichi Nuclear Power Station.*

References

Furukawa M, Takagai Y (2016) Split flow online solid-phase extraction coupled with inductively coupled plasma mass spectrometry system for one-shot data acquisition of quantification and recovery efficiency. Anal Chem 88:9397–9402

Furukawa M, Matsueda M, Takagai Y (2017) Internal standard corrected signal integration method for determination of radioactive strontium by online solid phase extraction/ICP-MS. Bunseki kagaku 66:181–187

Furukawa M, Matsueda M, Takagai Y (2018) Ultrasonic mist generation assist argon–nitrogen mix gas effect on radioactive strontium quantification by online solid-phase extraction with inductively coupled plasma mass spectrometry. Anal Sci 34:471–476

Takagai Y, Furukawa M, Kameo Y, Suzuki K (2014) Sequential inductively coupled plasma quadrupole mass-spectrometric quantification of radioactive strontium-90 incorporating cascade separation steps for radioactive contamination rapid survey. Anal Methods 6:355–362

Takagai Y, Furukawa M, Kameo Y, Matsueda M, Suzuki K (2017) Radioactive strontium measurement using ICP-MS following cascade preconcentration and separation system. Bunseki kagaku 66:223–231

www.ingramcontent.com/pod-product-compliance
Lightning Source LLC
LaVergne TN
LVHW082133200926
848484LV00008B/2049
* 9 7 8 1 0 1 3 2 7 4 2 7 5 *